Advanced Computer-Assisted Techniques in Drug Discovery

edited by Han van de Waterbeemd

VCH

Methods and Principles in Medicinal Chemistry

Edited by
R. Mannhold
P. Krogsgaard-Larsen
H. Timmerman

Volume 1
Hugo Kubinyi,
QSAR: Hansch Analysis and Related Approaches

Volume 2
Han van de Waterbeemd (ed.),
Chemometric Methods in Molecular Design

Volume 3
Han van de Waterbeemd (ed.),
Advanced Computer- Assisted Techniques in Drug Discovery

Advanced Computer-Assisted Techniques in Drug Discovery

edited by Han van de Waterbeemd

VCH

Weinheim • New York
Basel • Cambridge • Tokyo

Volume editor:
Dr. Han van de Waterbeemd
F. Hoffmann - La Roche Ltd.
Pharma Research New Technologies
CH-4002 Basel
Switzerland
Editors:
Prof. Raimund Mannhold
Biomedical Research Center
Molecular Drug Research Group
Heinrich- Heine-Universität
Universitätsstraße 1
D-40225 Düsseldorf
Germany

Prof. Povl Krogsgaard-Larsen
Dept. of Organic Chemistry
Royal Danish School of Pharmacy
DK-2100 Copenhagen
Denmark

Prof. Hendrik Timmerman
Faculty of Chemistry
Dept. of Pharmacochemistry
Free University of Amsterdam
De Boelelaan 1083
NL-1081 HV Amsterdam
The Netherlands

Published jointly by
VCH Verlagsgesellschaft mbH, Weinheim (Federal Republic of Germany)
VCH Publishers, Inc., New York, NY (USA)
Editorial Director: Dr. Thomas Mager
Production Manager: Dipl.-Ing. (FH) Hans Jörg Maier
Library of Congress Card No. applied for.
British Library Cataloguing-in-Publication Data: A catalogue record for this book is available from the British Library.
Deutsche Bibliothek Cataloguing-in-Publication Data:
Advanced computer assisted techniques in drug discovery / ed. by Han van de Waterbeemd. - Weinheim ; New York ; Basel ; Cambridge ; Tokyo : VCH, 1994
(Methods and principles in medicinal chemistry ; Vol. 3)
ISBN 3-527-29248-9
NE: Waterbeemd, Han van de [Hrsg.]; GT

Printed on acid-free and chlorine-free paper.

Composition:K+V Fotosatz GmbH, D- 64743 Beerfelden.
Printing: betz-druck gmbh, D-64291 Darmstadt.
Printed in the Federal Republic of Germany.

Distribution:
VCH, P.O. Box 10 11 61, D-69451 Weinheim (Federal Republic of Germany)
Switzerland: VCH, P.O. Box, CH-4020 Basel (Switzerland)
United Kingdom and Ireland: VCH (UK) Ltd., 8 Wellington Court, Cambridge CB1 1HZ (England)
USA and Canada: VCH, 220 East 23rd Street, New York, NY 10010-4606 (USA)
Japan: VCH, Eikow Building, 10-9 Hongo 1-chome, Bunkyo-ku, Tokyo 113 (Japan)

Preface

The main objective of this series is to offer a practice-oriented survey of techniques currently used in Medicinal Chemistry. Following the volumes on Hansch analysis and related approaches (Vol. 1) and multivariate analyses (Vol. 2), the present handbook focuses on some new, emerging techniques in drug discovery; emphasis is placed on showing users how to apply these methods and to avoid time-consuming and costly errors.

Four major topics are covered. The first centers on three-dimensional QSAR, and some of the enormous progress achieved in this field is summarized. Both the various 3D-QSAR methods available as well as the chemometric tools for handling the statistical problems involved in 3D-QSAR studies are covered.

Intimately coupled with 3D-QSAR is the current trend in pharmaceutical industry to establish chemical structure databases as a tool for identifying new leads. Correspondingly, in the second section, problems encountered in our understanding of molecular similarity and aspects of compound selection by clustering databases are treated.

The third section covers advanced statistical techniques in drug discovery. Inter alia the approach of Svante Wold to apply PLS to non-linear structure-activity relations deserves to be mentioned here.

Last but not least, the use of neural networks for data analysis in QSAR problems is discussed. Advantages and disadvantages are critically analysed by comparing networks versus statistics.

The editors would like to thank all contributors and VCH publishers for their fruitful cooperation.

Summer 1994

Düsseldorf — Raimund Mannhold
Kopenhagen — Povl Krogsgaard-Larsen
Amsterdam — Hendrik Timmerman

A Personal Foreword

It is no coincidence that the first three volumes of *Methods and Principles in Medicinal Chemistry* deal with computer-assisted medicinal chemistry. After the classical Hansch method in Volume 1 and applications of chemometric methods in Volume 2, the present volume of the series contains a number of emerging new techniques. Of course, all approaches using molecular modeling techniques, such as structure-based design and de novo design, rely on computers as well. These will be treated separately in a forthcoming volume.

This volume is a logical continuation of Volume 2. In fact, after analyzing the methods that have been developed following the Hansch method, we came to the conclusion that a number of these techniques have now matured, while others still require further developments. This criterion was used to select the chapters for Volumes 2 and 3.

In reviewing the contents of the first three volumes in this series, it is evident that highly specialized tools have become available for the analysis of complex biological and chemical data sets in order to unravel quantitative structure-activity relationships. It has not become easier for the bench chemist to select the ideal method for dealing with the analysis of structure-activity relationships using chemical and biological data. Specialist support is required to validate and apply statistical or chemometric and other computer-assisted tools. Volume 3 focusses very much on the newest methods employed by the chemometrician. We hope that, in an indirect way, some of the methods discussed will be of use to molecular design on a day to day basis.

I am grateful to, and would like to thank all the contributing authors for their efforts in compiling this volume.

February 1994, Basel — Han van de Waterbeemd

Contents

List of Contributors

Dr. Mohammed A-Razzak
Infolink
Decision Services Ltd.
9–11 Grosvenor Gardens
London SW1 W0BD, UK
Tel.: +44 71 233 7333

Dr. Laura Bonati
Dipartimento di Chimica
Fisica ed Elettrochimica
Università degli Studi di Milano
Via C. Golgi 19
20133 Milano, Italy
Tel.: +39 2 2660 3252
Fax: +39 2 7063 8129

Dr. Brian W. Clare
School of Mathematical
and Physical Sciences
Murdoch University
Murdoch, Perth
Western Australia 6150, Australia
Tel.: +61 9 360 6000
Fax: +61 9 360 2507

Prof. Sergio Clementi
Laboratorio di Chemiometria
Dipartimento di Chimica
Università di Perugia
Via Elce di Sotto 8
06123 Perugia, Italy
Tel. and Fax: +39 75 45646

Dr. Ugo Cosentino
Dipartimento di Chimica
Fisica ed Elettrochimica
Università degli Studi di Milano
Via C. Golgi 19
20133 Milano, Italy
Tel.: +39 2 2660 3252
Fax: +39 2 7063 8129

Dr. Gabriele Cruciani
Laboratorio di Chemiometria
Dipartimento di Chimica
Università di Perugia
Via Elce di Sotto 8
06123 Perugia, Italy
Tel. and Fax: +39 75 45646

Dr. Andrew M. Davis
Fisons PLC Research and
Development Laboratories
Bakewell Road
Loughborough LE11 ORH, UK
Tel.: +44 509 611011/44370
Fax: +44 509 236609

Dr. Jean-Christopher Doré
Chimie Appliquée
aux Corps Organisés
CNRS URA 401 & Muséum National
d'Histoire Naturelle
63, Rue de Buffon
75231 Paris Cedex 05, France
Tel.: +33 1 4079 3136
Fax: +33 1 4079 3147

Dr. Geoffrey M. Downs
Department of Information Studies
University of Sheffield
Western Bank
Sheffield S10 2TN, UK
Tel.: +44742825083
Fax: +44742780300

Dr. Martyn G. Ford
University of Portsmouth
School of Biological Sciences
King Henry Building
King Henry I Street
Portsmouth, Hants PO1 2DY, UK
Tel.: +44705842036
Fax: +44705842070

Dr. Elena Fraschini
Dipartimento di Chimica
Fisica ed Elettrochimica
Università degli Studi di Milano
Via C. Golgi 19
20133 Milano, Italy
Tel.: +39226603252
Fax: +39270638129

Dr. Robert C. Glen
Wellcome Research Laboratories
Department of Physical Sciences
Langley Court
Beckenham, BR3 3BS, UK
Tel.: +44816582211
Fax: +44816633788

Dr. Mark Johnson
Upjohn Laboratories
Computational Chemistry
The Upjohn Company
Kalamazoo, MI 49001-0199, USA
Tel.: +16163857830
Fax: +16163858488

Dr. Michael S. Lajiness
Upjohn Laboratories
Computational Chemistry
The Upjohn Company
Kalamazoo, MI 49001-0199, USA
Tel.: +16163857830
Fax: +16163858488

Dr. Marina Lasagni
Dipartimento di Chimica
Fisica ed Elettrochimica
Università degli Studi di Milano
Via C. Golgi 19
20133 Milano, Italy
Tel.: +39226603252
Fax: +39270638129

Dr. Paul J. Lewi
Information Science Dept.
Janssen Research Foundation
Janssen Pharmaceutica NV
Turnhoutseweg 30
B-2340 Beerse, Belgium
Tel.: +3214602111
Fax: +3214602841

Dr. David J. Livingstone
SmithKline Beecham Pharmaceuticals
The Frythe
Welwyn, AL6 9AR, UK
Tel.: +44438782088
Fax: +44438782550

Dr. Halliday J.H. MacFie
AFRC Institute of Food Research
Earley Gate
Whiteknights Road
Reading, RG6 2EF, UK
Tel.: +44734357172
Fax: +44734267917

Dr. Gerald M. Maggiora
Upjohn Laboratories
Computational Chemistry
The Upjohn Company
Kalamazoo, MI 49001-0199, USA
Tel.: +1 616 385 7830
Fax: +1 616 385 8488

Dr. Jonathan A. Malpass
School of Mathematical Studies
University of Portsmouth
Portsmouth, Hants PO1 2EG, UK
Tel.: +44 705 842036
Fax: +44 705 842070

Dr. David T. Manallack
SmithKline Beecham Pharmaceuticals
The Frythe
Welwyn, AL6 9AR, UK
Tel.: +44 223 420430
Fax: +44 223 420440

Prof. Dr. Raimund Mannhold
Biomedical Research Center
Molecular Drug Research Group
Universitätsstrasse 1
40225 Düsseldorf, Germany
Tel.: +49 211 311 2759
Fax: +49 211 312631

Dr. Henri Moereels
Janssen Research Foundation
Theoretical Med. Chem. Dept.
Janssen Pharmaceutica NV
B-2340 Beerse, Belgium

Dr. Joseph B. Moon
Upjohn Laboratories
Computational Chemistry
The Upjohn Company
Kalamazoo, MI 49001-0199, USA
Tel.: +1 616 385 7830
Fax: +1 616 385 8488

Dr. Giorgio Moro
Dipartimento di Chimica
Fisica ed Elettrochimica
Università degli Studi di Milano
Via C. Golgi 19
20133 Milano, Italy
Tel.: +39 2 2660 3252
Fax: +39 2 7063 8129

Dr. Tiiu Ojasoo
Groupe Cristallographie et Simulations
Interactives des Macromolécules
Biologiques
Université Pierre et Marie Curie (VI)
63, Rue de Buffon
75231 Paris Cedes 05, France
Tel.: +33 1 4079 3136
Fax: +33 1 4079 3147

Dr. James D. Petke
Upjohn Laboratories
Computational Chemistry
The Upjohn Company
Kalamazoo, MI 49001-0199, USA

Prof. Demetrio Pitea
Dipartimento di Chimica
Fisica ed Elettrochimica
Università degli Studi di Milano
Via C. Golgi 19
20133 Milano, Italy
Tel.: +39 2 2660 3252
Fax: +39 2 7063 8129

Dr. Douglas C. Rohrer
Upjohn Laboratories
Computational Chemistry
The Upjohn Company
Kalamazoo, MI 49001-0199, USA
Tel.: +1 616 385 7830
Fax: +1 616 385 8488

Dr. Valerie Sally Rose
Department of Physical Sciences
Wellcome Research
South Eden Park Road
Beckenham, BR3 3BS, UK
Tel.: +44 81 658 2211
Fax: +44 81 663 3788

Dr. Davis W. Salt
University of Portsmouth
School of Biological Sciences
King Henry Building
King Henry I Street
Portsmouth, PO1 2DY, UK
Tel.: +44 705 842036
Fax: +44 705 842070

Dr. Klaus-Jürgen Schaper
Forschungsinstitut Borstel
Parkallee 1–40
23845 Borstel, Germany
Tel.: +49 4537 10248
Fax: +49 4587 10245

Dr. Roberto Todeschini
Dipartimento di Chimica
Fisica ed Elettrochimica
Università degli Studi di Milano
Via C. Golgi 19
20133 Milano, Italy
Tel.: +39 2 26603252
Fax: +39 2 70638129

Dr. Han van de Waterbeemd
F. Hoffmann-La Roche Ltd.
Pharma Research New Technologies
CH-4002 Basel, Switzerland
Tel.: +41 61 688 8421
Fax: +41 61 688 1745

Prof. Dr. Peter Willett
University of Sheffield
Department of Information Studies
Regent Court, 211 Portobello
Sheffield, S10 2TN
Tel.: +44 742 768555
Fax: 44 742 780300

Dr. John Wood
Department of Physical Sciences
Wellcome Research
South Eden Park Road
Beckenham, BR3 3BS, UK
Tel.: +44 81 658 2211
Fax: +44 81 663 3788

Dr. E. Watcyn Wynn
School of Mathematical Studies
University of Portsmouth
Portsmouth, Hants PO1 2EG, UK

1 Introduction

Han van de Waterbeemd

Abbreviations

CFS	conformationally flexible searching
CLOGP	calculated log *P* values
CoMFA	comparative molecular field analysis
3D	three-dimensional
2D QSAR	traditional Hansch analysis
3D QSAR	quantitative models based on 3D superposition of molecules
GOLPE	generating optimal PLS estimations
MDL	Information Systems Inc.
MIC	minimum inhibition concentration
PLS	partial least squares projection to latent structures
SAR	structure-activity relationships
SPC	structure-property correlations
QSAR	quantitative structure-activity relationships

Symbols

log $1/C$	C is the molar concentration that produces a certain biological effect
log P	logarithm of the partition coefficient
E_s	Taft steric constant
σ	Hammett electronic substituent constant
IC_{50}	concentration at which 50% inhibition is observed

1.1 3D QSAR

Over the last two decades the art of drug discovery has changed dramatically with the introduction of new analytical tools. [1, 2] Analytical chemistry revolutionized both the analysis of chemical compounds and the study of biological processes. Today crystallography and NMR contribute significantly to biostructural research and have led to the unraveling of many details about the structure and function of macromolecules, such as nucleic acids and proteins. The second revolution, developed in parallel and which is now indispensable, concerns the use of computers in molecular design and in the lead discovery process.

The present series "Methods and Principles in Medicinal Chemistry" compiles the progress made in medicinal chemistry and illustrates the use of new methods and their limitations. It is no coincidence that the first two volumes involve the use of computers in molecular design, and that the present volume again discusses computer-assisted techniques. The development of the field quatitative structure-activity relationships (QSAR) and related topics has been covered in Volume 1 [3]. Traditionally, this approach, propagated by Hansch and Fujita since the 1960 s, employs mutiple linear regression techniques to obtain quantitative relationships [4, 5]. However, the statistical relevance of many a published equation may be disputed, or is simply non-existent. Modern statistical methods have been developed and are frequently used in data analysis problems, thus, a completely new discipline named *chemometrics* was born. Such statistical approaches are widely used in analytical chemistry and are also applied to quantitative molecular design. Many examples can be found in Vol. 2 [6]. Pattern recognition and regression using biological and chemical data are now widely employed in medicinal chemistry.

Chemical descriptors used in structure-property correlations (SPC) are often based on the lipohilic, electronic and steric nature of substituents [3, 6]. Although some of the steric descriptors, such as molar volume, encode some 3D information, molecular conformation has rarely been considered. The recent development of 3D QSAR are attempts to add this, a third dimension, to studies in quantitative molecular design. The first textbook on this relatively new subject appeared in 1993 [7]. The comparative molecular field analysis (CoMFA) method has been critisized and should still be considered as being in its infant years. The major problem being that CoMFA models are based on an alignment of compounds in a series, which is far from being a trivial problem [8]. Some progress has been made using genetic algorithms [9] and 3D ACC transforms (based on autocorrelation and cross-covariance of field descriptors [10].

In summary, computers in molecular design are used in the following ways:

- chemical information systems [11],
- computational chemistry [12–14],
- combinatorial chemistry, molecular diversity, molecular similarity [15–17],
- de novo design [18–21],
- molecular modeling [22],
- pharmacophore generation [23–26],
- property prediction [27–28],
- SPC, 2D QSAR [1–6],
- 3D QSAR, COMFA, GOLPE [7],
- synthesis planning, reaction databases [29].

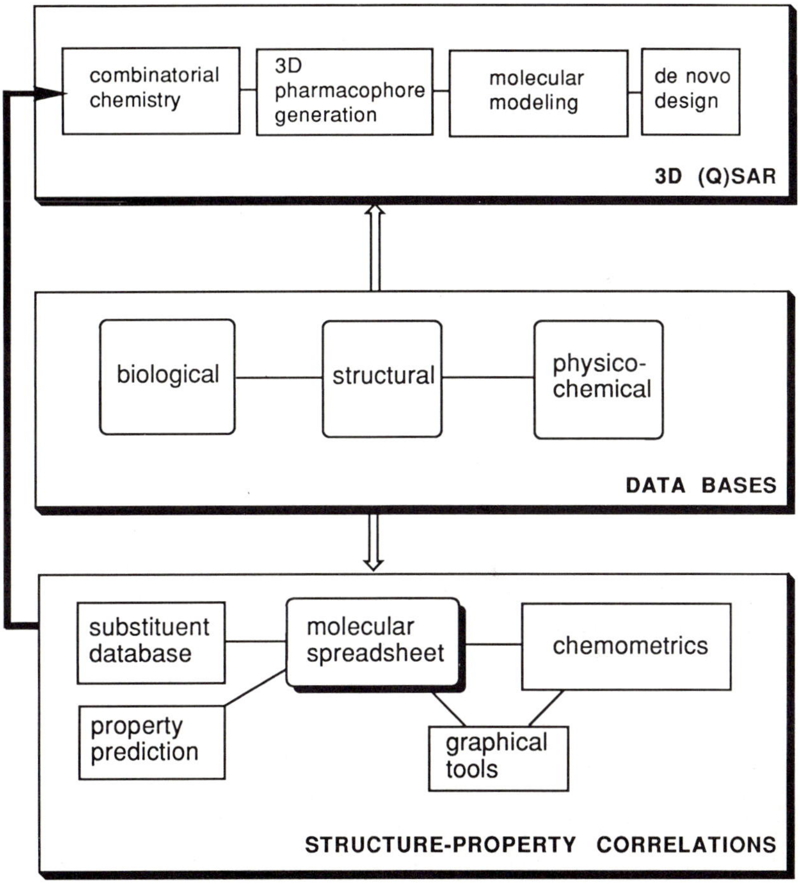

Figure 1. Important elements of computer-assisted medicinal chemistry.

Some confusion in semantics arises with the terms *computer chemistry* and *computational chemistry* [30–32]. For some authors, computational chemistry is just merely number crunching as, e.g. in quantum chemical or X-ray or NMR calculations, and computer chemistry relates to organic synthesis planning [32]. Others may understand computational chemistry as being equivalent to computer-assisted molecular design (CAMD).

In Fig. 1 a schematic representation of the main building blocks used in computer-assisted methods in medicinal chemistry is given. The core is formed by databases for in-house and external data collections. The different ways of looking at these data are the structure-property correlations approach and the 3D (Q)SAR approach. By the latter, we mean all methods looking at 3D structural data, thus, including molecular modeling and de novo design, pharmacophore generation tools and methods to screen 3D structural databases using conformationally flexible searching (CFS) strategies. Support for combinatorial chemistry or molecular diversity projects comes from a combination of 3D SAR and SPC techniques.

1.2 Databases

Large banks of chemical, biological and medical data are available and are potentially of interest to any drug discovery program. Chemical information systems and databases have become essential to handling such data [11]. Most pharmaceutical companies have used commercial software to store their in-house chemical information in database systems, e.g. MDL's MACCS-II, is widely used for structure handling. With increasing computational power and memory, as well as a huge storage potential, it has now become possible to create 3D versions of large chemical databases [33–36]. Recent software products include, e.g. MACCS-3D [37]. SYBYL-3D-UNITY [26], CATALYST/Hypo and CATALYST/Info [25], APEX [24], and RECEPTOR [23]. 3D Queries and semi-automatic pharmacophore generation using conformationally flexible searching (CFS) have increased the possibilities in rational lead finding for the medicinal chemist. Searching chemical databases using 3D (geometric), 2D (structural topology) and 1D (property) features and constraints are now within reach. Generation of new leads is an important aspect of preclinical research, and database searching is one approach, while blind and targeted screening with batteries of tests is another. Most compounds screened are taken from in-house depositories, which are growing at a phenomenal rate through combinatorial chemistry projects.

1.3 Progress in Multivariate Data Analysis

The quantification of electronic substituent effects by Hammett inspired Hansch and Fujita [38–41] to develop an analogous approach to define the contributions to the lipophilicity of an organic compound. Further studies on the role of lipophilicity in drug transport processes finally led to the introduction of quantitative models to describe relationships between biological effects and chemical structure [41]. These can be expressed by the Hansch Equation in the following form:

$$\log 1/C = a(\log P)^2 + b \log \mathrm{P} + cE_s + d\sigma + e \tag{1}$$

where C is the concentration of a standard response (e.g. an IC_{50} or MIC value), $\log P$ is the 1-octanol/water partition coefficient, E_s is Taft's steric descriptor and σ the well-known Hammett constant reflecting the electronic contributions of substituents. However, multiple linear regression, also called ordinary least squares, appears not always suited to deriving such quantitative models. In Vol. 2 [6], various alternatives have been discussed, particularly, partial least squares (PLS) regression which is the current standard for establishing quantitative models.

Various pattern recognition techniques have been developed to handle the problems of embedded data. This often occurs when active compounds are compared to inactive ones and the point here is that there are numerous reasons as to why a compound is inactive. Potentially important progress has been achieved with complex data sets using applications from the field of artificial intelligence. An increasing number of publications have appeared using neural network algorithms [42, 43]. These are well suited for pattern recognition applications using traditional molecular descriptors. Combinations of neural networks and molecular similarity matrices seem to be particularly promising [44] and other techniques of machine-learning are being explored [45].

1.4 Scope of this Book

Some of the above-mentioned topics are dealt with in this book, while other computer-assisted methods will be addressed in forthcoming volumes. In Sec. 2 we want to present some studies demonstrating how chemometric (statistical) methods can be combined with molecular modeling tools, an approach now called 3D-QSAR. Both the CoMFA and GOLPE methods are discussed within this context. Clustering of compounds and chemical descriptors can be accomplished very well with pattern recognition techniques, such as principal component analysis, cluster analysis and cluster significance analysis (see Vol. 2). In Sec. 3 of this book we deal with similarity criteria for rational clustering and searching through chemical databases. Furthermore, it is illustrated how clustering techniques can be used to extract information from protein sequence databases.

As stated above, recent developments in the understanding of certain data analysis problems may have applications in the field of molecular design. This involves, for example, the analysis of embedded data. A number of advanced statistical techniques are presented in Secs. 4 and 5 of this book. In Sec. 4 existing methods, have been developed further while Sec. 5 deals with new methods taken from the field of artificial intelligence (AI). It must be emphasized that the claims of many of these new methods in molecular design problems have yet to be verified and proven. However, this book illustrates the considerable efforts that are being made to broaden the scope of the methods employed to investigate the complex relationship between biological activity and molecular structure.

References

[1] Van de Waterbeemd, H., *Quant. Struct.-Act. Relat.* **11**, 200–204 (1992)

[2] Van de Waterbeemd, H., *Drug Des. Disc.* **9**, 277–285 (1993)

[3] Kubinyi, H. (ed.), *QSAR: Hansch Analysis and Related Approaches* (Methods and Principles in Medicinal Chemistry, Vol. **1**), VCH, Weinheim, 1993

[4] Tute, M. S., *History and Objectives of Quantitative Drug Design.* In: *Quantitative Drug Design* (Comprehensive Medicinal Chemistry, Vol. **4** Hansch, C., Sammes, P. G. and Taylor, J. B., eds. Pergamon Press, Oxford, 1990, p 1–31

[5] Topliss, J. G., *Perspect. Drug Disc. Des.* **1**, 253–268 (1993)

[6] Van de Waterbeemd, H., (ed.) *Chemometric Methods in Molecular Design* (Methods and Principles in Medicinal Chemistry, Vol. **2**), VCH, Weinheim, 1994

[7] Kubinyi, H., (ed.), *3D QSAR in Drug Design. Theory, Methods and Applications,* Escom, Leiden, 1993

[8] Klebe, G. and Abraham, U., *J. Med. Chem.* **36**, 70–80 (1993)

[9] Payne, A. W. R. and Glen, R. C., J. Mol. Graph. **11**, 74–91 (1993)

[10] Cruciani, G., Clementi, S. and Baroni, M., *Variable Selection in PLS Analysis.* In: *3D QSAR in Drug Design,* Kubinyi, H. (ed.), Escom, Leiden, 1993, p 551–564

[11] Dietrich, S. W., *Med. Chem. Res.* **2**, 127–147 (1992)

[12] Loew, G. H., Villar, H. O. and Alkorta, I., *Pharm. Res.* **10**, 475–486 (1993)

[13] Hyde, R. M. and Livingstone, D. J., *J. Comput.-Aid. Mol. Des.* **2**, 145–155 (1988)

[14] Saunders, M. R. and Livingstone, D. J., *Electronic Structure Calculations in Quantitative Structure – Property Relationships.* In: *Advances in Quantitative Structure – Property Relationships,* Charton, M., (ed.), JAI Press, Conneticut, 1994

[15] Johnson, M. A. and Maggiora, G. M., (eds.) *Concepts and Applications of Molecular Similarity,* Wiley, New York, 1990

[16] Simon, R. J., Martin, E. J., Miller, S. M., Zuckermann, R. N., Blaney, J. M. and Moos, W. H., publication submitted

[17] Moos, W. H., Green, G. R. and Pavia, M. R., *Ann. Rep. Med. Chem.* **28**, 315–324 (1993)

[18] Böhm, H.-J., *Ligand Design.* In: *3D QSAR in Drug Design,* Kubinyi, H. (ed.), Escom, Leiden, 1993, p 551–564

[19] Gillet, V. J., Johnson, A. P., Mata, P. and Sike, S., *Tetrahedron Comput. Meth.* **3**, 681–696 (1990)

[20] Tschinke, V. and Cohen, N. C. *J. Med. Chem.* **36**, 3863–3870 (1993)

[21] Cramer, R. D., *Chem. Design., Automat. News* **8**, 32–33 (1993)

[22] Cohen, N. C., Blaney, J. M., Humblet, C., Gund, P. and Barry, D. C., *J. Med. Chem.* **33**, 883–894 (1990)

[23] RECEPTOR, MSI, Molecular Simulations Inc., 200 Fifth Avenue, Waltham, MA 02154, USA

[24] APEX, Biosym, 9685 Scranton Road, San Diego, CA 92121-2777, USA

[25] CATALYST, MSI, Molecular Simulations Inc., 200 Fifth Avenue, Waltham, MA 02154, USA

[26] SYBYL-UNITY, Tripos Associates Inc., St Louis, MO 63144-2913, USA

[27] CLOGP, Daylight CIS, 18500 Von Karman Ave 450, Irvine, CA 92715, USA and BioByte Corp, PO Box 517, Claremont, CA 91711-0517, USA

[28] Bawden, D., *J. Chem. Inf. Comput. Sci.* **23**, 14–22 (1983)

[29] Kos, A. J. and Grethe, G., *Nachr. Chem. Tech. Lab.* **35**, 586–594 (1987)

[30] Ugi, I., *Topics Curr. Chem.* **166** (1993)

[31] Trinajstic, N., Book reviews, *Comput. Chem.* **4**, 405–406 (1993)

[32] Ugi, I., Stein, N., Knauer, M., Gruber, B. and Bley, K., *Topics Curr. Chem.* **166**, 199–233 (1993)

[33] Güner, O.F., Hughes, D.W. and Dumont, L.M., *J. Chem. Inf. Comput. Sci.* **31**, 408–414 (1991)

[34] Martin, Y.C., *J. Med. Chem.* **35**, 2145–2154 (1992)

[35] Humblet, C. and Dunbar, J.B., *Ann. Rep. Med. Chem.* **28**, 275–284 (1993)

[36] Willett, P., *Three-Dimensional Chemical Structure Handling*, Research Studies Press, Taunton, UK, 1991

[37] MACCS-3D, MDL Information Systems, Inc., San Leandro, USA

[38] Hansch, C., Maloney, P.P., Fujita, T. and Muir, R.M., *Nature* **194**, 4823–4825

[39] Hansch, C., Muir, R.M., Fujita, T., Maloney, P.P., Geiger, F. and Streich, M., *J. Amer. Chem. Soc.* **85**, 2817–2824 (1963)

[40] Hansch, C. and Fujita, T., *J. Amer. Chem. Soc.* **86**, 1616–1626 (1964)

[41] Fujita, T., Iwasa, J. and Hansch, C., *J. Amer. Chem. Soc.* **86**, 5175–5180 (1964)

[42] Salt, D.W., Yildiz, N., Livingstone, D.J. and Tinsley, C.J., *Pestic. Sci.* **36**, 161–170 (1992)

[43] Ajay, J., *Med. Chem.* **36**, 3565–3571 (1993)

[44] Good, A.C., So, S.-S. and Richards, W.G., *J. Med. Chem.* **36**, 433–438 (1993)

[45] Bolis, G., Di Pace, L. and Fabrocini, F., *J. Comput.-Aid. Mol. Des.* **5**, 617–628 (1991)

2 3D QSAR: The Integration of QSAR with Molecular Modeling

2.1 Chemometrics and Molecular Modeling

Demetrio Pitea, Ugo Cosentino, Giorgio Moro, Laura Bonati, Elena Fraschini, Marina Lasagni, and Roberto Todeschini

Abbreviations

5-HT	5-Hydroxytryptamine
Ah	Arylhydrocarbon
AII	Angiotensin II
CoMFA	Comparative Molecular Field Analysis
EC_{50}	50% Effective Concentration
GOLPE	Generating Optimal Linear PLS Estimations
HOMO	Highest Occupied Molecular Orbital
ID_{50}	50% Inhibitory Dose
K_d	Affinity constant
HMG-CoA	3-Hydroxy-3-methylglutaryl Coenzyme A
LDA	Linear Discriminant Analysis
LDCT	Linear Discriminant Classification Tree
LUMO	Lowest Unoccupied Molecular Orbital
MCDD	Monochlorinated Dibenzo-*p*-dioxins
MEP	Molecular Electrostatic Potential
PC	Principal Component
PCA	Principal Component Analysis
PCDD	Polychlorinated Dibenzo-*p*-dioxins
PES	Potential Energy Surface
PLS	Partial Least Squares
QSAR	Quantitative Structure-Activity Relationships
r^2	Squared Correlation Coefficient
r^2_{cv}	Cross-validated Squared Correlation Coefficient
RDA	Regularized Discriminant Analysis
SAR	Structure-Activity Relationships

SCF-HF	Self Consistent Field-Hartree Fock
SD	Standard Deviation
TCDD	Tetrachloro-dibenzo-*p*-dioxins
TrCDD	Trichloro-dibenzo-*p*-dioxins

2.1.1 Introduction

The role of computational simulations of molecular mechanisms in understanding biological processes is emphasized in this sentence by Weinstein [1]:

"The rapid growth in our mechanistic understanding of biological systems and processes combined with the recently developed technological capabilities of molecular biology, have engendered the promise that the essential modulatory processes, and the mode in which they are affected by environmental factors, can be understood at a discrete molecular level. For this promise to be realized fully, this new field requires detailed information, at the atomic level of resolution, about the structures and properties of the key molecular species involved in the underlying mechanisms. The approaches of physical chemistry and chemical physics are fundamental tools in this quest for essential information. Thus, structural information about biological systems can be obtained with the experimental methods of physical chemistry, especially X-ray crystallography and multidimensional NMR, but these approaches are still costly in time and resources. The theoretical aspects of physical chemistry, utilizing computational simulations of molecular mechanisms, support the experimental efforts and, in some cases, provide essential alternatives. Complementing both formal theory and direct experimentation, and resting on data and inferences from both molecular biology and structural biology, computational simulations of biological systems and mechanisms have become a major factor in modern research."

The main goals of Structure-Activity Relationship, SAR, studies are the rationalization of the activities observed in a specific class of compounds, finding a hypothesis for the molecular mechanisms and design of new molecular structures with a more specific and enhanced activity. One way to achieve these goals can be through the development of computational methodologies based on the combined use of molecular modeling and chemometrics. Molecular modeling provides considerable molecular information on the conformational and stereoelectronic properties of the molecular systems under study. Chemometrics is an efficient method to condense useful information form large data sets and to obtain reliable predictive models in classification or regression problems.

The activity of a drug can be considered as the end result of a series of complex phenomena: the methodologies involved in SAR studies are, generally speaking, focussed on the first steps of the interaction between the ligand, i.e. the drug, and the biological macromolecule, i.e. the receptor. The first step is the recognition process between the two species, and all subsequent processes leading toward the final biological response are generally not taken into account. In other words, SAR studies try to analyze and model the approach of the ligand to the binding site and the formation of the ligand-macromolecule complex. Therefore, the search for relationships between biological activity and molecular structure can be seen as the search for those stereoelectronic properties required for the recognition process to occur. The set of stereoelectronic requirements necessary for a family of compounds to elicit a certain activity is generally defined as the *pharmacophore.*

In most cases, only the structure of compounds characterized by a common biological activity is available, while structural information on the receptor active site is lacking. In these cases the pharmacophore can be determined by means of a comparative analysis of the physico-chemical properties of known ligands. This approach relies on the hypothesis that the recognition process between the ligand and the receptor is based on the spatial distribution of certain properties of the active site being complementary to those of the interacting ligands: the properties common to the ligands would provide the information about the stereoelectronic requirements of the receptor active site. This approach is defined as the *indirect approach* toward the rationalization of structure-activity relationships. Since the early 1980s, this approach has received huge impetus due to the development of both reliable computational methods and hardware technology, especially for powerful graphical representation. Nowadays, molecular modeling can provide many molecular stereoelectronic properties which can be employed as useful descriptors in SAR analysis; thus, the traditional two-dimensional description of molecular systems in SAR studies has now been replaced by a more detailed and appropriate three-dimensional representation offered by molecular modeling. Not only can a large number of stereoelectronic properties be calculated, but also each compound can show a large number of energetically accessible conformations. In QSAR analysis all these conformations should be taken into account, leading to data sets of ever increasing dimensionality. A large data set with several hundreds of objects and which is affected by noise or correlation between variables, requires the use and development, of computational techniques which are able to extract from the data set only the necessary information for solving the SAR problem. Chemometrics fulfills this requirement and is now becoming a common tool in SAR studies.

2.1.2 QSAR Methodology using Molecular Modeling and Chemometrics

The strategies generally adopted with the indirect approach assume, that in the first instance, the stereoelectronic property distributions of the ligands can be approximated by the spatial disposition of functional groups present in the ligands. Based on the hypothesis that ligands interact with the biological target in a specific conformation, usually called the *binding conformation,* the functional groups thought to be involved in the binding must show a common spatial disposition. Thus, the search for the pharmacophore corresponds to the search for geometric similarity among the conformations of different molecules, the similarity being defined through a common geometric disposition of selected functional groups. These groups must be present in all the examined compounds and must be equal or bioisosteric, i.e. they must all behave in the same way during interaction with the binding site. The approach based on the search for geometric similarity is commonly known as the *Active Analog Approach* [2]. This approach leads to a geometric model for the pharmacophore, i.e. the *geometric pharmacophore.*

To obtain quantitative correlations with the activity, a further step involving the calculation of stereoelectronic properties for the binding conformations, and the search for similarities in their spatial distribution is needed. In recent years many approaches have been developed to find correlations between the calculated molecular properties or their distributions and activities: Among these are the classical QSAR studies using molecular or atomic descriptors and CoMFA [3], (see Sec. 2.2) and GOLPE [4], (see Sec. 2.3) and methods that are able to localize points within the spatial distribution of properties which are strongly related to the activity. Comparison of molecular property distributions and the measurement of similarity between distributions can also be achieved by means of similarity indexes such as those developed by Carbò [5], Richards [6] and Sanz [7].

Our laboratories have developed a method for SAR analysis based on the combined use of molecular modeling and chemometrics. Molecular modeling is used to evaluate the minimum energy conformations of each compound from the data set and to calculate the stereoelectronic properties of the binding conformations. Chemometrics techniques are used to reduce the number of variables in the data set, select the geometric and stereoelectronic descriptors which contain useful information, and finally, ascertain the best predictive regression models for QSAR.

In our method a search for the geometric pharmacophore, formally derived from the Active Analog Approach, was initially performed by the following steps:

- evaluation of minimun energy conformations for each compound by means of Conformational Analysis;
- optimization of conformational descriptors, i.e. variable reduction by means of Principal Components Analysis (PCA);
- search for geometrical similarities between all accessible conformations to select the binding conformation by means of cluster analysis or classification methods.

Once the binding conformation has been defined the method involves:

- calculation of suitable stereoelectronic properties and their distribution for the binding conformation;

– variable selection to ascertain the best predictive regression model for the activity by means of cross-validated regression techniques.

2.1.2.1 Search for the Geometric Pharmacophore

All the approaches aimed at correlating molecular properties to activity are greatly dependent on the problem of choosing the binding conformation for which the molecular properties need to be evaluated. Therefore, the search for the geometric pharmacophore, i.e. the binding conformation, is a fundamental step in every SAR strategy.

Conformational Analysis

In our method the choice of the most suitable computational method for the conformational analysis, as well as the choice of an efficient method for complete sampling the conformational space of each compound, has assumed a particular relevance.

Due to the dimensions of the molecular systems involved in SAR studies, the use of approximated calculation methods is often required for modeling conformational properties. In general, the results obtained from quantum mechanical semi-empirical methods or molecular mechanics methods are unforeseeablely affected by parameterization. For this reason the reliability of approximated methods which determine conformational and stereoelectronic properties of a considered class of compounds must always be evaluated. In order to ascertain such a reliability, the results from the approximated methods must be compared with the available experimental evidence and/or results from ab initio calculations.

As far as conformational sampling problems are concerned, a systematic search is the only exhaustive method for sampling the conformational space. However, use of this straightforward method is considerably limited for highly flexible molecules due to the rapidly increasing number of relevant degrees of freedom and, as a consequence, of the total number of accessible conformations. To overcome this problem, many efficient methods, which are essentially based on Monte Carlo or Molecular Dynamics techniques, have been developed to sample the conformational space satisfactorily [8].

In most of the cases we analyzed the dimension of the molecular systems and performed conformational analysis by using a systematic search; we adopted the MULTICONF [9] option procedure included in the molecular modeling software MACROMODEL [10]. For each compound sampling was performed by employing a rigid rotor model and systematically increasing the conformationally relevant torsional angles to generate the starting geometries. Then each of these geometries was fully optimized, relaxing all the geometric degrees of freedom. All the stationary points localized on the Potential Energy Surface (PES) were characterized as minima or saddle points by a second derivative analysis. Finally, in the subsequent analysis consideration was given to all the minimum energy conformations within a predefined energy threshold, with respect to the global minimum (usually 6–8 kcal mol^{-1}).

Choice and Optimization of Conformational Descriptors

The pharmacophore model is defined by N atoms selected to represent the spatial distribution of the groups thought to be relevant for activity. Then, the variables used to describe each conformation are the $N(N-1)/2$ interatomic distances defined by these N atoms. The data set is, thus, a matrix containing the conformational minima of all compounds: each row represents a conformation and each column an interatomic distance. The data matrix is then autoscaled and subjected to Principal Component Analysis (PCA) [11]. The aim of PCA is to reduce the number of variables required to describe the system without a significant loss of information and to exclude redundant information. In general, few Principal Components (PCs) account for a high percentage of the total variance of the system. The projection of coefficients of the original variables onto these relevant PCs (loading projections) allows for the recognition of truly independent variables and, thus, an effective variable reduction.

In our case, the loading projections of the interatomic distances onto the relevant *PC*s highlight any correlation between distances and the selection of only those that are truly independent, i.e. the least number of variables which are able to accurately describe the total variance of the data set. Thus, each conformation is then represented by the values of these selected distances.

Selection of the Binding Conformation

During the development of our method two main strategies were adopted to ascertain the geometric pharmacophore. In the first, considerable attention was given to determining the geometric similarity between the conformations, and, thus, cluster analysis techniques were used. The second strategy, on the contrary, was based on the use of classification methods, thus providing, besides the binding conformation, a quantitative model that discriminates between different classes of biological activity.

Cluster Analysis Strategy

Cluster analysis looks for natural groups in data sets: each object within a cluster is more similar to the other objects in that cluster than to any object belonging to other clusters (see Chap. 3.2).

Among the different cluster analysis methods, the non-hierarchical Jarvis-Patrick method [12] seems to provide the most consistent level of predictive performance in classification problems]13]. This method attributes objects to clusters on the basis of common nearest neighbors, *K*. For each object the nearest neighbors, *K*, are evaluated, the *K* value being a parameter fixed by the user. Two objects A and B are then assigned to the same cluster if the following conditions are satisfied: i) A is among the nearest neighbors, *K*, of B; ii) B is among the nearest neighbors, *K*, of A; iii) common neighbors, *R*, are present among the nearest neighbors, *K*, of both A and B. The clustering procedure is dependent on the parameter *R*: increasing *R*, i.e. increasing the selectivity of the clustering procedure increases the total cluster number.

To select meaningful clusters, i.e. clusters that can be considered candidates as the geometric pharmacophore, the simultaneous presence of conformations of the maximum number of active compounds and the minimum number of poorly active and inactive compounds, is adopted as the leading criterion. Finally, it is possible to find one cluster in which the centroid corresponds to an area in the descriptor space where the highest "concentration" of conformations of active compounds and the lowest "concentration" of conformations of poorly or inactive compounds are present. This cluster is called the "active" cluster and its centroid defines the best geometric model for activity. Moreover, the lowest energy conformation of each compound present in the active cluster is assumed to be the binding conformation for that compound.

Classification Strategy

When different molecules are assigned to classes of different activity, classification methods can be employed to model the biological response. A necessary condition for the use of classification methods is the presence of at least two classes of activity: active and inactive. Each compound is defined as active or inactive on the basis of its measured activity value; then each conformation is initially assigned to the activity class of the corresponding compound. For instance, when two classes are present, in the initial estimate all conformations of the active compounds are assigned to the active class and all conformations of the inactive compounds to the inactive class. Starting from this initial estimate, a classification method is applied recursively, which finally results in the determination of one or more subsets of geometric pharmacophore models.

Among the classification methods available, we used Regularized Discriminant Analysis, RDA [14], and Linear Discriminant Classification Tree, LDCT [15].

RDA is used in an iterative procedure: at each step, the classification method is used to relocate the conformations in the class calculated by the discriminant function, i.e. misclassified conformations are moved to the corresponding calculated classes. The iterative process is stopped when the classification of the active conformations no longer changes in two successive steps. At the end of the process, the conformations assigned to the active class are subjected to cluster analysis to test the presence of more than one group of similar conformations, i.e. the presence of different models for the geometric pharmacophore within the active class.

LDCT consists in the combined use of Linear Discriminant Analysis, LDA, [16, 17] see Volume 2, and tree classification methods [18, 19]. The tree structure is composed of nodes and leaves, the starting node being the initial estimate. At each step, two new nodes can be generated from each node by binary splitting. When a node is no longer split, it becomes a leaf. At each splitting in our procedure, the conformations present in the node are partioned into two groups and classified using LDA, obtaining two new nodes. During this process, a node containing conformations of only one class, or a percentage of conformations of a particular class which is greater than a threshold value, is no longer split, i.e. it becomes a leaf: the conformations included in the leaf are all assigned to that class. At the end of the classification process, the obtained tree structure is validated by a leave-one-out cross-validation technique [20]. The final tree contains active, inactive and "fuzzy" leaves. Each leaf of the active class represents a model for the geometric pharmacophore.

2.1.2.2 Quantitative Correlation between Molecular Properties and Activity

Based on the hypothesis that ligands with a similar spatial distribution of stereoelectronic properties interact in the same way with the active site of the receptor in the recognition process, once the binding conformation has been defined, properties thought to be connected to the activity are calculated for this conformation.

Quantum mechanical calculations provide a considerable number of molecular descriptors that are useful for modeling the interaction process. Among these descriptors, global properties such as the dipole moment, HOMO and LUMO energies, polarizability and shape and volume parameters can be considered. Moreover, local properties such as electron density and properties derived from the latter (point charges, Molecular Electrostatic Potential, (*MEP*), Molecular Electrostatic Field (*MEF*)), can be considered. Among these local descriptors, particular attention has been given to *MEP*, the most effective descriptor of the long range intermolecular interactions involved in ligand-receptor recognition processes [21].

Molecular Property Calculations

As discussed above for conformational analysis, the stereoelectronic property calculations also require particular care in the choice of computational method. It must be emphasized that methods that give good molecular geometries, one cannot necessarily accurately calculate properties relating to the electron distribution. Therefore, when approximated calculation methods are necessary due to the molecular dimensions, a preliminary check should be made on their reliability by comparing the results against the available experimental data, or against the results derived from ab initio calculations at the highest possible level.

An analysis of the distribution of the local properties of the ligand can be performed by analyzing either the isovalue surfaces surrounding the molecules or two-dimensional maps, particularly when significant planes can be detected in the molecules.

When local property distributions are compared, a large number of variables must be handled: in fact, property values calculated for all points in the selected space surrounding the molecule constitute the set of variables. A variable reduction can be performed by selecting a few relevant points. In the case of *MEP* analysis, the minima are usually considered because of their physical significance and they represent the preferred molecular sites for electrophilic attack. In other cases, the information retained in the overall *MEP* distribution can be summarized by means of a few points representative of the topological characteristics of isopotential surfaces.

Optimized Regression Models

Once the molecular descriptors, which are useful for the QSAR analysis have been obtained, the variables representing redundant or useless information must be identified and discarded in order to obtain acceptable models: models with a high degree of correlation are expressed by good values of the cross-validated correlation coefficient, r^2_{cv}. A straightforward approach to variable selection can be performed by means of a systematic search for all possible models that can be derived from any possible combination of the variables. Unfortunately, this approach is not always feasible in practice, even with present day computational power. Different alternative strategies have been developed for variable selection: forward/backward methods [22], methods based on the use of regression techniques combined either with experimental design [4] or genetic algorithms [23].

To perform variable selection when a systematic search is not feasible, we adopted an iterative procedure based on the r^2_{cv} value calculated by the Partial Least Squares (PLS) method [24, 25]. At each step, variables representing either useless information or noise can be assessed by the value of their standardized regression coefficients when autoscaled variables are used. In other words, standardized regression coefficient values can be regarded as the weight of each variable in the model; thus, the elimination of the variable with the lowest coefficient improves the model. This improvement is quantitatively estimated by means of cross-validation techniques. In fact, elimination of these variables causes an increase in the r^2_{cv} value up to a maximum; after this point any further elimination causes a decrease in r^2_{cv} value and the iterative procedure is stopped.

2.1.2.3 Computer Programs

The molecular modeling calculations were performed using both ab initio and semi-empirical quantum mechanical methods as well as molecular mechanics methods. The following programs were used: Gaussian 90 [26] for the ab initio calculations; MOPAC [27] for the semi-empirical AM1 [28] and MNDO [29] calculations; MACROMODEL [10] and SYBYL [30] for the MM2 [31], AMBER [32] and OPLS [33] molecular mechanics calculations.

Chemometrics calculations were performed by the SCAN [34] program which contains the PCA, clustering, classification and regression methods employed. The LDCT calculations were performed by a non-commercial programm which is available upon request.

2.1.3 Illustrative Examples

To highlight different aspects of the proposed method, the results for five examples are shown, in which the search for the geometric pharmacophore and analysis of the stereoelectronic properties were undertaken.

The studies concerning a series of amnesia-reversal compounds [35] and a series of Angiotensin II receptor antagonists [36, 37] illustrate these two strategies which are based on cluster analysis and classification methods adopted in the search for the geometric pharmacophore. In the case of the HMG-CoA reductase inhibitors, [38–40] it has been shown that geometric similarity alone cannot always explain biological behavior and the stereoelectronic properties must be included in the model for a more appropriate representation of the recognition process. Finally, in the cases of antagonists at the 5-HT_3 receptor [41] and of Polychlorinated dibenzo-*p*-dioxins, PCDDs [42], the situation where the geometric approach is not fruitful is illustrated. Despite the fact that 5-HT_3 antagonists fulfill the geometric requirements, the latter exhibit different activities; on the other hand, PCDD isomers do not have conformational flexibility and exhibit quite different toxic effects. In the latter two examples our strategy was to perform the variable selection and to obtain the best predictive regression model.

2.1.3.1 Amnesia-Reversal Compounds

Primary degenerative dementia, also called Alzheimer's disease, is a clinical syndrome involving reduced intellectual functioning with impairment of memory, language and cognition. Research on new molecules, which are able to improve impaired cognitive functions, has led to the development of new compounds showing cognition-activating properties.

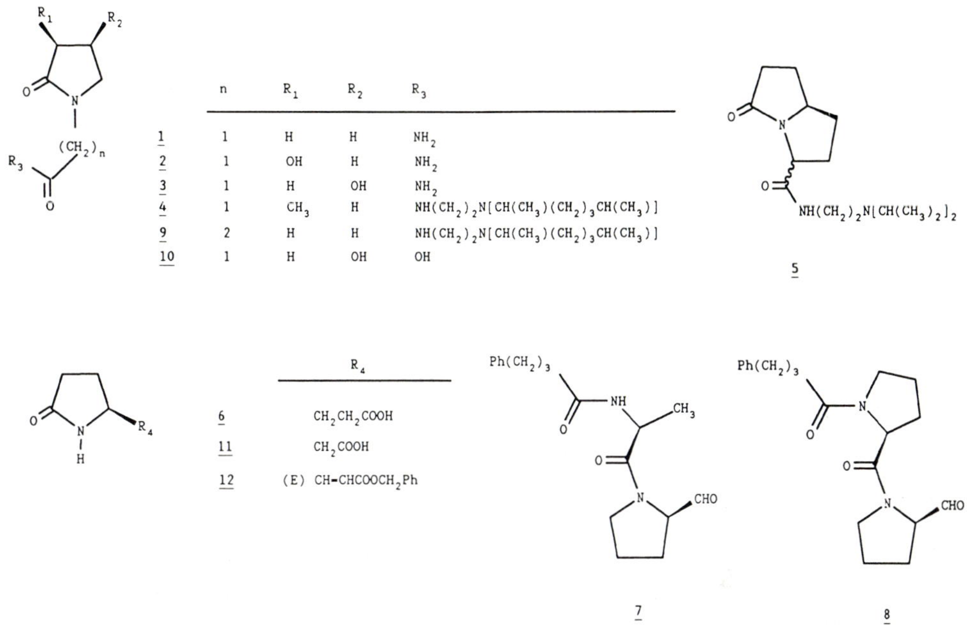

Figure 1. Investigated amnesia-reversal compounds. (Reprinted with permission from Ref. 35, copyright 1990, VCH).

This study [35] has analyzed the conformational features of twelve known amnesia-reversal compounds (Fig. 1), searching for common structural features which can explain the observed cognition-activating properties. Compounds **1**–**6** present a similar anti-amnesic effect; compounds **7** and **8** are the most active compounds in the series and compounds **9**–**12** are inactive.

The conformational analysis was carried out using the MM2 force-field, selected on the basis of an earlier study [43]: a comparison made with ab initio calculations has shown that this method is more reliable in the study of conformational properties of compounds containing a pyrrolidinonic ring than the AMBER and OPLS force-fields, or the quantum mechanical semi-empirical AM1 method.

A systematic search for minimum energy conformations was performed for each compound; all minimum energy conformations were retained and a total of 382 conformations was found.

All the compounds considered exhibit two polar functional groups, the N–C = O amide group and the X–C = O group, with X = O, N. These polar groups, which are presumed to be involved in the biological interaction, define our pharmacophore. In principle, the definition of the relative spatial disposition of the six atoms of the pharmacophore requires 15 interatomic distances (Fig. 2a), however, in the present case, six of the distances are almost constant for all the compounds as they are con-

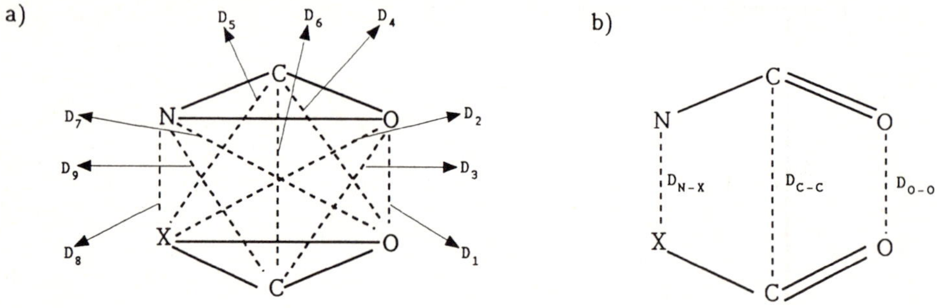

Figure 2. (a) Labels of the nine variable distances (dashed lines) of the pharmacophore; solid lines show the six distances constrained by bond lengths or angles. (b) Selected interatomic distances as conformational descriptors. (Reprinted with permission from Ref. 35, copyright 1990, VCH).

strained by bond lengths or bond angles. Thus, in the first instance, nine interatomic distances ($D_1 - D_9$) were thought to be necessary to completely describe our system.

These nine distances were evaluated for all the conformational minima; the correlation matrix of their autoscaled values was subjected to PCA. Two principal components, labeled $PC1$ and $PC2$, accounted for about 96% of the variance of the data set. The loading projections (Fig. 3) of the nine original variables onto the $PC1$ and $PC2$ plane show three groups in which the original variables present similar loading values; the distances between one atom of the N−C = O moiety and the three atoms of the X−C = O moiety are present in each group. Therefore, to represent the total geometric variability of the system, three interatomic distances, one for each group, suffice, thus, reducing a 9-dimensional problem to a 3-dimensional one. The three distances reported in Fig. 2b and labeled D_{O-O}, D_{C-C} and D_{N-X} were chosen.

In order to determine the binding conformation, the Jarvis-Patrick cluster analysis was performed on the conformational minima of compounds **1–12**, using the three selected interatomic distances to define the multivariate pattern space for clustering. In order to select meaningful clusters, the simultaneous presence of all the active compounds (**1–8**) was adopted as the leading criterion. Results (Table 1) show that only two clusters, referred to as A and B, fulfill our criteria. Therefore, by means of

Table 1. Amnesia reversal compounds. Total conformations included (Conf.), molecules in the cluster (labeled Y), centroid values (Å) and cluster Standard Deviation, *SD*, (Å) for the two selected clusters A and B.

	Conf.	Molecules												D_{O-O}	D_{C-C}	D_{N-X}	*SD*
		1	**2**	**3**	**4**	**5**	**6**	**7**	**8**	**9**	**10**	**11**	**12**				
A	42	Y	Y	Y	Y	Y	Y	Y	Y	–	Y	–	–	3.57	3.14	3.07	0.28
B	24	Y	Y	Y	Y	Y	Y	Y	Y	–	–	–	–	3.65	3.24	3.75	0.19

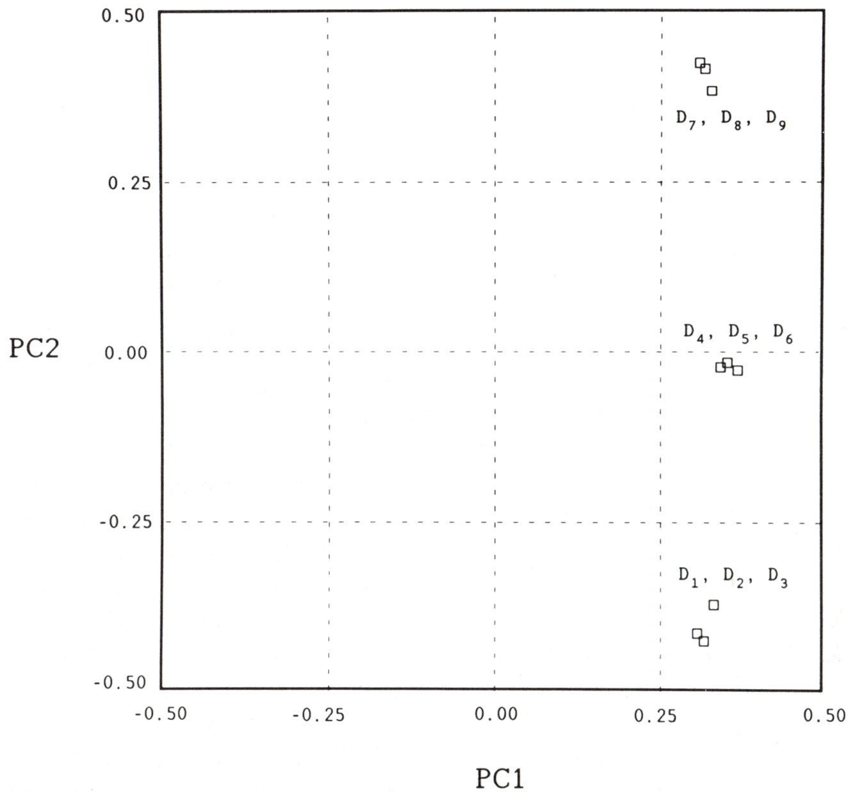

Figure 3. Loading projections of the nine original variables onto the first two principal components *PC*1 and *PC*2 plane. (Reprinted with permission from Ref. 35, copyright 1990, VCH).

cluster analysis two sets of values for the interatomic distances were selected, which defined two possible pharmacophore models for the group of compounds examined. Fig. 4 shows the superimposed conformations of compounds **5** and **6** present in the two clusters A and B with respect to compound **8**.

The study of amnesia-reversal compounds provides an illustrative example of applying certain aspects of the method relating to the search for the geometric pharmacophore: in this class of compounds, a geometric model appears to suffice in rationalizing the observed activities.

2.1.3.2 Non-Peptide Angiotensin II Receptor Antagonists

Linear octapeptide Angiotensin II, AII, is a powerful endogenous vasosuppressor. Antagonists to the AII receptor have been shown to be effective in treating human hypertension.

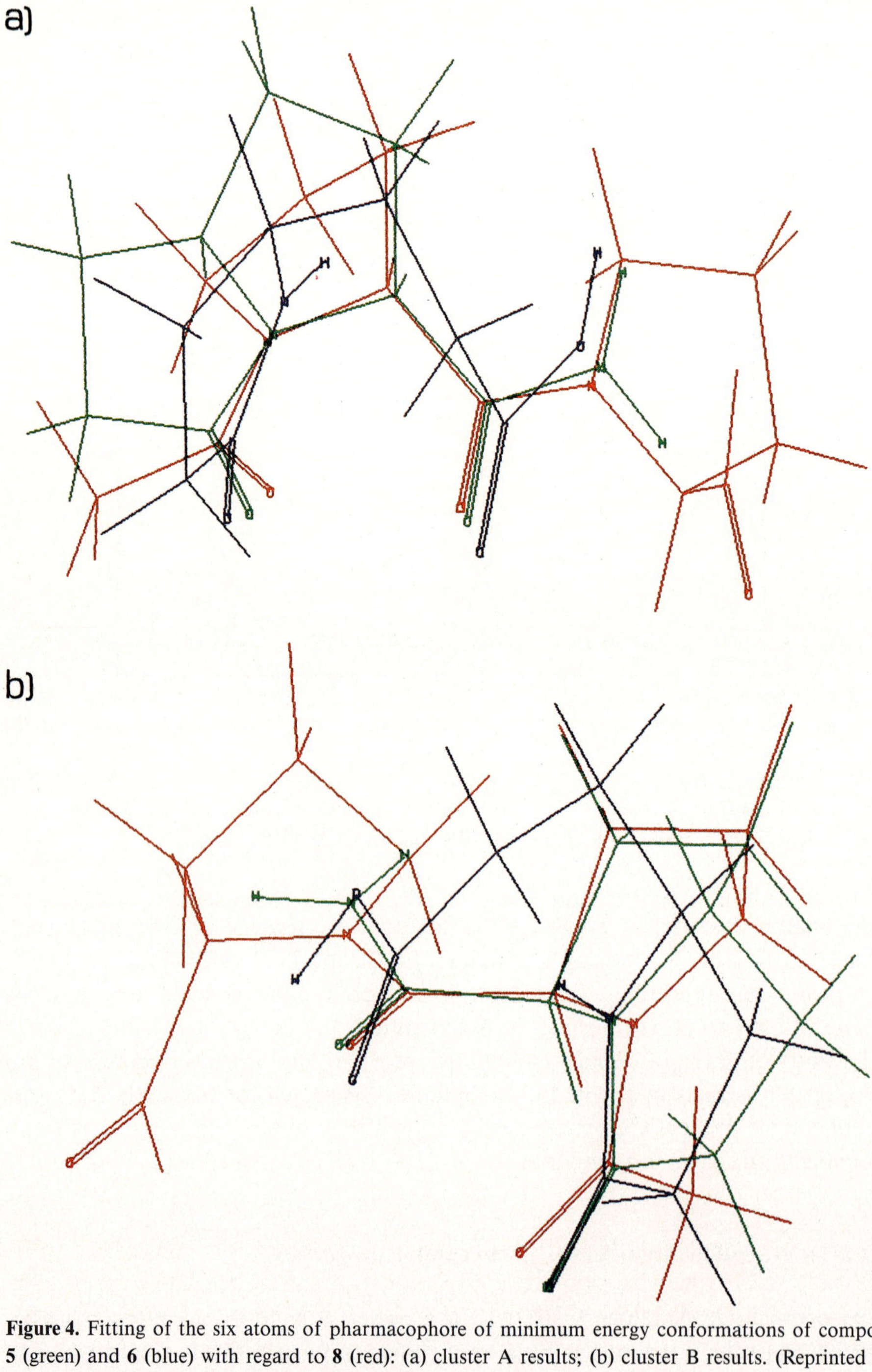

Figure 4. Fitting of the six atoms of pharmacophore of minimum energy conformations of compound **5** (green) and **6** (blue) with regard to **8** (red): (a) cluster A results; (b) cluster B results. (Reprinted with permission from Ref. 35, copyright 1990, VCH).

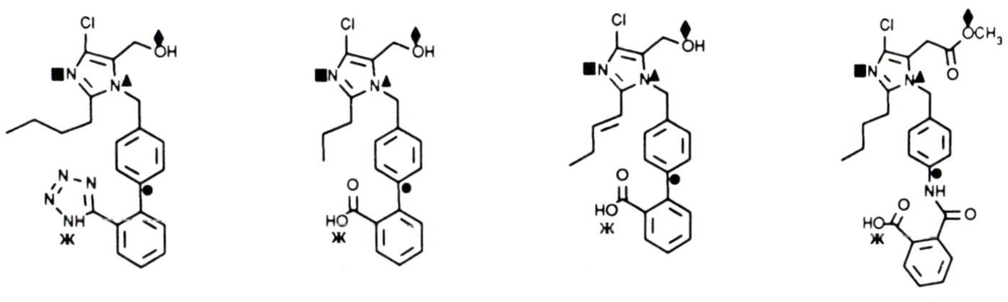

A1 (0.019) A2 (0.16) A3 (0.08) A4 (0.14)

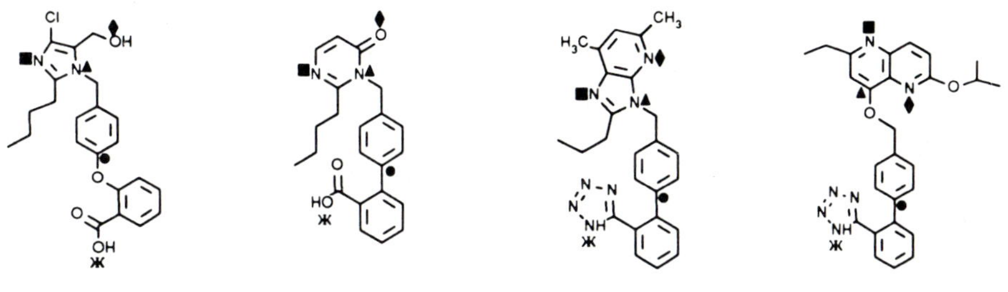

A5 (0.4) A6 (< 1) A7 (0.0003) A8 (0.006)

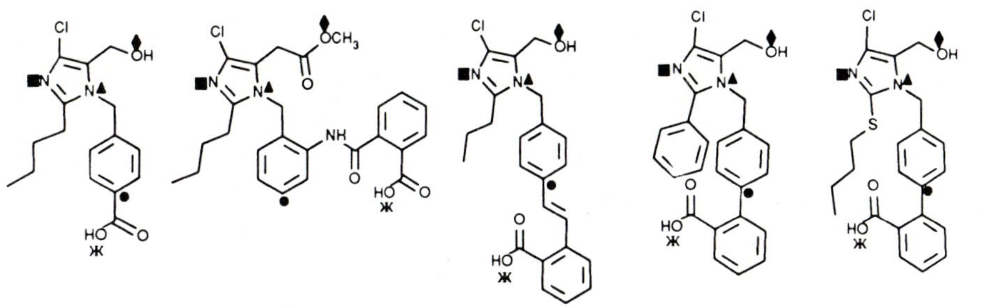

I1 (1.7) I2 (100) I3 (> 5) I4 (24) I5 (> 5)

Figure 5. Structure and binding affinity (IC_{50}, μM) of the investigated non-peptide AII receptor antagonists. Labeled atoms are considered in the pharmacophore definition. (Reprinted with permission from Ref. 37, copyright 1993, ESCOM).

After a preliminary study [36], 8 active and 5 inactive non-peptide (Fig. 5) were examined [37]. For each compound, a random search for the minimum energy conformations was performed; a large number of minimum energy conformations, within 8 kcal mol^{-1} with respect to each global minimum, was found for a total of approximately 9000 conformations. Each conformation was described by ten interatomic distances defining the relative spatial disposition of the key structural elements which constitute the pharmacophore [37].

To reduce the total number of conformations, cluster analysis employing the hierarchical Unweighted Averaged Linkage method [13], was performed separately on the conformations of each molecule described by the ten selected distances. For each compound, cluster analysis highlights a certain number of clusters whose centrotypes are conformations representative of the total accessible conformational space; as a result, a total number of 734 conformations were retained. PCA was then performed on these conformations, allowing for the recognition of correlated distances and the reduction of the number of variables from ten to eight. Thus, the use of cluster analysis and PCA enabled a reduction in the complexity of the original data set in terms of samples and variables, with only a slight loss of useful information.

LDCT was then used to model the biological activity of the AII receptor antagonists. In the initial estimate all the conformations (449) of active compounds were assigned to the active class and all the conformations (285) of inactive compounds to the inactive class. As a result of the final cross-validation process, LDCT provides 4 active, 1 inactive and 2 fuzzy leaves (Fig. 6) and thus, four geometrical models for the pharmacophore were obtained.

Of the four active leaves, only one leaf contains only active conformations with no inactive conformations present. Nevertheless, none of the four pharmacophore

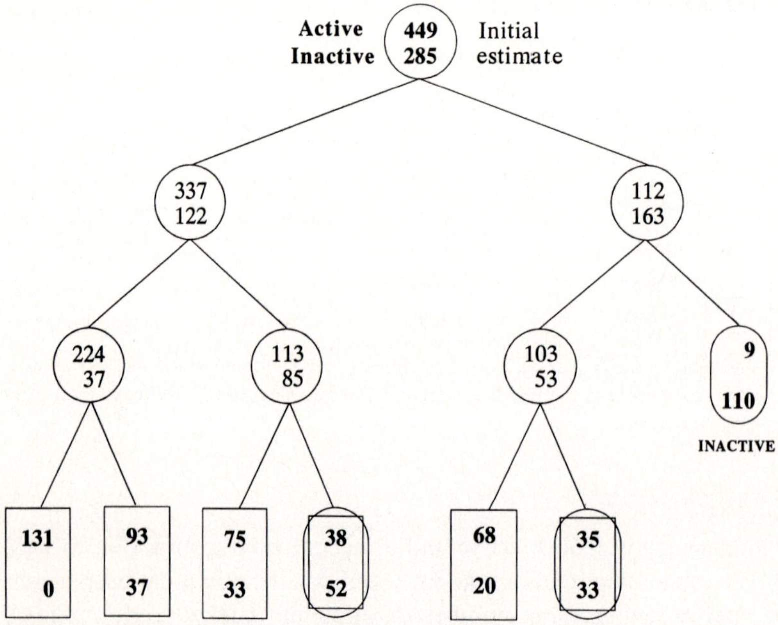

Figure 6. LDCT models for AII receptor antagonists.

geometric models contains conformations of all the active compounds, thus, it is difficult to suggest a unique set of distances defining the geometric pharmacophore, and an extension of the indirect approach to other molecular descriptors seems necessary to better rationalize the activity of this set of AII receptor antagonists.

2.1.3.3 HMG-CoA Reductase Inhibitors

The biosynthetic pathway for cholesterol involves more than 25 different enzymes and the major rate-limiting step in this pathway is regulated by the 3-hydroxy-3-methylglutaryl Coenzyme A (HMG-CoA) reductase, the enzyme that catalyzes the conversion of HMG-CoA to mevalonic acid.

In this work [38–40], eleven inhibitors (Fig. 7), classified as active (**1–5**), poorly active (**6–8**) and inactive (**9–11**) according to the literature activity data, were examined. Conformational analysis of compounds **1–11** was undertaken using the MM2 force field; its reliability in determining molecular geometries and conformational relative energies was first confirmed using model compounds and comparing the results with semi-empirical and ab initio calculations. For each compound a systematic search for minimum energy conformations was performed: all minimum energy conformations within 6 kcal mol^{-1} above each global minimum were retained furnishing a total of 432 conformations.

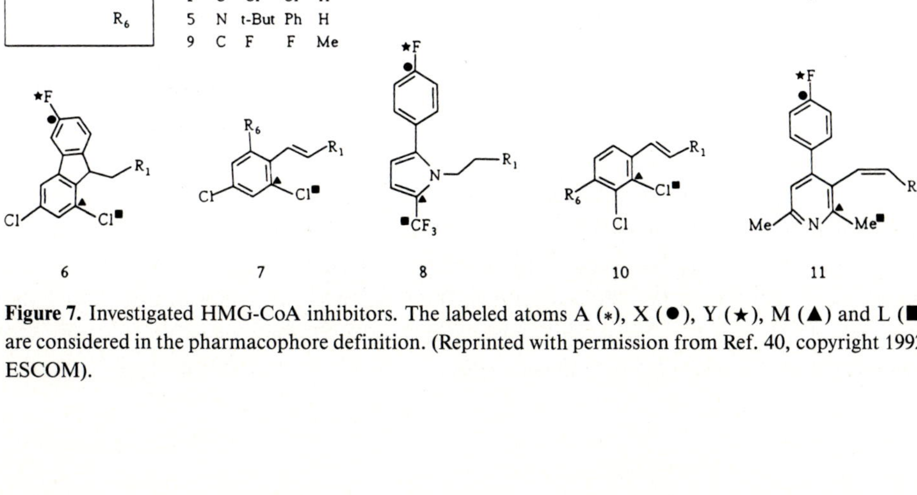

Figure 7. Investigated HMG-CoA inhibitors. The labeled atoms A (∗), X (●), Y (★), M (▲) and L (■) are considered in the pharmacophore definition. (Reprinted with permission from Ref. 40, copyright 1992, ESCOM).

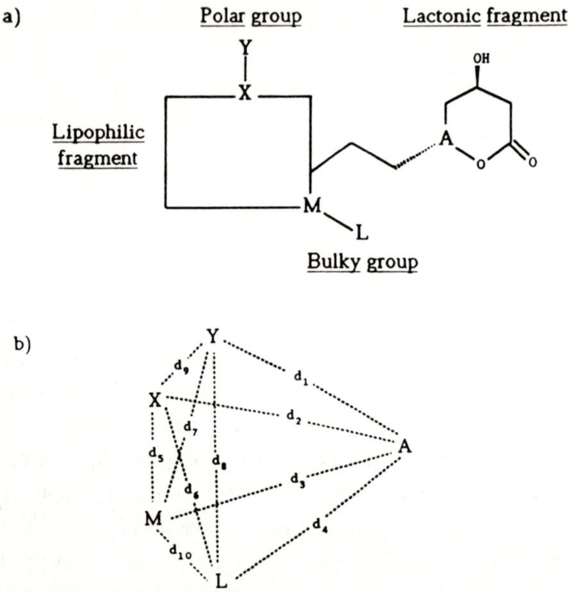

Figure 8. a) Main structural features connected with the activity of HMG-CoA inhibitors and atoms included in the pharmacophore. b) Interatomic distances considered as conformational descriptors. (Reprinted with permission from Ref. 40, copyright 1992, ESCOM).

The knowledge to date of the properties in this class of compounds prompted us to select atoms A, M and X, L and Y (Fig. 8a) as representative of the spatial disposition of the lactonic (A), lipophilic (M and X), bulky (L) and polar (Y) groups respectively, i.e. of the main structural features relating to the activity. The ten interatomic distances defined by the five atoms (Fig. 8b) were initially considered as conformational descriptors and were evaluated for all the conformational minima. The correlation matrix of their autoscaled values was subjected to PCA. The first three components account for approximately 85% of the total variance of the data set. From the loading projections of the ten original variables, three main groups, in which the original variables contained the same information, could be highlighted. Thus, only three interatomic distances, one from each group, was sufficient to represent the variability of the system and the distances d_1, d_4 and d_5 were chosen as conformational descriptors.

In the search for the geometric pharmacophore, the conformational minima were subjected to the Jarvis-Patrick cluster analysis (Table 2): only one cluster fulfills the criterion for cluster relevance, i.e. the cluster contains conformations of the maximum number of the active compounds **1**–**5**. Moreover, this cluster contains not only the greatest number of active compounds, but also the lowest number of poorly active and inactive compounds. This cluster represents the best possible solution and was defined as the active cluster. Nevertheless, it should be noted that the active com-

Table 2. HMG-CoA reductase inhibitors. Total number of conformations (Conf.), molecules included (Y), centroid value (Å) and cluster standard deviation, *SD*, (Å) for the selected cluster.

Conf.	Molecules											d_1	d_4	d_5	*SD*
	1	**2**	**3**	**4**	**5**	**6**	**7**	**8**	**9**	**10**	**11**				
64	Y	Y	–	Y	Y	–	–	Y	Y	–	–	6.03	5.14	6.50	0.50

pound **3** is not present in this cluster, while the poorly active **8** and the inactive **9** are in this cluster.

Thus, it seems that, in this case, geometric similarity is an insufficient criterion to rationalize the biological behavior of these compounds, and other properties have to be included in the model for a more appropriate description of the recognition process. For these reasons the *MEP* distributions of the lowest energy conformation for each compound present in the active cluster were calculated from the MNDO semi-empirical wave function in the plane of the lipophilic fragments (Fig. 9). The

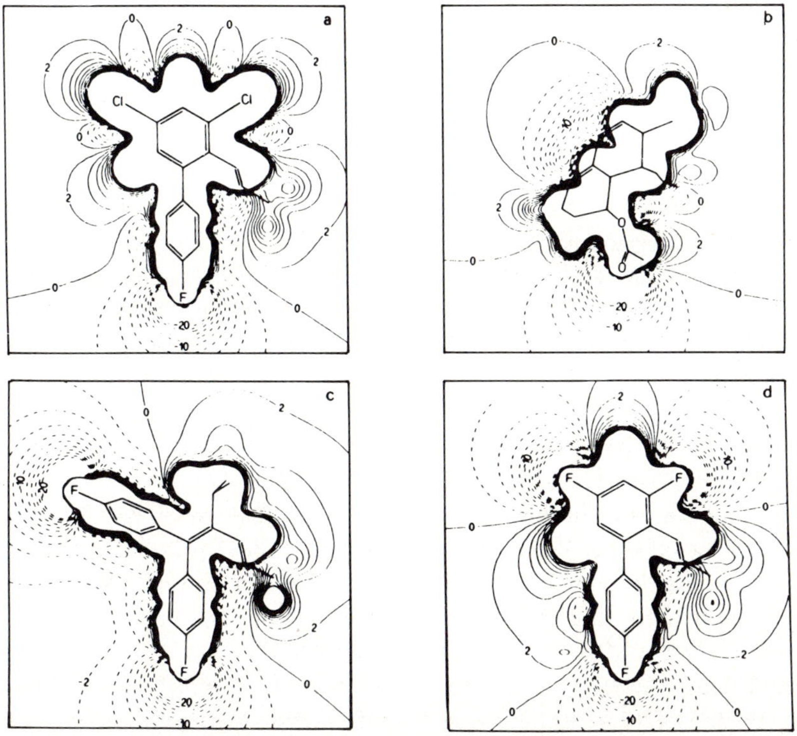

Figure 9. Molecular Electrostatic Potential maps in the lipophilic plane for compounds **1** (a), **(3)** (b), **4** (c), and **9** (d). Solid and dashed lines correspond to positive and negative values, respectively. Isocontour levels every 2 kcal mol^{-1}. (Reprinted with permission from Ref. 40, copyright 1992, ESCOM).

upper right zone of the maps seems to be the most discriminating as regards to activity: in fact, all the active compounds show positive *MEP* values in this zone, while all the poorly active and inactive compounds show negative values.

Indeed, compounds **8** and **9**, although present in the active cluster, have different characteristics in their *MEP* distributions in this zone and thus, can be differentiated from the active compounds. On the other hand, compound **3**, which had no conformations present in the active cluster, shows all the *MEP* distribution characteristics of the active compounds. Moreover, the *MEP* minimum located in the lower zone of the map is lower than the corresponding minima of the other active compounds. The carbonyl moiety present in this compound facilitates a stronger interaction with the secondary binding site, unlike compounds which have a fluorine atom as the polar group. This interaction can partially overcome the geometric differences between the conformations of compound **3** and the centroid of the active cluster.

In conclusion, the case of the HMG-CoA inhibitors presents an interesting SAR problem because both the geometric and electronic properties of these compounds have to be taken into consideration in order to model the activity. The search for geometric similarity, although inadequate, is a necessary step in the SAR analysis, whereby a reasonable model for the binding conformation is formulated. The search for similarities in electronic distribution in the active conformations revealed further analogies between compounds which are useful for rationalizing activity.

2.1.3.4 Antagonists at the 5-HT$_3$ Receptor

In this study ten benzimidazolone derivatives (Fig. 10), antagonists at the 5-HT$_3$ receptor, were investigated [41]. In the last few years, several geometric pharmacophore models for this class of compounds have been proposed [44–46]. All include an aromatic ring, a hydrogen bond acceptor linking group (acyl or heterocyclic) coplanar to the aromatic moiety, and a hydrogen bond donor or a positively charged center, such as an amino nitrogen. A specific spatial disposition of such groups is a requisite for the effectiveness of compounds as antagonists. According to these models, the active conformation consists of acyl groups in an *anti*-periplanar (*app*) conformation.

Interestingly, compounds **1–10** showed a wide spectrum of activity in both in vitro and in vivo tests (Table 3, see p. 30), even though all fulfill the requirements for the geometric pharmacophore model. The aim of this study was to ascertain whether additional descriptors relating to conformational and electronic properties could explain the different degrees of activity of these molecules.

The conformational analysis of compounds **1–10** was performed by the AM1 method, where the reliability in predicting the conformational features of compound **1** had been tested by comparing the results with the available experimental evidence: the X-ray structure [47] and the structural information provided by IR spectroscopy [48].

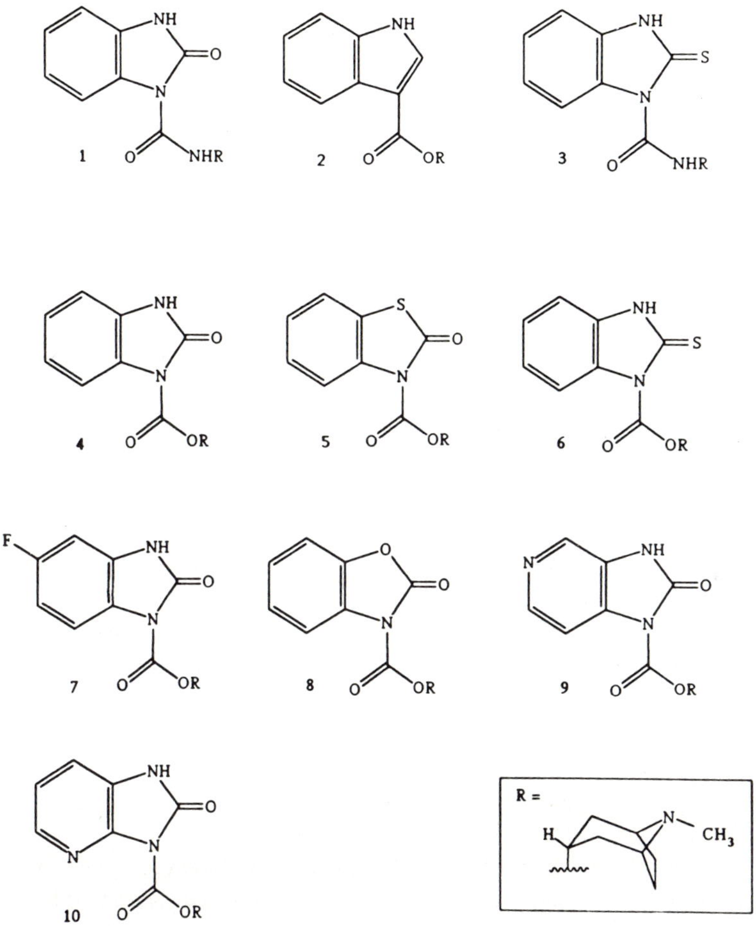

Figure 10. Chemical structure of the considered 5-HT_3 receptor antagonists (Reprinted with permission from Ref. 41, copyright 1993, Elsevier).

All the compounds exhibit two minima in which the endocyclic and exocyclic carbonyl or thiocarbonyl groups are in a periplanar (*pp*) or *anti*-periplanar conformation (*app*), respectively; in all compounds the *app* conformation is the global minimum. On the basis of energy differences ($\Delta E = \Delta H_f^{pp} - \Delta H_f^{app}$) between the relative and global minima, the relative population of the *app* conformation at 37 °C, n_{app}, has been calculated (Table 3).

For the *app* conformation of each compound, i.e. for the binding conformation, the following molecular descriptors were calculated by the AM1 method: atomic charges, dipole moment, HOMO and LUMO energies, polarizability, a shape parameter derived from the moment of inertia [49], the relative population of the *app* con-

Table 3. Biological activities, expressed as inverse of the logarithms of the 50% Inhibitory Dose, pID_{50}, and of the binding affinity constant, pK_d, of the 5HT$_3$ receptor antagonists **1**–**10**. The relative population of the *app* conformation (n_{app}) and *MEP* minimum values (kcal mol^{-1}) above the aromatic ring are reported.

Compd	pID_{50}	pK_d	n_{app}	MEP
1	9.05	9.08	1.000	−19.4
2	8.73	9.10	0.619	−27.8
3	8.07	8.89	1.000	−16.7
4	7.75	8.16	0.934	−17.1
5	7.41	8.32	0.934	−15.7
6	6.91	8.85	0.969	−12.5
7	6.55	8.06	0.902	−9.8
8	6.34	7.77	0.876	−10.6
9	5.96	6.38	0.909	−6.6
10	5.57	6.61	0.742	−6.4

formation, the electrophilic and nucleophilic superdelocalizability indexes, as well as the corresponding frontier superdelocalizability indexes [50]. Moreover, for the *app* conformation of each compound, the *MEP* distribution was calculated at the ab initio SCF-HF level (3–21G basis set) in a 0.3 Å spaced grid-point of the plane 1.70 Å above the molecular plane. All compounds exhibited a negative zone above the aromatic system, with the *MEP* minimum located on the benzene ring. The *MEP* minimum value above the aromatic system was included in the molecular descriptors (Table 3).

The theoretical descriptors considered are not all necessarily suitable for defining the QSAR model. They can be correlated with each other and/or can contain useless information or noise. Thus, a selection of uncorrelated variables related to the in vitro and in vivo activities must be performed. The procedure discussed in Sec. 2.1.2.2 was adopted. The PLS calculations were performed iteratively on the autoscaled variables, discarding the variable with the lowest coefficient value in each step, i.e. the variable containing useless information, until the maximum value of r^2_{cv} was obtained. All the PLS calculations were performed separately on pK_d and on pID_{50} with a "leave-one-out" cross-validation technique.

At the end of the procedure, different acceptable mathematical models were obtained. To select the most significant QSAR models, both physical interpretability and the value of r^2_{cv} were considered. From the possible choices, one model was selected for the in vivo (model **Ia**) and one for the in vitro (model **Ib**) activities (Table 4). Both models contain the *MEP* and the n_{app} variables with very similar values for the regression coefficients: both activities increase when the *MEP* minimum decreases and the *app* population increases. Calculated values vs experimental values of pK_d and pID_{50} obtained with the two models are shown in Figure 11.

Table 4. $5HT_3$ receptor antagonists. Least squares models[a] for pK_d and pID_{50}. Standard errors are reported in parentheses.

	Model **Ia**	Model **Ib**
Activity	pK_d	pID_{50}
intercept	3.301 (±1.382)	1.712 (±0.787)
MEP	−0.135 (±0.026)	−0.180 (±0.015)
n_{app}	3.258 (±1.390)	3.324 (±0.792)
r^2	0.804	0.956
r^2_{cv}	0.678	0.873
SDEC[b]	0.407	0.232
SDEP[c]	0.522	0.395
std. err.	0.487	0.277
$F_{calc.}$	7.379	24.12
degrees of freedom	7	7

[a] In these models the *PLS* and the least squares results are coincident as the number of PLS components equals the number of variables.

[b] $SDEC = \sqrt{\frac{RSS}{n}}$ and *RSS* is the Residual Sum of Squares.

[c] $SDEP = \sqrt{\frac{PRESS}{n}}$ and *PRESS* is the Predicted REsidual Sum of Squares.

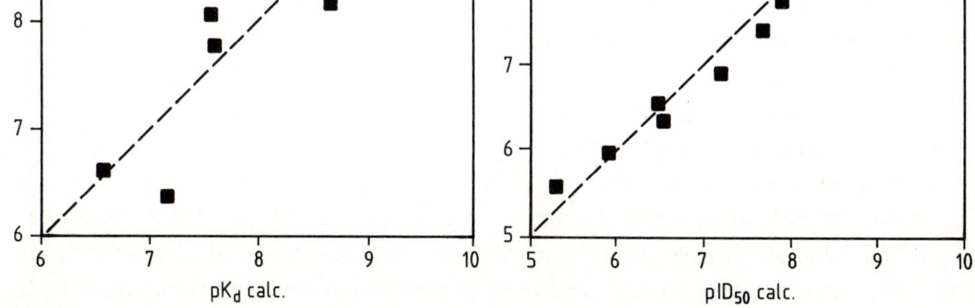

Figure 11. Calculated vs experimental values of pK_d and pID_{50} obtained with models **Ia** and **Ib**. (Reprinted with permission from Ref. 41, copyright 1993, Elsevier).

Thus, the relationship between the n_{app} and the activities quantitatively supports the hypothesis that the two acyl groups must be *anti*-periplanar. On the other hand, the relationship between the *MEP* minima values and the activities suggests that the presence of electron-rich aromatic fragments enhances the activity. As a consequence, the π electron distribution seems to be directly involved in receptor recognition. Moreover, the same model accounts for the activities observed both in vitro and in vivo. In conclusion, the mathematical model obtained provides a straightforward physical interpretation and leads to a better insight into the interaction process at the molecular level.

2.1.3.5 Polychlorinated Dibenzo-*p*-dioxins

Polychlorinated dibenzo-*p*-dioxins (PCDDs) are a group of chemicals that produce a broad pattern of toxic and biological effects, most of which are mediated by binding to the Ah (Aromatic hydrocarbons) receptor. As the molecular structure of the receptor is still unknown indirect approaches that compare ligand properties are needed.

The study of PCDD activity is an example where the geometric pharmacophore is unequivocally defined owing to poor conformational flexibility. Therefore, electronic property distributions must be considered.

We focussed our attention on the first recognition step of the PCDD-Ah receptor interaction, seeking patterns in the *MEP* of PCDDs which could be related to their binding affinities. A series of 14 PCDD isomers, which showed a significant range of binding affinity values, were analyzed and their molecular structures and experimental EC_{50} values are reported in Fig. 12.

On the basis of preliminary studies [51, 52], the *MEP* was obtained by SCF-HF ab initio calculations with the 3−21G basis set. Analysis of the *MEP* distribution [42, 53] in two-dimensional maps for 8 of the considered PCDD isomers showed that the *MEP* minima are more affected by the degree of substitution than by the differences in substitution patterns which have important consequences for biological activity. On the contrary, visual analysis of the *MEP* isopotential surfaces highlighted some electrostatic properties required for high affinity, i.e. a strong concentration of the valence electron charge at both lateral sides of the principal molecular axis and a charge depletion over the oxygen atom region. On this basis, we proposed that the relevant information retained in the overall *MEP* distribution could be summarized by a few descriptors, i.e. the *MEP* values in points properly located around the molecules. Four points were proposed (Fig. 13), α and β, to model the electrostatic accessibility of the central region of the molecule toward an electron-rich site of the receptor, γ and δ, to describe the possibility of a favorable interaction with electrophilic sites in the lateral positions. The linear correlations between the *MEP* values at these points and the negative logarithm of the experimental EC_{50} binding affinities

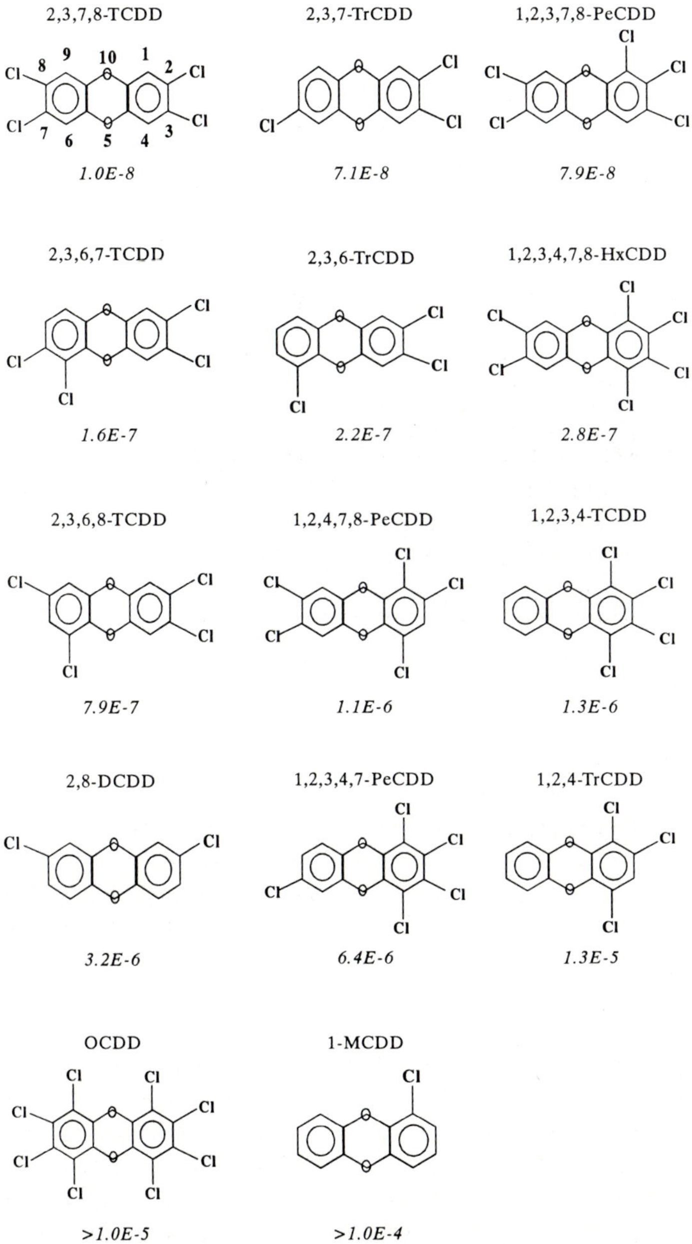

Figure 12. Molecular skeletons of the analyzed PCDD isomers and experimental binding affinity (EC_{50}, M) values.

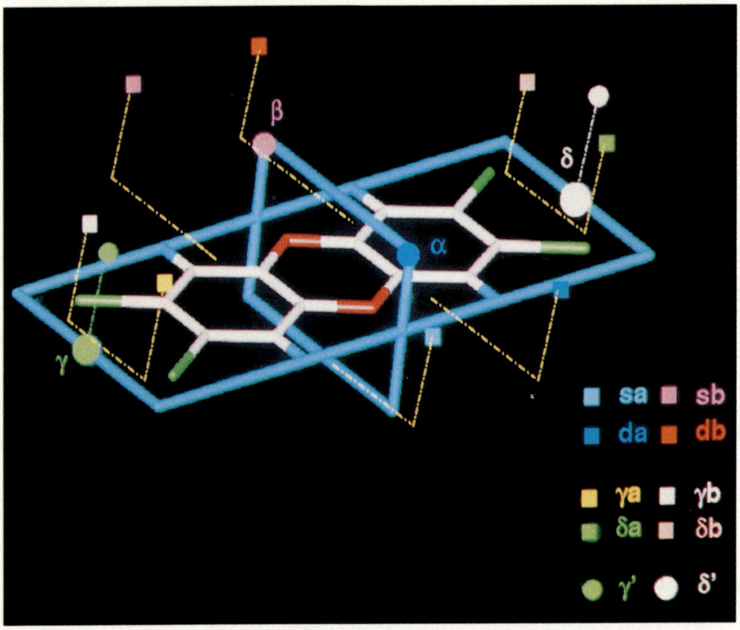

Figure 13. Location of the significant points around 2,3,7,8-TCDD.

(pEC_{50}) were calculated by the linear PLS method and a good correlation was found ($r^2 = 0.981$; $r^2_{cv} = 0.854$) [42].

Our recent results, not yet published, show that the four descriptors do not correlate with the affinities of the whole PCDD series. A PCA analysis has enabled us to differentiate two subsets of isomers on the basis of the values for the four proposed descriptors. All the isomers belonging to the first set (Set 1) are characterized by a typical electronic polarization along the principal molecular axis toward both lateral sides. Set 2 contains less highly chlorinated isomers with an uneven substitution pattern with respect to the sides of the principal molecular axis (2,3,6-TrCDD, 1,2,3,4-TCDD, 1,2,4-TrCDD and 1-MCDD). Only the affinities of isomers belonging to the Set 1 have been well accounted for by the previously proposed model.

To obtain a global model that takes into account the electrostatic characteristics of the whole PCDD series, other points have been added and selected in three-dimensional *MEP* distributions (Fig. 13): γ', δ', γa, γb, δa and δb are useful for improving the description of the lateral zones, *sa, sb, da, db* in order to sample the oxygen atom zone. A variable selection has been performed by means of a systematic search for all the possible models associated with any combinations of these descriptors. The models with the highest degrees of correlation which we obtained are reported in Table 5 alongside the r^2 and r^2_{cv} values. This procedure indicates that the most significant decriptors of the PCDD binding affinities are the *MEP* values in δ, γ' and δ'. The same procedure, which was performed separately for Sets 1

Table 5. Standardized PLS models for the PCDD binding affinities.

No. of isomers	Variables	PLS model (standardized coefficients)[a]	r^2	r^2_{cv}
14	δ, γ', δ'	$pEC_{50} = 4.09\,MEP(\delta) - 0.96\,MEP(\gamma') - 4.39\,MEP(\delta')$	0.874	0.759
10 (Set 1)	γ', δ'	$pEC_{50} = -0.71\,MEP(\gamma') - 0.76\,MEP(\delta')$	0.950	0.898
4 (Set 2)	δ	$pEC_{50} = 0.99\,MEP(\delta)$	0.991	0.956
	γ	$pEC_{50} = -0.99\,MEP(\gamma)$	0.984	0.906

[a] Models with only one variable have been obtained by the least squares method.

and 2 (Table 5), makes it possible to rationalize the global model on the basis of the characteristics of the two sets. The best models for Set 1 take into account the *MEP* features in both the lateral regions, as summarized by γ' and δ' and high affinity values are related to negative *MEP* values at both these points. The best models for Set 2 are given solely by the δ or γ variable: binding affinity increases as the *MEP* value in δ increases (or as the *MEP* value in γ decreases). A similar trend in affinity with respect to the *MEP* values in γ' and δ', or in δ is maintained in the global model. On this basis, it can be inferred that the potential in the lateral regions is the most relevant as regards to affinity: high affinity is related to high negative *MEP* values in at least one of the lateral zones, or in both, if the molecule is characterized by a balanced electron polarization with respect to the principal molecular axis.

In conclusion, a good quantitative model obtained by using descriptors derived from the *MEP* distributions confirms that the selected points are physical meaningful descriptors of the electrostatic properties recognized by the receptor and provides some insight into the nature of the interaction.

2.1.4 Conclusions

An understanding of complex biological systems and processes at the molecular level requires instruments which are able to give a detailed description of molecular structures and properties, as well as techniques suitable for handling large amounts of information. We have demonstrated that molecular modeling and chemometrics can be the right tools for such purposes. In fact, from our experience, summarized in the reported examples, we infer that methods based on the combined use of molecular modeling and chemometrics can give quantitative and predictive models of activity. It is important to stress that the models obtained are not only mathematically acceptable, but can also allow a straightforward physical interpretation of the ligand-active site interactive process. Thus, a mechanistic hypothesis of the activity can be derived from modeling ligand properties.

Of course, as in every modeling problem, only experimental validation of the hypothesis derived from the models can provide real growth in the knowledge and understanding of the problem.

Acknowledgements

All the reported works were supported financially by the Italian National Research Council (Grants CNR CT92.00040.12 and Progetto Finalizzato Chimica Fine II) and the Italian Ministry of Scientific and Technological Research (Grants MURST 40% and Programma Nazionale di Ricerca per i farmaci CITFI).

References

[1] Weinstein, H., *Computational Simulations of Molecular Structure, Dynamics and Signal Transduction in Biological Systems: Mechanistic Implications For Ecological Physical Chemistry.* In: *Trends in Ecological Physical Chemistry.* Bonati, L., Cosentino, U., Lasagni, M., Moro, G., Pitea, D. and Schiraldi, A., (eds.) Elsevier Science Publishers B.V., Amsterdam, 1993, p 1–16

[2] Marshall, G., *Annu. Rep. Med. Chem.* **15**, 267–276 (1980)

[3] Cramer III, R.D., Patterson, D.E. and Bunce, J.D., *J. Am. Chem. Soc.* **110**, 5959–5967 (1988)

[4] Baroni, M., Costantino, G., Cruciani, G., Riganelli, D., Valigia, R. and Clementi, S., *Quant. Struct. Act. Relat.* **1**, 9–20 (1993)

[5] Carbò, R., Leyda, L. and Arnau, M., *Int. J. Quant. Chem.* **17**, 1185–1189 (1980)

[6] Hodgkin, E.E. and Richards, W.G., *Int. J. Quant. Chem., Quant. Biol. Symp.* **14**, 105–110 (1987)

[7] Manaut, F., Sanz, F., Josè, J. and Milesi, M., *J. Comput.-Aided Mol. Des.* **5**, 371–380 (1991)

[8] Leach, A.R., *A Survey of Methods for Searching the Conformational Space of Small and Medium-Sized Molecules.* In. *Reviews in Computational Chemistry,* Vol. **2**, VCH Publishers, New York, 1991, p 1–55

[9] Lipton, M. and Still, W.C.J., *Comput. Chem.* **9**, 343–355 (1988)

[10] MACROMODEL: Still, W.C., Columbia University, N.Y

[11] Wold, S., Esbensen, K. and Geladi, P., *Chemometrics and Int. Lab. Syst.* **2**, 37–52 (1987)

[12] Jarvis, R.A. and Patrick, E.A., *IEEE Trans. Comput.* **C22**, 1025–1034 (1973)

[13] Willett, P., *Similarity and Clustering in Chemical Information Systems,* Research Studies Press, Letchworth, 1987

[14] Friedman, J.H., *J. Am. Stat. Ass.* **84**, 165–175 (1989)

[15] Todeschini, R. and Marengo, E., *Chemometrics Intell. Lab. Syst.* **16**, 25–35 (1992)

[16] Mardia, K.V., Kent, J.T. and Bibby, J.M., *Multivariate Analysis,* Academic Press, London, 1979

[17] James, M., *Classification Algorithms,* Collins, London, 1985

[18] Breiman, L., Friedman, J.H., Olsen R.A, and Stone, C.J., *Classification and Regression Trees,* Wadsworth & Brooks, Monterey, 1984

[19] Bratko, I. and Lavrac, N., (eds.) *Progress in Machine Learning,* Sigma, Wilmslow, 1987

[20] Efron, B., *The Jackknife, the Bootstrap and other Resampling Plans,* Society for Industrial and Applied Mathematics, Bristol, 1982

[21] Politzer, P. and Truhlar, D.G., (eds) *Chemical Applications of Atomic and Molecular Electrostatic Potential,* Plenum, New York, 1981

[22] Lanteri, S., *Chemometrics Intell. Lab. Syst.* **15**, 159–169 (1992)

[23] Leardi, R., Boggia, R. and Terrile, M., *J. Chemometrics* **6**, 267–281 (1992)

[24] Wold, S., Geladi, P., Esbensen, K. and Oehman, J., *J. Chemometrics* **1**, 41–56 (1987)

[25] Lorber, A., Wangen, L.E. and Kowalski, B.R., *J. Chemometrics* **1**, 19–31 (1987)

[26] *GAUSSIAN 90, Revision H:* Frisch, M.J., Head-Gordon, M., Trucks, G.W., Foresman, J.B., Schlegel, H.B., Raghavachari, K., Robb, M., Binkley, J.S., Gonzalez, C., Defrees, D.J., Fox, D.J., Whiteside, R.A., Seeger, R., Melius, C.F., Baker, J., Martin, R.L., Kahn, L.R., Stewart, J.J.P., Topiol, S. and Pople, J.A., Gaussian, Inc., Pittsburgh PA, 1990

[27] *MOPAC 5.0 ESP:* Merz, K.M. and Besler, B.H., *QCPE Bull.* **10**, 589 (1990)

[28] Dewar, M.J.S., Zoebish, E.G., Healy, E.F. and Stewart, J.J.P., *J. Am. Chem. Soc.* **107**, 3902–3909 (1985)

[29] Dewar, M.J.S. and Thiel, W., *J. Am. Chem. Soc.* **99**, 4899–4907 (1977)

[30] *SYBYL Molecular Modeling Software:* TRIPOS Associates, Inc., St. Louis, MO, USA

[31] Burket, U. and Allinger, N.L., *Molecular Mechanics.* American Chemical Society, Washington, 1982

[32] Weiner, S.J., Kollman, P.A., Nguyen, D.T. and Case, D.A., *J. Comput. Chem.* **7**, 230–252 (1986)

[33] Jorgensen, W.L. and Tirado-Rives, J., *J. Am. Chem. Soc.* **110**, 1657–1666 (1988)

[34] *SCAN*: Todeschini, R., Frank, I.E., Moro., G. and Cosentino, U., JerIl Inc., 790 Esplanada, Stanford, CA, USA

[35] Cosentino, U., Moro, G., Pitea, D., Todeschini, R., Brossa, S., Gualandi, F., Scolastico, C. and Giannessi, F., *Quant. Struct.-Act. Relat.* **9**, 195–201 (1990)

[36] Belvisi, L., Salimbeni, A., Scolastico, C., Todeschini, R. and Vulpetti, A., *Pharm. Pharmacol. Lett.* **1**, 57–60 (1991)

[37] Belvisi, L., Bravi, G., Scolastico, C., Vulpetti, A., Salimbeni, A. and Todeschini, R., *J. Comp.-Aided. Mol. Des.* **8**, 211–220 (1994)

[38] Cosentino, U., Moro, G., Pitea, D., Scolastico, S., Todeschini, R. and Scolastico, C., *Molecular Modeling and Chemometrics for Pharmacophore Identification in a Series of HMG-CoA Reductase Inhibitors.* In: *QSAR: Rational Approaches to the Design of Bioactive Compounds.* Silipo, C. and Vittoria, A., (eds.) Elsevier Science Publishers B.V., Amsterdam, 1991, p 323–326

[39] Cosentino, U., Moro, G. and Pitea, D., *J. Chim. Phys.* **88**, 2639–2644 (1991)

[40] Cosentino, U., Moro, G., Pitea, D., Scolastico, S., Todeschini, R. and Scolastico, C., *J. Comp.-Aided Mol. Des.* **6**, 47–60 (1992)

[41] Cosentino, U., Moro, G., Gil Quintero, M., Giraldo, E., Rizzi, C.A., Schiavi, G.B. and Turconi, M., *J. Mol. Struct. (THEOCHEM)* **286**, 275–291 (1993)

[42] Bonati, L., Fraschini, E., Lasagni, M. and Pitea, D., *J. Mol. Struct. (THEOCHEM)* **303**, 43–54 (1994)

[43] Cosentino, U., Scolastico, C., Moro, G., Morosi, G., and Pitea, D., *J. Mol. Struct. (THEOCHEM)* **201**, 199–212 (1989)

[44] Hibert, M.F., Hoffmann, R., Miller, R.C. and Carr, A.A., *J. Med. Chem.* **33**, 1594–1600 (1990)

[45] Swain, C.J., Baker, R., Kneen, C., Herbert, R., Moseley, J., Saunders, J., Seward, E.M., Stevenson, G.I., Beer, M., Stanton, J., Watling, K. and Ball, R.G., *J. Med. Chem.* **35**, 1019–1031 (1992)

[46] Youssefyeh, R.D., Campbell, H.F., Airey, J.E., Klein, S., Schnapper, M., Powers, M., Woodward, R., Rodriguez, W., Golec, S., Studt, W., Dodson, S.A., Fitzpatrick, L.R., Pendley, C.E. and Martin, G.E., *J. Med. Chem.* **35**, 903–911 (1992)

[47] Collin, S., Gil Quintero, M., Moureau, F., Turconi, M., Giraldo, E., Vercauteren, D.P., Evrard, G. and Durant, F., *Eur. J. Med. Chem.,* in press

[48] Turconi, M., Nicola, M., Gil Quintero, M., Maiocchi, L., Micheletti, R., Giraldo, E. and Donetti, A., *J. Med. Chem.* **33**, 2101–2108 (1990)

[49] Cartier, A. and Rivail, J.L., *Chemometrics and Intell. Lab. Systems* **1**, 335–347 (1987)

[50] Brown, R.E. and Simas, A.M., *Theor. Chim. Acta* **62**, 1–16 (1982)

[51] Bonati, L., Fraschini, E., Lasagni, M. and Pitea, D., *J. Chim. Phys.* **88**, 2631–2638 (1991)

[52] Bonati, L., Cosentino, U., Fraschini, E., Moro, G. and Pitea, D., *J. Comput. Chem.* **13**, 842–850 (1992)

[53] Bonati, L., Fraschini, E. and Pitea, D. On the Relationships Between the Biological Activity and the Molecular Electrostatic Potential Distribution of Tetrachlorodibenzo-*p*-dioxin isomers. In: *Trends in Ecological Physical Chemistry.* Bonati, L., Cosentino, U., Lasagni, M., Moro, G., Pitea, D. and Schiraldi, A., (eds.) Elsevier Science Publishers B.V., Amsterdam 1993, 27–38

2.2 3D QSAR Methods

Andrew M. Davis

2.2.1 Introduction

The measured biological activity of a drug, which is active at a particular receptor, is a highly complex number indeed. It encodes the energetics of drug transport from the site of administration to the target receptor. It encodes for partitioning of the drug from the bulk phase into the receptor. Finally, it encodes the complementarity of three dimensional electrostatic fields, hydrogen bonding, hydrophobicity and shape of drug and receptor. Since the pioneering work of Hansch [1], QSAR methods have been used to "decode" the importance of such interactions to the observed biological activity of drugs binding to their receptors. The success of this approach is obviously dependent on being able to ascertain the physico-chemical properties that could be used to accurately describe these molecular properties.

Traditional QSAR studies have used descriptors based on experimentally derived 1-octanol-water partition coefficients to model the "hydrophobic effect" and Hammett substituent constants to model electronic effects. The influence of molecular shape has always been difficult to describe, and a wide range of descriptors, from simple molecular weights to complex topological indices, have been employed to model steric interactions [2, 3].

In recent years the growth in importance of approaches employing computational chemistry has provided a plethora of molecular and atom-based descriptors that can be and have been employed in QSAR studies. These include descriptors derived from individual atomic partial charges, HOMO/LUMO energies, nucleophilic/electrophilic superdelocalizabilities etc. [4].

In general, these descriptors only describe the magnitude of particular physical properties and not any directional preferences that those properties may have. The CoMFA approach of Cramer, Patterson and Bunce though, looked at molecules in 3-dimensions, from the viewpoint of the "receptor", and described the magnitude and directional preferences of electronic and steric interactions [5, 6]. This technique measured the interaction energies between a small probe atom or group at a series of regular grid positions around and through the series of molecules. The molecules were previously overlaid/aligned to occupy the same position in space. At each grid point, the steric and electrostatic interaction energies between the probe and each molecule in the series are recorded. This set of numbers becomes a new steric and electrostatic descriptor in a QSAR analysis (Fig. 1). Therefore, many hundreds or even thousands of descriptors are generated which can be employed in a QSAR anal-

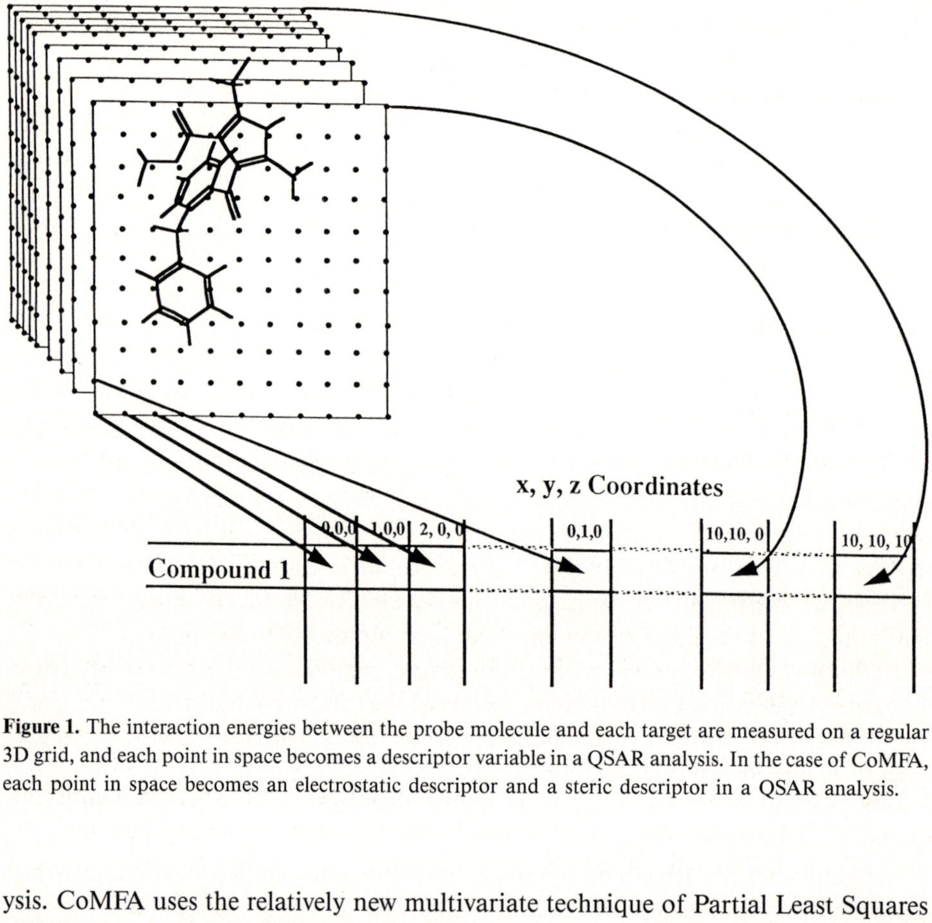

Figure 1. The interaction energies between the probe molecule and each target are measured on a regular 3D grid, and each point in space becomes a descriptor variable in a QSAR analysis. In the case of CoMFA, each point in space becomes an electrostatic descriptor and a steric descriptor in a QSAR analysis.

ysis. CoMFA uses the relatively new multivariate technique of Partial Least Squares (PLS) to ascertain predictive relationships between CoMFA fields and biological activity. The advantage of the CoMFA approach is that the result of the analysis can be mapped back into 3D space. This provides a three dimensional picture of the electrostatic and steric forces which are important for controlling biological activity in the series of molecules under consideration. The sole commercial implementation of the CoMFA procedure is in the SYBYL molecular modeling package [7], but similar types of analyzes can be performed using proprietary "off-the-shelf" packages. We decided to adopt the latter route, as we wanted to learn the advantages and pitfalls of the 3D QSAR method [38].

We employed the GRID force-field [8 – 11] to compute the interaction energy between a series of target molecules and a probe atom or group, over a regular 3D grid both around and through the target molecules. GRID calculates the total energy of interaction, which is the sum of electrostatic, steric and hydrogen bonding terms. The probe can be chosen from a wide choice of predefined probe molecules. The force-field was originally developed to probe the interior of proteins for interaction sites useful for drug design. GRID has been used successfully to predict binding sites of small ligands in proteins [12], and has been extended to evaluate the properties of small molecules, when the receptor is unknown. The table-based statistical package, RS/1 [13], was employed to compile the generated grids into a QSAR table ready for statistical analysis. Statistical analysis was undertaken employing the multivariate technique of partial least squares as implemented in the QSAR package SIMCA [14], and the results were displayed in Chem-X [15].

2.2.2 3D QSAR of a Series of Calcium Channel Agonists

We will illustrate the 3D QSAR method with an analysis of the molecular features which control the observed activity of a set of calcium channel agonists [16]. The modulation of transmembrane calcium movement is an important area of current pharmacological research with applications in many therapeutic areas. The compounds were tested for their ability to increase cardiac contractility. The inotropic potency of the compounds was expressed as the concentration of drug which increased the tension developed to 50% of the isoprenaline maximum in guinea pig atria paced at 1 Hz. The y descriptor used in the QSAR analysis was $-\log EC_{50}$, expressed relative to the standard calcium channel agonist Bay K 8644 (Fig. 2). Early lead optimization in this series was directly guided by a linear regression model [16], which showed the importance of lipophilicity and steric size for the observed activity, and this led to the synthesis of FPL64176, (R = benzyl). Thus, this data set provided a good vehicle for our study of the usefulness of GRID and SIMCA in identifying structural features which can be used in the rational design of new compounds. We hoped that inclusion of compounds synthesized more recently would provide a greater insight into the physico-chemical factors which control activation of the calcium channel by this class of compounds. The data set is shown in Table 1.

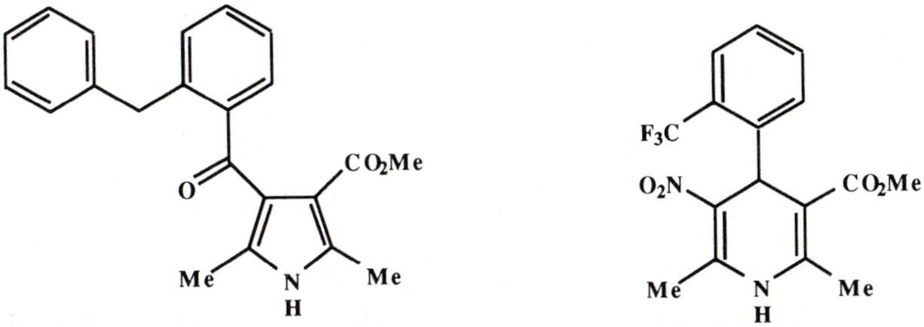

FPL 64176 (13) Bay K 8644

Figure 2. The structures of FPL64176 and Bay K 8644. The inotropic potencies were expressed relative to the standard calcium channel agonist Bay K 8644.

Table 1. CLOGP, CMR and force of contraction, EC_{50}, measured relative to Bay K 8644 for 36 compounds used in the QSAR analysis

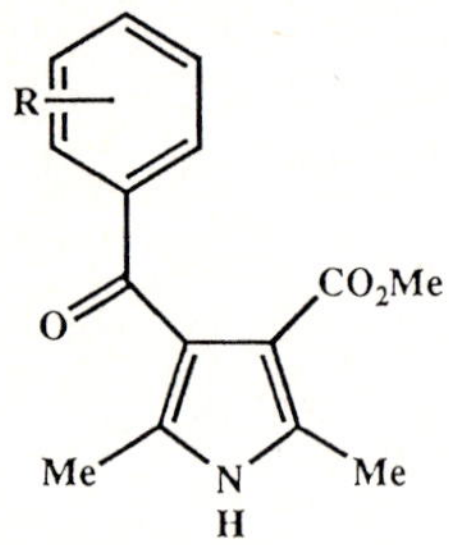

Compound	R	CLOGP	CMR	Relative force EC_{50}
1	2-Cl	2.63	7.58	0.0943
2	2-CF_3	3.09	7.60	0.27
3	2-OCH_3	2.03	7.70	0.0053
4	2-H	2.18	7.08	0.059
5	2-OCO-(2′-OH-C_6H_5)	4.10	10.40	0.34
6	2-CH_3	2.67	7.58	0.14
7	2-F	2.34	7.10	0.0093
8	2,4-Cl_2	3.35	8.07	0.33
9	2-I	3.04	8.39	0.22
10	2-Br	2.78	7.86	0.15
11	2-OCH_2Ph	3.80	10.21	1.13
12	2-Cl,4-NO_2	2.41	8.30	0.16
13	2-CH_2Ph	4.09	10.06	35.5
14	2-Ph	4.06	9.60	0.174
15	2-SCH_2Ph	4.61	10.87	2.89
16	2-$SOCH_2$Ph	2.41	10.90	0.312
17	2-SO_2CH_2Ph	2.16	10.93	0.021
18	2-CH_2CH_2Ph	4.62	10.52	8.00
19	2-CH_3,4-CH_3	3.17	8.01	0.0568
20	2-SPh	4.62	10.40	2.57
21	2-SOPh	2.18	10.44	0.34
22	2-NH-Ph	4.79	9.96	18.91
23	2-CH_2-(4′-NO_2-Ph)	3.84	10.79	4.31
24	2-CH_2-(2′-NO_2-Ph)	3.56	10.78	2.90
25	2-S-(4′-NO_2-Ph)	4.46	11.13	1.24
26	2-O-(4′-NO_2-Ph)	4.13	10.47	0.96
27	2-CH_2-(4′-NH_2-Ph)	2.87	10.43	0.0457
28	2-OSO_2-(4′-Me-Ph)	3.06	10.62	0.0072
29	2-OPh	4.21	9.75	2.90
30	2-NH-pyrid-2-yl	3.94	9.75	7.70
31	2-CH_2-C_6H_{11}	5.32	10.15	27.6
32	2-NH-C_6H_{11}	4.84	10.06	19.8
33	2-Br-4-F	2.93	7.88	0.220
34	2-CH_2-(4′-F-Ph)	4.24	10.08	14.0
35	2-CH_2Ph,4-F	4.25	10.08	19.0
36	2-CH_2(4′-F-Ph),4-F	4.40	10.09	19.0

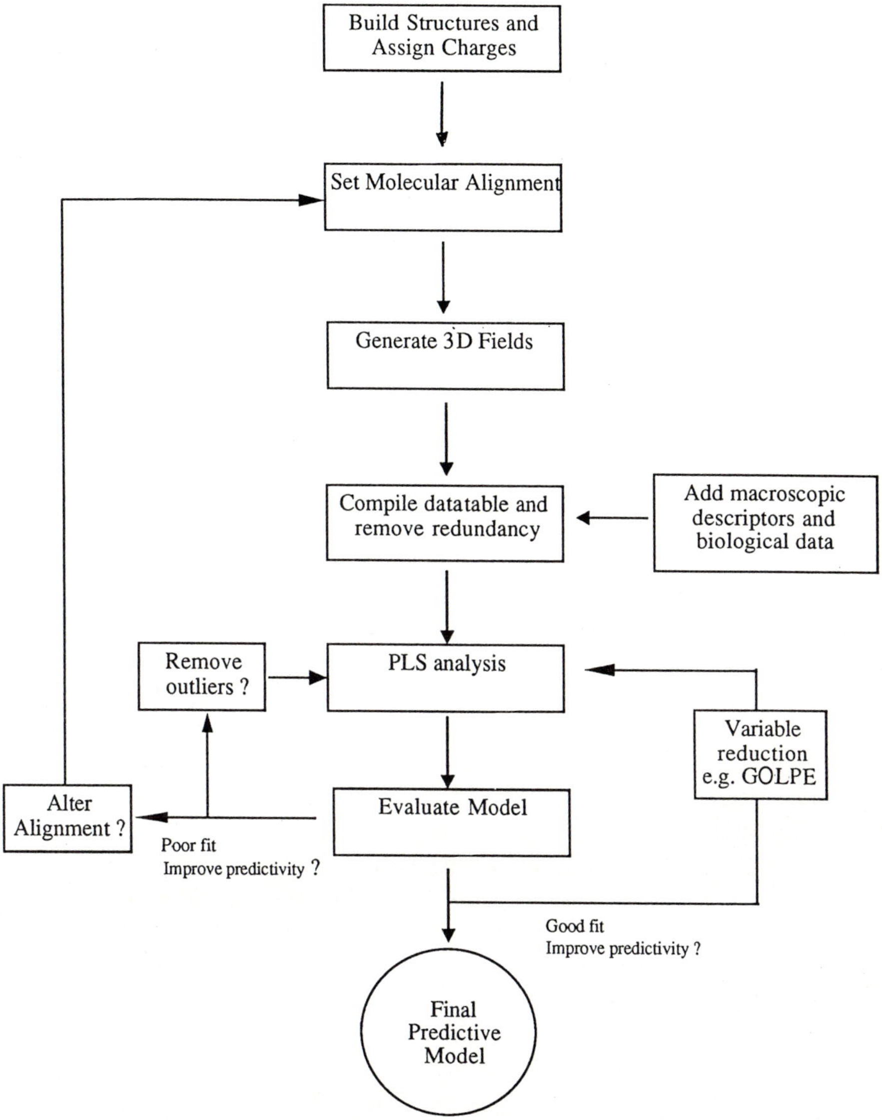

Figure 3. Flow diagram showing the 3D QSAR procedure.

The procedure for a 3D QSAR analysis can be summarized in Fig. 3. Each of the steps in the flow chart is important and can affect the predictability of the resulting model.

2.2.2.1 Molecular Alignment

Molecular alignment is probably the most crucial element of the analysis, as a poor alignment can result in an inadequate statistical model, if none at all. If the set of molecules contains an important binding region that is invariant in that molecular set, then the problem is considerably easier to deal with. The perfect overlay of this would then be the basis of the alignment. An underlying assumption in QSAR analyses, and not just in CoMFA, is that all molecules in the data set showing high activity bind to their receptor in a similar way; inactive or poorly active compounds can be poor for many reasons, and may bind in different ways. Therefore, one should choose one of the most active compounds in the data set to define the molecular alignments. This is termed the "active-analog alignment" and is similar in concept to the molecular modeling procedure often termed the "active analog approach" [17], where the conformation which is adopted by all active compounds is sought after. The underlying assumption again, being that all active compounds must adopt a similar conformation. Which conformation one chooses is, in our view, arbitrary and the phrase "bioactive conformation" is often employed favourably in this context. Usually, this is impossible to deduce, but lack of this information has not hampered 3D QSAR analyses published to date! This is highlighted in a recent study by Klebe and Abraham [18] of inhibitors of Thermolysin and human Rhinovirus 14. Although crystallographic data of the protein-ligand complex provided information on the true binding conformation of the ligand, alignments based on a theoretical binding conformation gave CoMFA models as having equal, or superior predictive power, to those based on experimentally determined binding conformations [18].

In our study, the conformation deduced from the X-ray data for FPL 64176, one of the most active compounds in the data set, was used as the starting point for the construction of 3D structures of the 36 compounds. Substituent variations were built in Chem-X using standard bond lengths and angles. The structures were not fully optimized. Full optimization would have introduced small differences in the bond angles, bond lengths and torsion angles of the common portions of the molecules in the test set, and this would have given rise to "noise" in the GRID analysis. In this case, all the molecules shared a common molecular fragment, a dimethyl substituted pyrrole ring, which is known to be important for binding. Structural variation was introduced on the phenyl ring at the *ortho* position with respect to the linking keto group adjoining the pyrrole ring. Initial molecular alignment involved overlaying the pyrrole ring of each structure, followed by conformational analysis of the side chain. Here we fitted all the side chains to the conformation adopted by the benzyl side chain of FPL64176, since this was a low energy conformation for the compound in question. Since it was not possible to deduce the bioactive conformation of FPL64176, we selected an arbitrary conformation, i.e. the one deduced from the X-ray.

If the molecules in the data set do not contain an obviously similar binding region, then the alignment procedure becomes more difficult. One would still choose one of the most active compounds to set the alignment. Intuitively, one should base the alignments upon similarities in the 3D interaction fields. The CoMFA option in SYBYL contains a "FIELD-FIT" procedure, which does just that, and the program ASP was written to align molecules based upon their electrostatic isopotential/field [19, 20].

Other workers are investigating the use of GRID fields and 3D autocorrelation functions to deduce an alignment [21]. A number of other packages are available to deduce molecular alignment including RECEPTOR [17, 22], DISCO [23] and APEX-3D [24].

It has been suggested that the lack of a good statistical model could be an indicator of a poor initial alignment [22]. Thus, the CoMFA procedure could be employed as an iterative method, trying different alignment procedures until a satisfactory and sensible alignment is obtained. We suggest further that statistical outliers in an analysis might be compounds that adopt a different binding conformation at the receptor binding site.

2.2.2.2 Charges

The quality of the charge scheme chosen can influence the quality of the model derived in a number of ways. One CoMFA study has shown that the use of ab initio charges derived from ab initio wave functions (6–31G*) give an improved cross-validated r^2 compared with Mulliken semi-empirical charges [25]. Also noted was an increased contribution of the electrostatic term to the overall regression. It was suggested by the authors that this could be due to the increased quality of the charges. The correlation of ab initio charges with semi-empirical charges is not clear, and in some cases the charge on a given atom can change sign! The charge scheme should be able to identify inductive and resonance effects of substituent variations. In CoMFA work it is recommended that charges should be calculated by at least one MNDO method [22]. The GRID program assigns its own charges while calculating GRID fields. The GRID-defined atomic charges were obtained from a look-up table, and were assigned according to atom-types. The GRID-defined charges are insensitive to changes in structure in small molecules, e.g. changing a substituent on a phenyl ring does not change the charges on the ring atoms. Therefore, in our work the GRID charges were replaced with MNDO/PM3 Mulliken charges as calculated using MOPAC 5.0. We felt that this level of approximation was an adequate balance between the quality of the charges and the speed of calculation.

2.2.2.3 Generating 3D Fields

The CoMFA option within SYBYL calculates the electrostatic fields and steric fields around the molecules in the data set and treats them separately. The probe most often used in CoMFA is a methyl probe with +1.0 charge. Other probes can be selected. One criticism of the CoMFA approach is that hydrogen bonding cannot be explicitly defined. However, hydrogen bonding is considered to some extent for two probe types, a proton with +1.0 charge identifies hydrogen bonding acceptor groups and a hydroxyl anion with −1.0 charge identifies hydrogen bond donating groups.

GRID calculates fields in a conceptually different manner from CoMFA and calculates the total interaction energy between a probe and the molecules in the data set. This is calculated as the sum of electrostatic, steric, and hydrogen bonding interaction energies at that point in space. The parameterization of the force-field has been made to reproduce experimentally determined binding energies. Therefore, the total interaction energy is an experimentally based balance between steric and electrostatic forces. We viewed this as an advantage of GRID fields. In our view, the use of a separate *x*-block describing electrostatic interaction, because of the scaling that has to be applied compared with the steric block (see later), increases the likelihood of overemphasis of electrostatic forces on any resulting model.

Other types of 3D field can also be used, for example, electrostatic isopotentials, MOPAC-derived HOMO fields [26], and so-called "lipophilicity potentials" [27] such as those generated by HINT [28].

The size of the 3D grid is an important consideration. First, the 3D grid should be large enough to contain all the molecules and extend far enough away from all the molecules in *x*, *y*, and *z* directions to allow the interaction energies to fall away to zero. This ensures fields are not truncated and allows room for larger compounds to be included in future analyses. The spacing of the grid points, if the data is handled appropriately, should not affect the results in an adverse way [29]. A grid spacing of 2 Å is often used in CoMFA analyses. A grid spacing of 1 Å gives a much better defined mapping of the results obtained from the analysis, but one should consider the increase in the number of grid points as this could be problematic. A $20 \times 20 \times 20$ Å grid would generate 1000 grid points at 2 Å spacing, but 8000 at 1 Å spacing. It has been mentioned that the PLS routine in CoMFA appears to be slow (compared to SIMCA or GOLPE) when handling large numbers of GRID columns [30].

For our work an alkyl hydroxyl probe was selected as the probe molecule since this would provide information on electrostatic interactions, hydrogen bond donor and acceptor ability, and steric effects due to its size. We decided that the nature of the probe was unimportant as long as it could interact by all mechanisms. It is possible that a probe also bearing an overall formal charge would place more emphasis on electrostatic interactions, and GRID affords the possibility of defining custom probes, if necessary.

During the GRID calculations the bulk dielectric constant was set to 4.0, representing the estimated dielectric constant of the active site of a receptor. In preliminary work, we used a bulk dielectric of 80.0, but we decided that using the more realistic lower value would give a better representation of hydrophobic effects. Setting the dielectric constant to 4.0 would also increase the contribution of the electrostatic term, and provide a good compromise between electrostatic and steric terms. If the dielectric constant had been set to lower than four, then the electrostatic term would have become a dominating factor. We do not know the effect of changing the dielectric constant in 3D QSAR analysis. As changing the dielectric constant would have resulted in a consistent change across the series, we did not feel it would have a major impact upon the quality of any derived model.

2.2.2.4 Compilation of GRID Maps

In CoMFA, a set of 3D data is represented as only one column in the resulting database table, the number displayed in each cell being roughly proportional to the volume of that compound. The actual energy values at each grid point in space remain hidden in the software.

In our analysis, the GRID maps were compiled into a table in RS/1, each column representing a point in space, and each row a compound in the test set, generating a 36 row by 15625 column table. The *x, y, z* coordinates of the GRID points were written as the column titles of this table. The column titles are the means to collapsing the dimensionality of the GRID block, removing redundant information, and regenerating the original GRID later in the analysis for the display of results. The negative energy values generaly ranged from 0 to −9, but the positive values from 0 to +50.0 (the cut-off value is set by GRID). As extraction of PLS components is scale-dependant, this would unduly bias the analysis towards the steric terms [29]. This is because each column in the PLS analysis is usually centered upon the mean (mean subtracted from every value), and a greater range associated with the repulsive energies would bias the position of centering, and hence, PLS component extraction. We, therefore, scaled all positive energies by 12.5, so they would only cover the range 0 to +4.0. CoMFA also gives the opportunity to change the default cut-off value of the steric repulsive energies, and this is recommended.

To analyze the information content of our RS/1 map data table, a table was constructed showing the distribution of column/GRID point ranges (Table 2). Analysis of this distribution table demonstrated that the compiled GRID map data table contained many GRID points/columns at which the probe showed little or no variation in interaction energy across the set of test compounds. This was because of the following:

- a very large grid was used, therefore, many grid points were so far away from all the molecules that the interaction energy between the probe and all molecules was zero or nearly zero kcal/mol.
- common parts of the molecule provide a constant interaction with the probe.
- as part of the molecular volume is common to the whole set, there are regions of space where the probe is inside the van der Waals surface of the whole set, so interaction energies were constant at +4.0 (after scaling).

Inclusion of these redundant columns would grossly affect the chance of extracting a useful PLS model. A table was constructed in RS/1 that was a subset of the 15625 master table that contained columns/GRID points where the range of energy values ($E_{max} - E_{min}$) was greater than 0.2 kcal/mol, generating a 1842 column table. This cut-off was arbitrary, and we could have equally used a higher cut-off, e.g. in the 0.3 to 0.4 kcal/mol range without losing too much *x*-block information. Thus,

Table 2. Distribution of the interaction energy ranges between the hydroxyl probe and the 36 compounds at each GRID point in space, which are each columns in the compiled RS/1 table.

Range Intervals ($E_{max} - E_{min}$ in each column) kcal/mol	Number of columns/GRID points with range in the interval
0 up to 0.1	12818
0.1 up to 0.2	965
0.2 up to 0.3	196
0.3 up to 0.4	94
0.4 up to 0.5	96
0.5 up to 1.0	427
1.0 up to 1.5	193
1.5 up to 2.0	141
2.0 up to 3.0	165
3.0 up to 4.0	66
4.0 up to 5.0	146
5.0 up to 6.0	172
6.0 up to 7.0	81
7.0 up to 8.0	42
8.0 up to 9.0	16
9.0 up to 10.0	7
Total	15625

only around 10% of the data contained potentially useful information. This column filtering is similar to the MINIMUM_SIGMA option in CoMFA, which ignores columns with a standard deviation smaller than a user defined cut-off.

2.2.2.5 Inclusion of Macroscopic Descriptors with 3D Field Data

We wanted to include the macroscopic descriptors CLOGP and CMR with the GRID information in the analysis. GRID (and CoMFA) only considers the enthalpic component in drug-receptor interactions. GLOGP and CMR are calculations of log *P* (log of the octanol/water position coefficient) and the molecular refractivity respectively, and are obtained from the MEDCHEM software [37]. Molecular refractivity describes molecular volume essentially. The importance of log *P* is that it can be employed to model the large entropic component for drug-receptor interaction. One of the main contributions to the free energy of partitioning of lipophilic compounds from water to a non-polar receptor phase, is the favourable gain in entropy [31]. This arises because a lipophilic solute in water disrupts the random hydrogen bonding network in bulk water, and causes the ordering of water molecules around the van der Waals surface of the solute. On partitioning out of water, the random hydrogen bonding network can reform, which is an entropically favourable process.

CLOGP and CMR were added to the data table with the activity data to generate a 1845 column by 36 row data table for analysis.

2.2.3 Statistical Analysis

The use of so many variables in a regression analysis dictates the use of a multivariate statistical technique, and partial least squares is the technique of choice for 3D QSAR analyses. The partial least squares method identifies summary variables in the x-descriptors that are correlated with y as much as possible. In our analysis of the calcium channel agonist data set, the PLS routine implemented in SIMCA (Version 4.4) was employed. In the version available to us, up to a 60 compound by 5600 variable matrix could be analyzed. After each component was extracted, the significance of that component to the model, and the overall significance of the model was tested by cross-validation. Cross-validation tests the uncertainty in prediction of the derived model. Normally, the model is derived several times, and at each stage groups of compounds are left out. Overall, all compounds are left out only once. Each time the model is used to predict those compounds left out, and the difference between the observed and predicted y is used to generate a predictivity statistic, the prediction of the sum of squared deviation from the correlation, *PRESS*.

$$PRESS = \Sigma i(Yi - yi)^2$$

For each PLS component the *PRESS/SS* is calculated, where *SS* is the residual sum of squares of the previous dimension. When the *PRESS/SS* (total or for any dimension) is smaller than a significance LIMIT (5% level), then the tested dimension is considered significant.

In our analysis of the calcium channel agonist data set, using as default 7 groups with 36 cases approximates to leaving 5 out at a time. The "leaving-groups-out approach" is recommended over the "leave-one-out approach", unless the compounds have been selected for the training set by an experimental design procedure which would ensure no clustering. If the compounds are clustered, leaving out only one at a time can give an over optimistic view of the model predictivity. This is because the model is still rigid enough to give a good prediction for the compound left out [30]. Clustering can be examined by plotting the scores for successive PLS dimensions or the scores from a PCA analysis, against each other in a 2D plot.

The use of a cross-validation technique to test significance of the model has many advantages over using distribution-based tests for significance such as F-tests. Cross-validation always tests the model for predictivity and as we wish to use the model to guide the design of new compounds, then this is preferable. Also the use of cross-validation does not impose any assumptions upon the distribution of errors in the model. Such assumptions may not be valid with this type of data [32]. Most statistical tests of significance assume that errors follow a normal distribution pattern.

PLS analysis is sensitive to the scaling of the x-block descriptors. Because the units of the GRID columns are identical, i.e. kcal/mol, the GRID columns were not scaled. Autoscaling, which sets the variance of each column to unity, would place undue

Table 3. PLS Regression Models for the Full 36 Compound Data-set.

Block Variances	PLS 1[a]	PLS 2	PLS 3	PLS 4	overall r^2
CLOGP = 1.0 act = 1.0	$r^2 = 0.69$				
GRID = 1458 act = 1.0	$r^2 = 0.42$	n/s	n/s	n/s	0.42
GRID = 1458 CLOGP = 1 CMR = 1 act = 1	$r^2 = 0.42$	n/s	n/s	n/s	0.42
GRID = 1 CLOGP = 1 CMR = 1 act = 1	$r^2 = 0.60$	$r^2 = 0.71$	$r^2 = 0.77$ n/s	$r^2 = 0.86$	0.86

n/s not significant by cross validation.
[a] First PLS component

weight on columns containing little variation in interaction energy over the test set of compounds. Autoscaling may have some advantage, if in the data-preprocessing all descriptors which are not relevant for the PLS model were discarded, for instance after GOLPE variable reduction. A consequence of not autoscaling the field data is that the importance of a particular point in space to the overall regression is weighted by the variance of energies observed at that particular point in space.

Inclusion of one whole molecule descriptor such as CLOGP along with hundreds or thousands of columns of GRID information requires careful attention to scaling. Table 3 shows the effect of changing the relative scaling of the variance of the GRID block to CLOGP and CMR column variances.

Inclusion of CLOGP and CMR with the 1842 columns of GRID information without blockscaling has no effect upon the model obtained when compared to the model extracted from just the GRID information alone. Although CLOGP alone describes 60% of the *y*-block variation, without blockscaling, the variable does not contribute significantly to the model. But when the GRID block variance is scaled to give the total variance of all columns as 1.0, the same as the CLOGP column, the overall model now accounts for 86% of the activity data in 4 PLS components. This shows that in this data set where lipophilicity is known to be important in controlling the observed inotropic potency, the best PLS model can only be identified after the appropriate scaling. A similar approach has recently been employed by Silipo [33], McFarland [34], and Hansch [35] for the inclusion of macroscopic descriptors with CoMFA data. The scaled model described the data set more adequately than using either GRID information or the bulk descriptors separately. Kim [36] has demonstrated that, in some cases, lipophilic effects can be parameterized directly from the molecular field of CoMFA. But for this data set where lipophilicity is known to be important in controlling biological activity, the best model can only be extracted by explicitly including the macroscopic descriptors CLOGP and CMR with the GRID data with appropriate block-scaling. Blockscaling is available in CoMFA, it is the default option when 3D field data is included in an analysis. This scaling is applied to the electrostatic block and steric block, and to any macroscopic descriptors included, so each has an equal chance of contributing to the model.

2.2.3.1 Results of the Analysis

The results of the PLS analysis most often used to interpret 3D QSAR analyses, is a regression equation, where biological activity is expressed as the sum of contributions from every variable in the model. The size of the coefficient for each variable underlies it's importance in describing activity, although the original variance of that variable also modulates the coefficient as previously discussed. As each variable represents variation in interaction energies at a defined point in 3D space, the regression

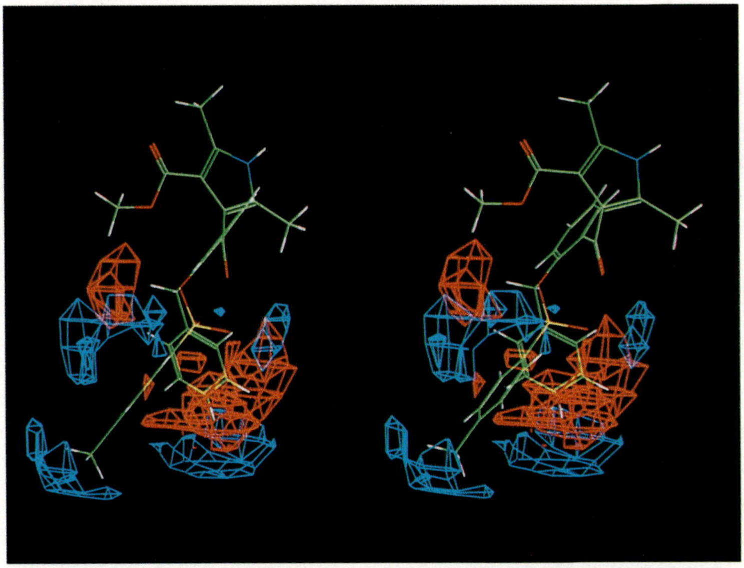

Figure 4. Regression map showing the overall 4 component PLS model for the calcium channel data set. Red regions (positive coefficients) are favourable for steric bulk and unfavourable for electrostatic/hydrogen bonding interactions. Blue regions (negative coefficients) are favourable for electrostatic interactions/hydrogen bonding interactions and unfavourable regions to place steric bulk. The map is displayed over the structures of the benzyl and *p*-toluene sulphonyl substituents, representatives of the most and least active compounds, respectively.

coefficients can be mapped back onto the initial *x, y, z* coordinates of the variables, generating the 3D regression map. In CoMFA, the default display is actually the coefficient multiplied by the standard deviation of energies at that point in space, to overcome the modulating effect of the columns variance on PLS extraction. This, with a plot of *y* predicted vs *y* observed is usually the basis of most CoMFA interpretation published. One usually looks for how compounds showing high and low biological activity, and not outliers on the *y* vs y_{pred} plot, interact with the regression maps. The contribution of a particular grid point to activity is calculated from the product of the regression coefficient and the interaction energy between the probe and target molecule at that point. Interpretation of the electrostatic field regression maps, therefore, requires consideration of the sign of the charge on the probe. So in a CoMFA analysis, taking a methyl probe with +1 charge as an example, the more positive the *y* descriptor, the more active the compounds. Highly active compounds should be those that have:

- a negative charge near regions of negative electrostatic regression mapping, since a negative coefficient multiplied by a favourable negative interaction energy equates to an increase in *y*,
- a positive charge near regions of positive regression mapping,
- steric bulk in regions of positive steric mapping,
- no steric bulk in regions of negative steric mapping.

The opposing arguments apply to poorly active compounds, but far more information can be extracted from the PLS analysis. One of the problems in interpreting the regression maps is identifying how many of the mapped regions contain useful information that can be interpreted. The questions are how many, and which, are the mapped regions which offer independent and useful information, and which of the mapped regions contain common information. This problem becomes complicated when more than one 3D field source is used in combination as in CoMFA, where the electrostatic and steric information are shown on two separate maps.

The number of components in the PLS model, in fact, provides this information. For instance a 4 component PLS model indicates that the statistical analysis has identified 4 underlying unique properties of the molecules in the data set, which are important for describing biological activity. Each PLS component identifies a separate "underlying property" that is important in determining biological activity. The *x*-components are extracted so each is orthogonal, i.e. are not correlated to those previously extracted. Variables that are weighted heavily on a particular component are important in defining that component. Other variables that are also weighted heavily on that component contain similar information. One can construct 3D weightings maps showing how each of the grid variables are weighted onto each PLS component, similar to the construction of regression maps. Therefore, all mapped regions that are weighted onto a single PLS component should have a single statistical/physical interpretation. The interpretation of the PLS weightings maps was aided by examining them against a plot of the scores, commonly denoted as "*t*" in SIMCA terminology, of the compounds in the PLS *x* component vs the scores of *y* on the PLS *y* component (*u*). The *x* and *y* scores for a particular component are the "projections" of each compound in the data set onto the new physical summary descriptors extracted. The *t* vs *u* plot shows the "inner relationship", or inner correlation between the *x*-PLS component and *y*-PLS component for a particular dimension. The weightings maps and scores plots contain complementary information. Compounds that appear at the positive and negative ends of the *t*/*u* axis, are those whose GRID fields are most important in defining that component. The weightings maps were displayed over the structures of the two compounds with the most negative *t*/*u* and the most positive *t*/*u* values, i.e. compounds that are the most influential in defining that component.

These methods were applied to the analysis of the 4 component PLS model extracted from our calcium channel data set. The overall regression map and y vs y_{pred} plot are shown in Figs. 4 and 5, respectively. The regression map is overlaid on the

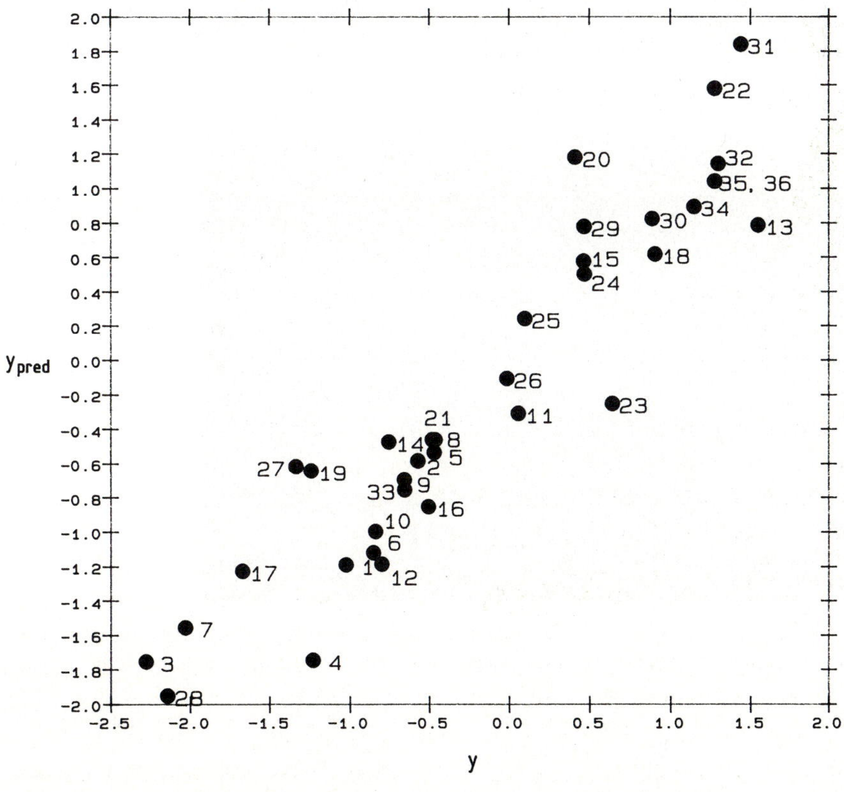

Figure 5. A plot of y predicted vs y observed for the overall 4 component model.

structures R = benzyl (**13**) and R = *p*-toluenesulphonyl (**28**), representing a highly active and poorly active compound. In this GRID analysis with a negatively charged alkyl hydroxyl probe, highly active compounds are those that have:

negative charge or hydrogen bonding groups near regions of negative regression coefficient/weighting,
steric bulk in regions of positive regression coefficient/weighting.

The opposing arguments apply to weakly active compounds. The interpretation was aided by examining the weightings maps together with *t* vs *u* plots for each PLS component. PLS 1 (the first PLS component) described 61% of the variance in biological activity, and 84% of the variance of the CLOGP descriptor and 67% of CMR weight onto this component. Fig. 6 shows the weightings of each GRID point that are also loaded onto PLS 1, which was interpreted in conjunction with a plot of *t*1 vs *u*1 (Fig. 7). The mapped GRID regions, therefore, showed points in space which are

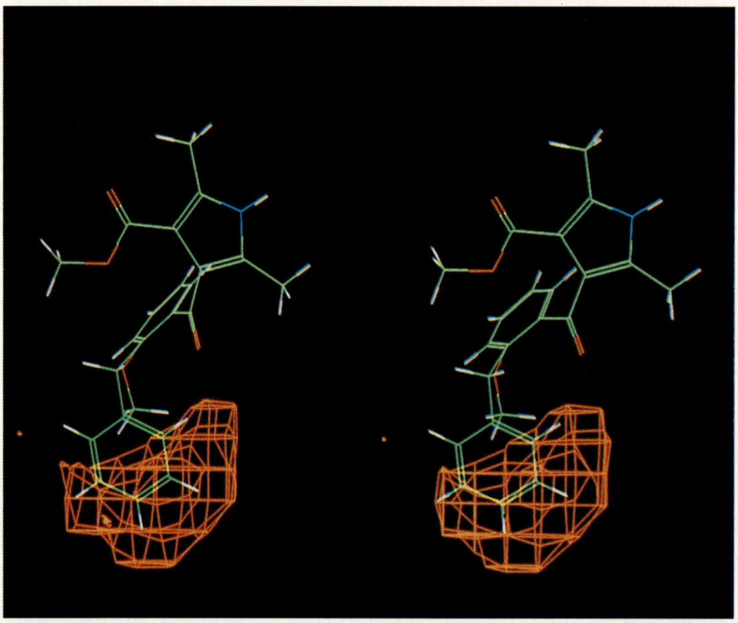

Figure 6. Map showing the weightings of the original grid points onto PLS 1. The positive and negative weightings are displayed at the same contouring level. This component is dominated by positive weightings, at this contouring level no negative weightings are observed. Also 84% of the variance of CLOGP and 67% of CMR weightings are loaded onto this component. The positive weighting regions, therefore, indicate where it is best to place lipophilicity. The weightings map is displayed over the structures of the benzyl substituent and the methoxy substituent, compounds that define the positive and negative ends of the inner relation for PLS 1, see Fig. 7.

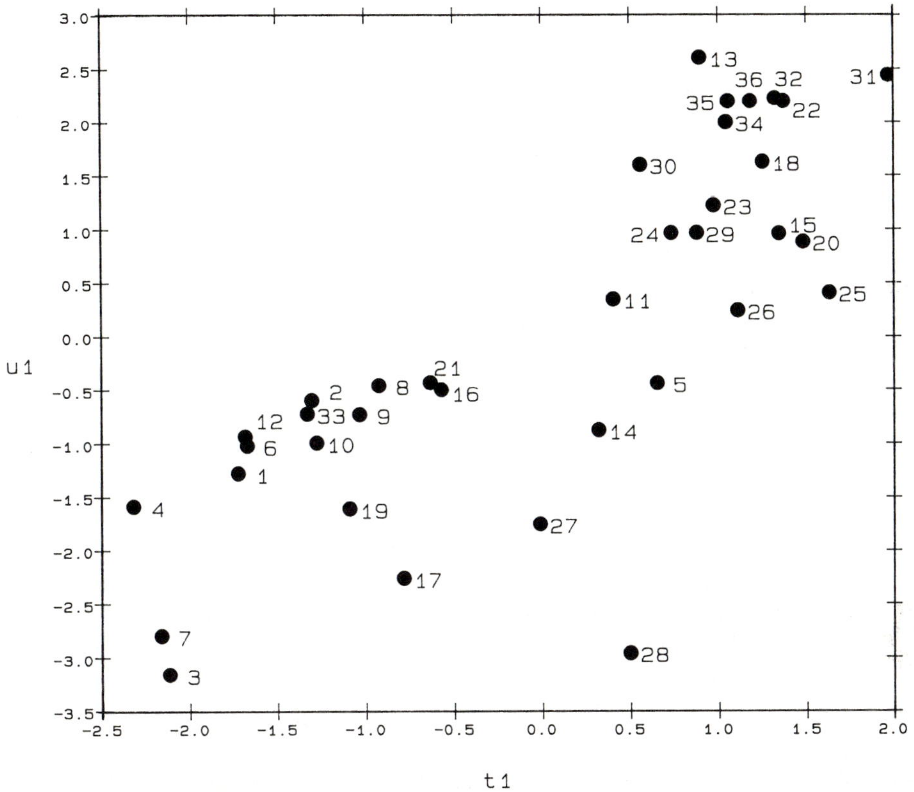

Figure 7. Plot of $t1$ vs $u1$, the inner correlation between the x-PLS component extracted and that proportion of y described for the first PLS dimension.

favourably occupied by a bulky lipophilic substituent. PLS 1 is dominated by regions of positive coefficients, regions in space from which the –OH probe is repelled, correlating with high biological activity.

Fig. 8 shows the weightings of GRID points onto PLS 2, and Fig. 9 a plot of $t2$ vs $u2$. The remaining 11% of the CLOGP and 30% of CMR that are loaded onto this component is the CMR term with a negative weighting. This component shows that too large a substituent can be detrimental to activity and negative PLS weights dominate this component.

Fig. 10 shows the weightings of GRID points onto PLS 3, and Fig. 11 a plot of $t3$ vs $u3$. This shows that benzyl substituents and their isosteres have favourable positive contours around the region of space they occupy, while the region of the aromatic ring of phenyl and phenethyl isosteres is filled with negative contours, which is unfavourable for this steric interaction. PLS 4 which describes only a further 7% of y (not shown) shows that benzyl substituents and phenethyl isosteres containing p-substituents appear to have less of an unfavourable effect upon biological activity.

Published CoMFA analyses so far have not been analyzed by inspecting individual PLS weightings maps with corresponding scores plots. But this information can be obtained from CoMFA, though scores plots are difficult to generate in CoMFA.

Inspection of weightings maps in 3D QSAR can be highly illuminating, especially when more than one field of information is being used in the analysis, for instance, as in CoMFA where the electrostatic and steric forces are separated. For any one PLS component, the electrostatic weightings map and steric weightings maps must have a common interpretation, as they show electrostatic and steric descriptors that are weighted onto the same PLS component.

2.2.3.2 Testing the Model

The only way to truly test any QSAR model is to use it to predict the activities of compounds that have not been included in deducing the model. The validation procedure used in PLS does test the model by leaving compounds out and predicting their activities, so one could argue that the model has been sufficiently tested. However, cross-validated predictions are not the same as true predictions and this is because, in deriving the PLS model, the properties of all the training set compounds supervise the PLS component extraction process. In 3D QSAR this also includes the initial preselection of variables based on minimum standard deviation, and variable reduction as achieved with GOLPE. It is possible that when the true predictivity of a model is tested, then the optimum number of components in the true prediction may be different from the optimum number derived by cross-validation on the training set [30].

As an example, we divided the calcium channel agonist data set into two groups, an 18 compound training set and an 18 compound test set. In this particular subset, we found that 1536 points in space showed a range >0.2 kcal/mol, and these were used with CLOGP and CMR in a PLS analysis. PLS analysis extracted 2 significant components. Table 4 shows the statistics of the model prediction of the 18 test set compounds for 1 to 4 components.

Table 4. Predictivity of an 18 compound training subset in predicting remaining 18 compounds.

Model dimensionality	Predicted vs observed r^2 36 compounds training set + test set	Predicted vs observed r^2 18 compounds test set only
1 PLS component	0.58	0.35
2 PLS components	0.71	0.54
3 PLS components	0.74	0.54
4 PLS components	0.76	0.54
regression model in log *P* alone	0.60	0.40

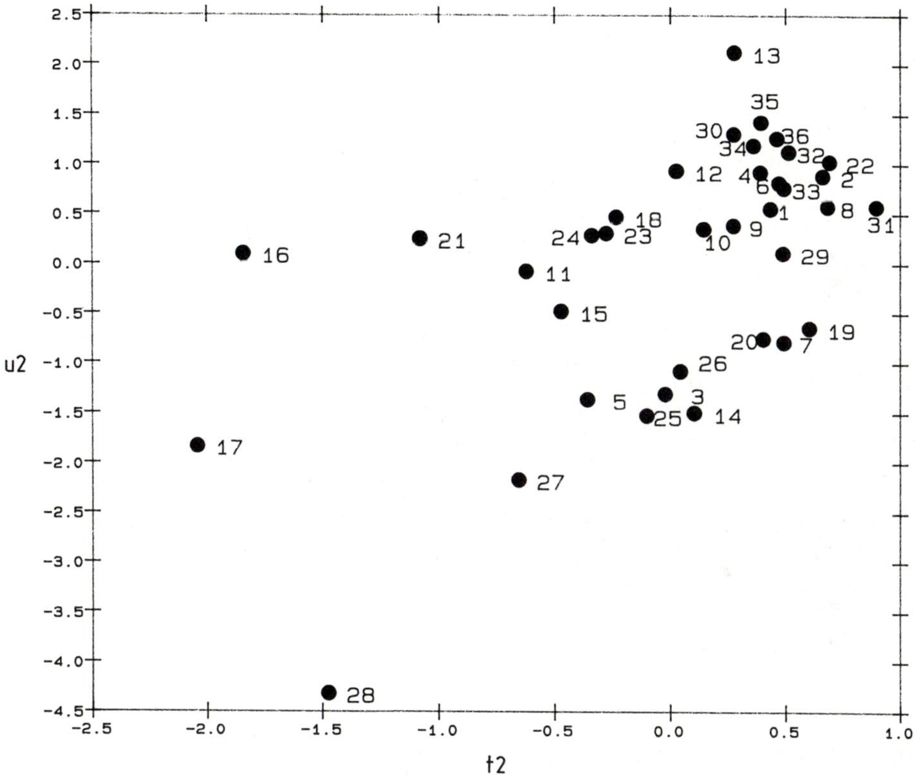

Figure 9. Plot of *t*2 vs *u*2, the inner correlation for the 2nd PLS dimension.

2.2.4 Conclusions

The CoMFA and 3D QSAR methods have at last brought together molecular modeling and traditional QSAR approaches. One of the problems with molecular modeling has always been that more data would always be generated than could be quantitatively analyzed, and so often the data was interpreted in only a semi-quantitative way. One of the problems of QSAR methods has always been the lack of suitable and relevant descriptors. Now these 3D methods are paving the way forward. The "glue" that binds these two previously separate disciplines together is of course PLS. Although these techniques are easy to perform (especially in CoMFA) and the main output is pictorial, they are in fact advanced statistical analyses, and are prone to the same pitfalls and errors as are all statistical analyses. Of particular importance, is the choice of the training set of compounds on which the model was developed in the first place, the detection of outliers in *x* space, *y* space and in the *x*–*y* correlation, and their careful use in prediction.

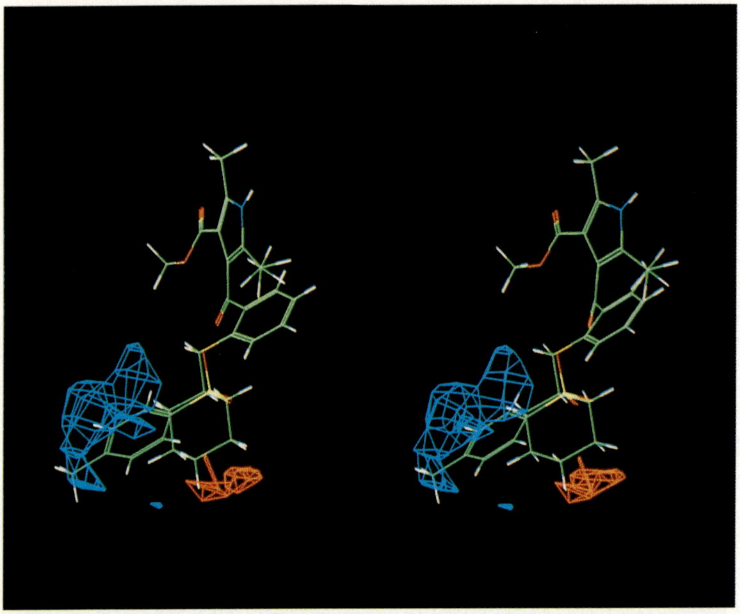

Figure 8. Map showing the weightings of the original grid points onto PLS 2. The map is displayed over the structures of the cyclohexylmethyl compound and the *p*-toluenesulphonyl substituted compound, compounds that define the positive and negative ends of PLS 2 inner correlation as shown in $t2$ vs $u2$ plot.

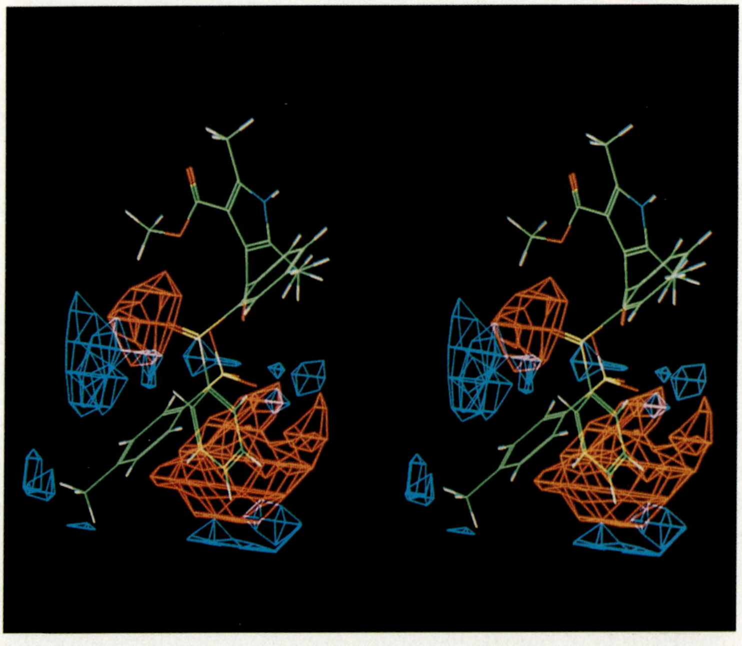

Figure 10. Map showing the weightings of the original grid points onto PLS 3. The map is displayed over the structures of compound **21** and compound **28**, which define the positive and negative ends of PLS 3 inner correlation as shown in $t3$ vs $u3$ plot, respectively.

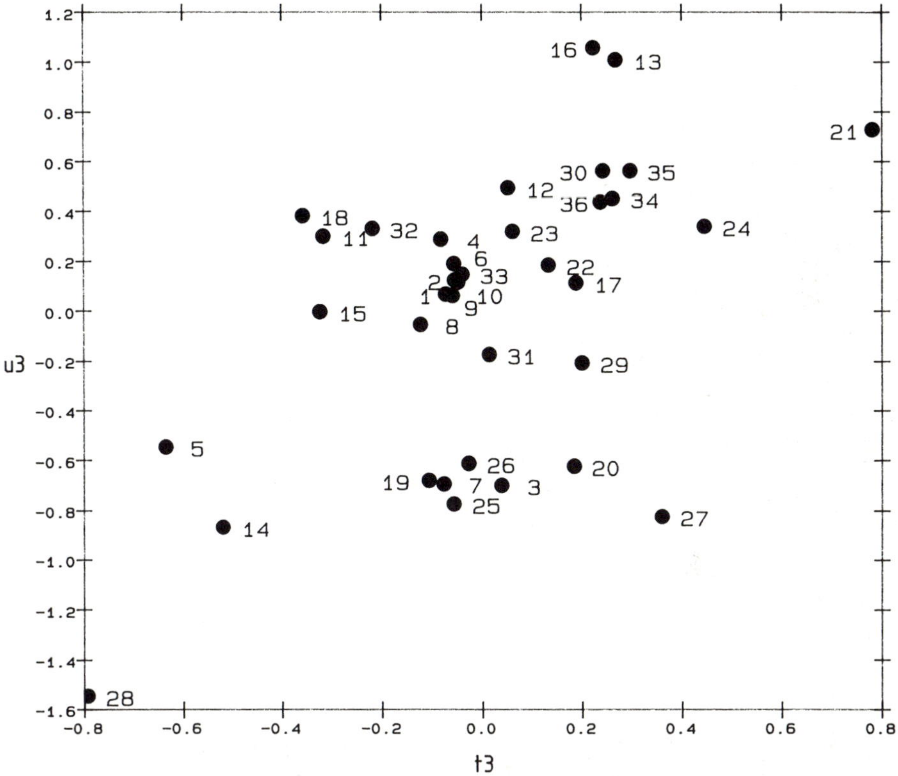

Figure 11. Plot of $t3$ vs $u3$, the inner correlation for the 3rd PLS dimension.

References

[1] Hansch, C. and Fujita, T., *J. Amer. Chem. Soc.* **86**, 1616–1626 (1964)

[2] Verloop, A., Hoogenstraaten, W. and Tipker, J., *Development and Application of New Steric Substituent Parameters in Drug Design.* In: *Drug Design,* Vol **VII**, Ariens, E. J., ed., Academic Press, New York, 1976

[3] Balaban, A. T., Chiriac, A., Motoc, I. and Simon, Z., *Steric Fit in Quantitative Structure-Activity Relations,* Springer-Verlag, Berlin, 1980

[4] Boel, M., *Theoretical Investigation on Steroid Structure and QSAR.* In: *Molecular Structure and Biological Activity of Steroids,* Duax, W. L., eds., CRC Press, Boca Raton, 1992

[5] Cramer, R. D., Patterson, D. E. and Bunce, J. D., *J. Amer. Chem. Soc.* **110**, 5959–5967 (1988)

[6] Cramer, R. D., Patterson, D. E. and Bunce, J. D., *Prog. Clin. Biol. Res.* **291**, 161–165 (1989)

[7] Tripos Associates Inc., 1699 South Hanley Road, Suite 303, St. Louis, Missouri, 63144, USA

[8] Boobyer, D. N. A., Goodford, P. J., McWhinnie, P. M., and Wade, R. C., *J. Med. Chem.* **32**, 1083–1094 (1989)

[9] Goodford, P. J., *J. Med. Chem.* **28**, 849–857 (1985)

[10] Wade, R.C., Clark, K.J. and Goodford, P.J., *J. Med. Chem.* **36**, 140–147 (1993)
[11] Wade, R.C. and Goodford, P.J., *J. Med. Chem.* **36**, 148–156 (1993)
[12] Itzstein, M., Yang, W.W., Kok, G.B., Pegg, M.S., Dyason, J.C., Jin, B., Phan, T.V., Smythe, M.L., White, H.F., Oliver, S.W., Colman, P.M., Varghese, J.N., Ryan, D.M., Woods, J.M., Bethell, R.C., Hotham, V.J., Cameron, J.M. and Penn, C.R., *Nature* **363**, 418 (1993)
[13] *RS/1, BBN Software Products Corporation,* 10 Fawcett Street Cambridge, MA. 02238, USA
[14] *SIMCA,* developed and distributed by Umetri A.B., Umea, Sweden
[15] *Chem-X,* developed and distributed by Chemical Design Ltd., Oxford, England
[16] Baxter, A.J.G., Dixon, J., Ince, F., Manners, C.N. and Teague, S.J., *J. Med. Chem.* **36**, 2739–2744 (1993)
[17] Dammkoeler, R.A., Karasek, S.F., Shands, E.F.B. and Marshall, G.R., *J. Comput-Aided Mol. Des.* **3**, 3–21 (1989)
[18] Klebe, G. and Abraham, U., *J. Med. Chem.* **36**, 70–80 (1993)
[19] *ASP,* distributed by Oxford Molecular Limited, Magdalen Centre, Oxford Science Park, Oxford, England
[20] Burt, C., Richards, W.G. and Huxley, P., *J. Comput. Chem.* **11**, 1139–1146 (1991)
[21] Clementi, S., Cruciani, G., Riganelli, D., Valigi, R., Costantino, G., Baroni, M. and Wold, S., *Pharm. Pharmacol. Lett.* **3**, 5–8 (1993)
[22] *SYBYL Version 6.01 Release Notes,* Tripos Associates, Inc. 1993, p. 2225–2277
[23] Martin, Y.C., Bures, M.G., Danaher, E.A., DeLazzer, J., Lico, I. and Pavlik, P.A., *J. Comp. Mol. Des.* **7**, 83–102 (1993)
[24] Golender, R.J. and Rozenblit, A.B., *Logical-Structural Approach to Computer-Assisted Drug Design.* In: *Drug Design* Vol. **10**, Academic Press, 1980
[25] Allen, M.S., LaLoggia, A.J., Dorn, L.J., Martin, M.J., Costantino, G., Hagen, T.J., Koehler, K.F., Skolnick, P. and Cook, J.M., *J. Med. Chem.* **35**, 4001–4010 (1992)
[26] Waller, C.L. and Marshall, G.R., *J. Med. Chem.* **36**, 2390–2403 (1993)
[27] Audrey, E., Dubost, J.P., Colleter, J.C. and Dallet, P., *Eur. J. Med. Chem.* **21**, 71–72 (1986)
[28] Kellog, G.E., Semus, S.F. and Abraham, D.J., *J. Comput-Aided Mol Des.* **5**, 545–552 (1991)
[29] Cocchi, M. and Johansson, E., *Quant. Struct.-Act. Relat.* **12**, 1–8 (1993)
[30] Baroni, M., Constantino, G., Cruciani, G., Riganelli, D., Vilagi, R. and Clementi, S., *Quant. Struct.-Act. Relat.* **12**, 9–20 (1993)
[31] Fersht, A., *Enzyme Structure and Mechanism,* 2nd edn., W.H. Freeman and Company, New York, 1985, p 294–301
[32] Wold, S. *Technometrics* **20**, 397–404 (1978)
[33] Greco, G., Novellino, E., Silipo, C. and Vittoria, A., *Quant. Struct.-Act. Relat.* **11**, 461–477 (1992)
[34] McFarland, J.W., *J. Med. Chem.* **35**, 2543–2550 (1992)
[35] Hansch, C., Debnath, A.K., Kim, K.H. and Martin, Y.C., *J. Med. Chem.* **36**, 1007–1016 (1993)
[36] Kim, K.H., *Med. Chem. Res.* **1**, 259–264 (1991)
[37] MEDCHEM, version 3.54; Daylight CIS: USA, 1993
[38] Davis, A.M., Gensmantel, N.P., Johansson, E. and Marriott, D.P., *J. Med. Chem.* **37**, 963–972 (1994)

2.3 GOLPE: Philosophy and Applications in 3D QSAR

Gabriele Cruciani and Sergio Clementi

Abbreviations

ACC	Auto and Cross Covariance
ACE	Alternating Conditional Expectations
APOLLO	Automated PharmacOphore Location Through Ligand Overlap
CoMFA	Comparative Molecular Field Analysis
CoMPA	Comparison of Molecular Potentials and Analysis
FFD	Fractional Fractorial Design
GOLPE	Generating Optimal Linear PLS Estimations
GPb	Glycogen Phosphorylase b
HINT	Hydrophobic INTeractions
LOO	Leave-One-Out
PCA	Principal Component Analysis
PCs	Principal Components
PLS	Partial Least Squares
PRESS	Predictive REsidual Sum of Squares
QSAR	Quantitative Structure-Activity Relationship
SDEP	Standard Deviation of Error of Predictions
SIMCA	Soft Independent Modeling of Class Analogy
SSY	Sum of Squares of response value

2.3.1 Introduction

Quite a number of chapters in the first two volumes of the series *Methods and Principles in Medicinal Chemistry* (VCH, Weinheim) illustrate several aspects which we consider appropriate for introducing a detailed account of Generating Optimal Linear PLS Estimations (GOLPE) [1]. In particular we wish to draw the reader's attention to the chapters on Principal Component Analysis (PCA) [2], Partial Least Squares (PLS) [4], Design [5], Comparative Molecular Field Analysis (CoMFA) [6], three-dimensional quantitative structure-activity relationships (3D QSAR) [6] and others. GOLPE is, in fact, a chemometric procedure, which is based on an advanced PLS method, aimed at obtaining models whith highly reliable predictivity by means of variable selection criteria. The procedure, developed by our group in Perugia over a period of five years, is oriented towards the research requirements of 3D QSAR, and is implemented in a computer program, which complements rather then rivals methods such as SIMCA and CoMFA.

In a brief introduction to this chapter we wish to state how we believe GOLPE should rank among the different chemometric tools used in QSAR, outline the philosophy which led us to develop such a procedure, and give a number of examples of some of the possible applications in conjunction with some comments and criticisms that have been raised so far.

GOLPE is one of the new chemometric tools that have been suggested for 3D QSAR modeling, and in our opinions, falls completely within the QSAR tradition started by Hansch some thirty years ago. In fact the Hansch approach, expanding Hammett and Taft, allowed analogy models to be established which used "constants" for the varying fragments (substituents, amino acids, etc.) around a common skeleton framework. We have always considered the Hansch approach as being highly appropriate within this context. Chemometricians, especially those with a background in physical organic chemistry, have suggested that there should be an update of the chemometric tools used in 3D-QSAR modeling. That is, to use PLS instead of ordinary multiple regression, design criteria in latent variables (principal properties), to select the least number of the most informative structures, to use validated models and to avoid using indicator variables. A review on some of these topics can be found in Vol. 2 of the present series [7].

However, the traditional approach may be considered to have some limitations. Besides the obvious requirements of the additional thermodynamic relationship, where only series of compounds with a common skeleton framework should be considered, conformational equilibria are not taken into account, and, in general, information on the 3D-structure is not employed at all.

On the other hand, molecular modeling techniques have become extremely popular, especially because of increasing computation. These methods, are aimed at calculating the energy of a number of conformations for each molecule at different levels of approximation, and then to study the possible interactions between the molecule and its binding site. It became possible from this approach to describe each molecule/conformation by a series of theoretically computed parameters, some of which are 3D in nature.

A 3D QSAR is, therefore strictly speaking, a QSAR relationship in which the structural descriptors have 3D nature: several compounds are studied at the same time within the framework of a regression model, with the objective of ascertaining which structural features significantly affect the biological response. Notably, these 3D descriptors are usually derived from the different modeling techniques.

At present, the best example of state-of-the-art 3D QSAR is given by the CoMFA procedure [8]. Molecules are first represented by a long vector of interaction energies with a probe situated at regular intervals in three dimensions, and subsequently aligned according to some fitting criterion. The chemometric method used is PLS, because of the largely greater number of variables (descriptors) over the number of objects (compounds), and the PLS models are validated by cross-validation techniques. The results are shown in terms of coefficients of a pseudo-regression equation with the original variables, here as locations in the 3D space, and are represented by 3D graphics.

Once again, it seems likely that the original idea of describing molecules in 3D by their interaction energies with different probes might be improved in the future by the application of more suitable chemometric strategies and/or newly developed tools.

2.3.1.1 3D Molecular Descriptors and Chemometric Tools

When a probe is moved around in a rectangular box of grid points and through a target molecule, it produces a three-dimensional box of interaction fields. Depending on the computational procedure used, these fields may represent total interaction energies (GRID) [9], steric or electrostatic fields (CoMFA) [8], molecular electrostatic potential fields (CoMPA) [10], hydrophobic interactions (HINT) [11], electron densities etc. These fields may be used as point descriptors of the 3D molecular structure and physico-chemical behaviour of the target molecule. Moreover, a graphical analysis allows a simple interpretation of the fields such as the visualization of the regions where the probe interacts most strongly with the target either by attraction or repulsion.

However, problems arise when a number of molecules are studied at the same time. In this case, a simple graphics analysis and visualization is not sufficient to provide the necessary information in order to understand the observed trend in the biological properties of a series of compounds. In such a case, appropriate chemometric tools may be extremely useful in order to condense and extract hidden information.

Principal Component Analysis (PCA) and Partial Least Squares (PLS) are statistical multivariate techniques for extracting and rationalizing the maximum amount of common information from a multivariate description of a biological system.

PCA is a projection method that provides an approximation of a matrix $\boldsymbol{X}$, in this case the descriptor matrix in terms of the product of two smaller matrices $\boldsymbol{T}$ and $\boldsymbol{P'}$ (Eq. (1)). The matrices $\boldsymbol{T}$ and $\boldsymbol{P'}$ extract the essential information and patterns from $\boldsymbol{X}$. By plotting the columns of $\boldsymbol{T}$, a picture of the dominant "object pattern" of $\boldsymbol{X}$ is obtained and, by analogy, plotting the rows of $\boldsymbol{P'}$ shows the complementary "variable pattern". The number of statistically significant dimensions for the PCA model is determined by cross-validation as implemented in the NIPALS algorithm [12].

The Principal Components (PCs) are linear combinations of the original variables, and in the 3D QSAR context can be regarded as the important 3D regions distinguishing between the characterized subsystems. Moreover, PCs are orthogonal to each other, so that they represent independent effects. Examples on how to use PCA in 3D QSAR studies are given in Sec. 2.3.3. However, PCA is not a tool for handling relationships between two different blocks of variables, and the relationship between the biological activity and the structural descriptors, described here in terms of 3D fields, must be modeled by PLS.

$$X = \bar{x} + TP' + E \tag{1}$$

$$Y = \bar{y} + UQ' + F \tag{2}$$

$$U = BT + H \tag{3}$$

$$Y = \bar{y} + BTQ' + F^* \tag{4}$$

The two-block PLS model relates matrix X (chemical description) to matrix Y (biological activities) with the purpose of predicting Y from X. The X-block model is the same as in PCA. The Y-block matrix is modeled in a similar manner (Eq. (2)), and the maximum correlation between the X and Y block models is obtained using a PLS-weight matrix W' [12]. The relationship between U and T, the "inner relationship", can therefore be modeled by Eq. (3), where B is a diagonal matrix and H is a residual matrix. In the case of a single biological response, Eqs. (2) and (3) are substituted by Eq. (4). Similarly to PCA, the statistical significance for each model dimension is determined by cross-validation [12, 13].

Predictions of y values for new compounds are obtained from the data of these compounds inserted into the PLS model in the sequence: $x \rightarrow t \rightarrow u \rightarrow y$. As the principal components in PCA, the PLS latent variables are linear combinations of all the original variables. In 3D QSAR, these PLS latent variables take into account the important 3D regions which results in a better model for the relationship between the X and Y matrices. Moreover, in PLS it is more appropriate to study the PLS-weights as the "variable pattern" instead of the P' loadings used in PCA.

2.3.1.2 Unfolding Three-way Matrices

Ordinary PCA and PLS methods require a two-way table of objects and variables. An object is often a physically distinguishable entity, such as target molecule or a probe, while a variable represents the results of an observation, or a measurement, or a computation undertaken with this object. However, 3D QSAR methods produce three-way matrices of molecular descriptors. In order to transform three-way matrices into a two-way table, an unfolding procedure and a simple reorganization of the data is required. The unfolding procedure is a method for transforming a multi-way matrix into a one-dimensional vector of numbers (Fig. 1), while a data reorganization is a procedure for organizing one-dimensional vectors into a two-way data table.

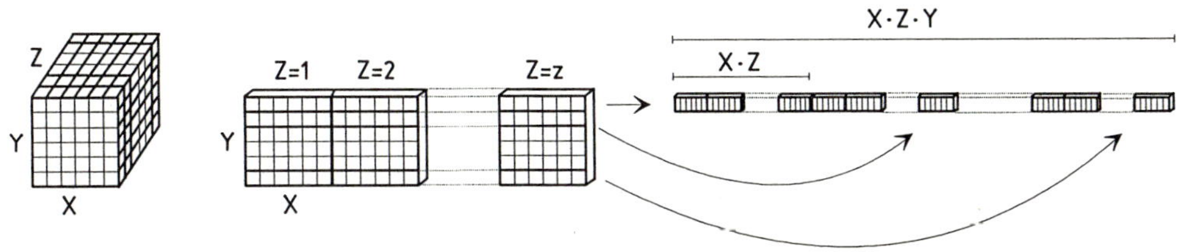

Figure 1. The unfolding procedure for a three-dimensional data array *X*.

The interaction energies between a target molecule and a probe produced by traditional 3D QSAR methods may be viewed as descriptors of the probe, or of target behavior. The twofold interpretation of descriptors leads to different methods for producing the two-way data reorganization. Usually a 3D descriptor matrix for a single molecule is organized as a three-way table where the rows, the columns and the sheets are variables; the table itself represents the object and this three-way table can be easily rearranged as a one-dimensional vector (Fig. 1).

In the presence of several molecules the procedure should be repeated for all the molecules and the vectors of variables assembled together in a two-way table in order to obtain a target matrix [14]. Thus, the target matrix will contain the interaction energies between all the molecules and one specific interacting group.

With only one target molecule, many different computations may result by varying the probe, and, in this case, a probe matrix is obtained [14]. The probe matrix contains information about the interactions of different chemical groups with the same target molecule. With such a problem, multivariate statistics may be used to select the most suitable probes in order to design selective target ligands.

Target matrices can be combined, thus, obtaining only one larger matrix. In the CoMFA procedure two probes are employed as blocks of descriptors and the resulting two target matrices are combined to form a unique matrix containing the same number of objects and twice the number of variables. Similarly, by using GRID, several probe matrices can be combined by keeping the number of variables constant and increasing the number of objects. Clearly, the choice of using either the target or the probe or combined matrices for individual studies depends on what is to be deduced from the data, and is closely related with the problem in question.

2.3.2 The GOLPE Philosophy

Since data matrices in 3D QSAR are characterized by a huge number of variables (usually thousands) and a relatively low number of molecules (usually tens), the requirements of an appropriate chemometric tool should involve a sound validation method and a reliable variable selection procedure. In fact, nowadays, it is clear that predictions should be made only on points excluded from the modeling phase, and that when there are many variables, keeping less important variables in the model can be detrimental to its predictivity. A thorough discussion on validation and variable selection can be found elsewhere [7], and is only briefly illustrated here.

The standard reference for measuring the predictive power of a model with the given complexity of the data set in cross-validation is *PRESS* (Eq. (5)), which is defined as the total sum of squares of predictions minus observation, and therefore, contains one term for each molecule.

$$PRESS = \Sigma(y - y_{\mathrm{PRED}})^2 \tag{5}$$

$$Q^2 = 1 - PRESS/SSY = R^2_{\mathrm{cv}} \tag{6}$$

$$SDEP = [\Sigma(y - y_{\mathrm{PRED}})^2/N]^{1/2} \tag{7}$$

If the PRESS value is transformed into a dimensionless term by relating it to the initial sum of squares (Eq. (6)) one obtains Q^2, i.e. the complement to one of the fraction of unexplained variance over the total variance. However, for purposes of the end-user, the square root of *PRESS/N*, which we suggested should be called *SDEP* [15] (Eq. (7)), seems to be more directly related to the uncertainty of the predictions, since it has the same units as the actual y values. Either *PRESS*, Q^2, or *SDEP* can be used to check the predictivity of regression models, including PLS.

However, the meaning of any statistical parameter depends on the way it is computed. In our case, for instance, the parameters depend upon the way points are held out in the cross-validation phase. We have shown that the *SDEP* parameter decreases on increasing the number of cross-validation groups [16]. Consequently, if one decides to use the maximum number of groups possible, i.e. with a leave-one-out (LOO) procedure, a much better result (a higher predictivity) will always be obtained than by using a smaller number of groups. Therefore, although a LOO procedure is computationally simpler and faster, we should be aware that it gives an overoptimistic estimation of predictivity, either simply on numerical grounds, or because of the clusters of structures we often find in a 3D problem because of the discrete nature of organic molecular systems. Furthermore, it has also been stated recently that the use of groups is better than a LOO procedure, also for theoretical reasons [17].

Clearly, the way in which one decides to compute the statistics depends on how the results are to be presented. Usually, one should be pleased when reliable estimations of the model parameters are obtained, so that reliable uncertainties are obtained for future predictions. Quite often, however, theoreticians seem to be more concerned about illustrating how good their models are solely in terms of R^2, R^2_{cv} or Q^2.

The GOLPE procedure was developed in order to improve as much as possible the reliability of future predictions. Consequently, even if the method is aimed at finding

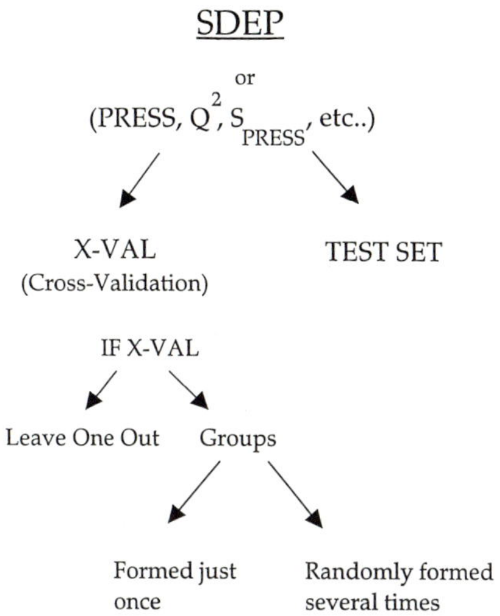

Figure 2. Validation Scheme.

the highest Q^2, or the lowest *SDEP*, in principle, one should not expect that GOLPE will give, as such, Q^2 values which are higher than those derived by other chemometric techniques. If in the eventuality it does so, then this is due to the variable selection procedure. In principle, the statistics produced by GOLPE are less optimistic and, seem to be closer to the experimental values obtained from QSAR studies and is, thus, a true reflection of the validity of the model.

We should also draw attention to the fact that no validation procedure can give an objective and absolute criterion for estimating the validity of a model. In fact, validation procedures work either in terms of self-consistency of a data set or on an external test set, and the same parameters (*PRESS, SDEP*, etc.) can be used for any of these ways of formulating the problem. In the former case, one implicitly assigns to the whole model the predictivity of a number of reduced models derived from the whole model. In the latter case, unless one has a designed data set which automatically is defined as the test set of all the molecules outside the design data set, results will depend upon the selection of the test set, and there is no unique way of determining this selection. This problem is highlighted in Fig. 2. When data are grouped, as in QSAR problems, a LOO procedure leads to an overpredictivity.

Accordingly, in the GOLPE procedure we proposed to use a smaller number of groups (e.g. 5) instead of a LOO procedure, but we proposed that the group formations were repeated several times e.g. 100) in a random way in order to avoid the results being dependent on onc single computational grouping. Consequently, *SDEP* is defined as the mean value of 100 individual "*sdep*" values, each obtained on predicting one fifth of the points at a time, with the five groups being randomly formed 100 times from the data set.

2.3.2.1 Variable Selection

Although the need for variable selection has increased sharply over the last few years and several different strategies have been suggested [7], the method implemented in GOLPE appears to be the only one which is really aimed at evaluating the effect of each individual variable on model predictivity. The other methods of selecting variables are based on their importance in the validated models.

It is impossible to check the predictivity of all possible combinations of variables and we have selected the most appropriate approximation. Finding an efficient way of selecting the best combination of vairables is a typical design problem and the design matrices used in fractional factorial designs (FFDs) provide a suitable tool [18]. The strategy was based on using combinations of variables according to a FFD where each of the two levels (plus and minus) corresponds to the presence and absence of the variable. The design matrix proposed to test the prediction ability of these reduced models involve a different combination of variables which include only the "plus" and exclude only the "minus" variables. For each such combination, the prediction ability of the corresponding PLS model can be evaluated by means of *SDEP* (Fig. 3).

In order to estimate the significance of a single variable effect on predictivity, a number of dummy (or ghost) variables were introduced into the design matrix.

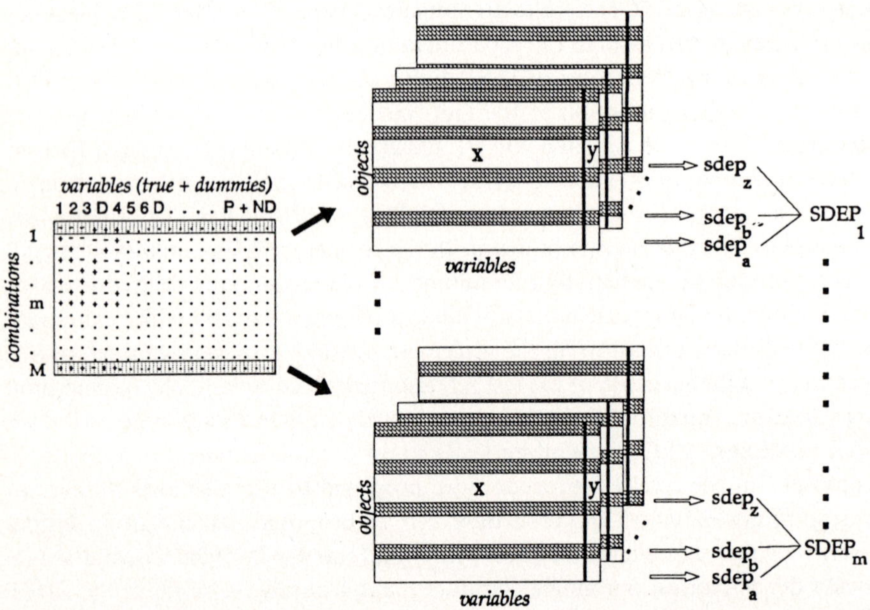

Figure 3. Variable selection procedure. For each variable combination, suggested by the design matrix, the model predictivity is evaluated dividing the objects set into five groups and repeating the formation of groups several times.

It is worth mentioning that these dummy variables are not numbers: we define dummy variables as some specific columns in the design matrix, say one over three or four. Since they are not true variables, the dummies are not used in the combinations of variables which evaluate the predictivity of each row of the design matrix. However, they are used to compute the effects on predictivity for a ghost variable, so that the positive or the negative effects of true individual variables can be ascertained on the basis of a Student-*t* tailoring.

The introduction of these dummy variables into the design matrix allows for a comparison between the effect of a true variable and the average effect of the dummies. If we assume that the variable selection works in an iterative manner, we can always retain the variables which have a positive effect on predictivity of the model, while variables with a negative effect can be excluded. This facilitates an increase in the stability of the results: the reduction of the number of variables leads to an increase in the degrees of freedom and, therefore, to a decrease of the critical value of the Student-*t* function besides a better control of variable combinations. The iterative process stops when no more variables are fixed or excluded.

However, as the reliable strategy just outlined would have been impractical in 3D QSAR problems, we, therefore, had to find from the beginning an alternative strategy providing a reduced number of variables with which we could then apply this fixing/excluding procedure. The most efficient way was to select variables in the loading space according to a D-optimal design for the purpose of this preliminary selection. The information is largely redundant for many of the variables and D-optimality appears to be an appropriate criterion for selecting variables in such a way that most of the redundant information is discarded while still retaining sufficient collinearity as required by the PLS algorithm.

Consequently, the first step of the GOLPE procedure in 3D QSAR is a normal linear PLS model with all variables, followed by the variable preselection according to a D-optimal design in the loading space. The selection results depend upon the dimensionality of the PLS model and are more stable for low PLS dimensions. Moreover, it is recommended that the D-optimality criterion is used in an iterative manner, so that no less than a half of the variables are kept for each run. The preselection phase should be stopped when the predictivity of the reduced model changes significantly from the previous run: this indicates that we have finished discarding redundant information before beginning to discard important information.

The GOLPE procedure appears, therefore, to be a powerful and efficient tool for variable selection. However, we should note that it can only be properly applied provided that the regression model on the whole data set has at least some predictive ability in the first whole model. If this is not the case, variable selection can still be undertaken provided that there is sufficient structure in the *X* data, implying that the dimensionality of the problem is lower than the number of variables [1].

2.3.3 Applications

Because of these peculiar choices in developing GOLPE, both in the validation approach and in the variable selection procedure, it seems to be appropriately designed to handle 3D QSAR matrices. In particular, it appears that it can be used profitably in CoMFA studies, not as an alternative method, but as an appropriate tool for exploring the structural space and to ascertain the grid locations which exert the greatest effect on the biological response and which can be used in connection with the QSAR/CoMFA module of SYBYL. [NOTE: A UNIX version of the GOLPE package for the SGI environment was developed in C at M.I.A. (Perugia) by Massimo Baroni and is distributed in collaborative agreement with Tripos.]

In principle, however, GOLPE could be used with any group of molecular 3D descriptors, which have been either calculated in-house or produced by commercially available software. Among them it seems that GRID, developed by Goodford [9], is particularly suitable either for the variety of probes it offers and for the reliability of its force field. On the other hand, the search for extremely precise theoretical calculations in terms of charges and energies, say, for instance, by molecular dynamics or ab initio methods, appears to be worthless in this context because of the overwhelming number of approximations employed in chemometrics. At present, the potential of combining GRID, CoMFA and GOLPE seems to provide the best possible working medium.

Before illustrating a few examples on how to use GOLPE, a brief discussion on data pretreatment and other parameters that can be selected in the procedure seems appropriate, since different choices lead to different results. Besides data scaling, results depend upon the alignment criterion, the cut-off values of the fields, the cut-off value of the standard deviation, the grid spacing, the number of PLS components, etc. We wish to state, however, that it is probably impossible, and perhaps not correct, to furnish a set of general rules which should always be used in 3D QSAR studies. We understand that this may be a requirement of end-users who apply chemometric tools merely in a procedural manner; we should, on the contrary, strongly support the use of chemometrics in a more active and interactive way for a better understanding of the problem under investigation.

Data pretreatment constitutes one of the subjects under discussion: depending on the required sensitivity, the raw field or energy data can be used either as such (but this might prevent ranking of the importance of locations), or autoscaled (but this blows up irrelevant variables), or blockscaled [13] (this cannot, however, solve the problem of the relative importance of individual properties, such as log *P*, in an objective manner). Either blockscaling and the standard CoMFA scaling are aimed at assigning equal importance initially to the different "logical" effects in the analysis. The results of the full paper from which example 3.3 has been taken [19] suggest that the final variable selection should be performed on autoscaled data, whereas the D-optimal preselection should be performed on non-scaled data.

Grid spacing and contour plots merit a few more words. On decreasing the grid spacing the number of grid nodes, and, therefore, of variables increases dramatically, for instance, by a factor of eight from 1000 to 8000. It is generally accepted that the large collinearities between variables introduce considerable noise. On the contrary, in our experience, at least with GRID probes, on using the variable selection criteria suggested in GOLPE, a grid spacing of 1 Ångstrom always gives better results than with 2 Ångstroms. The final warning takes into account the stability of the regression coefficients. When the PLS loadings are transformed into the coefficients rotated back into the original variable space they may give rise to misleading interpretations, if they are not derived from autoscaled models, and if they change significantly, as they sometimes do, with model dimensionality.

2.3.3.1 PCA on the Target Matrix

The GRID force field and Principal Component Analysis (PCA) have been used in order to predict the interactions of small chemical groups with all 64 different sequences of *beta* DNA [14]. In this example the target matrix contains 64 objects (the triplets) and 9510 variables which represent the interaction energies between the amide multiatom probe and each triplet, calculated at each grid point in the minor groove space.

The amide probe can accept two hydrogen bonds at the carbonyl oxygen atom, and donate two hydrogen bonds from the NH_2 group. The probe's ability to accept and donate hydrogen bonds at the same time allows it to define a high number of interaction sites with all of the 64 DNA triplets. GRID shows that the probe can interact in the minor groove forming hydrogen bonds with the O-2 oxygen of thymine and cytosine, N-3 nitrogen of guanine and adenine, O-4* oxygen of deoxyribose rings, O-P oxygen of phosphate groups and N-2 nitrogen of guanine. Moreover, many combinations of two, three or four hydrogen bonds are possible at these sites according to the geometric and energetic characteristics of the triplets and of the probe. The aliphatic amide probe can, therefore, from multiple hydrogen bonds with each triplet giving high energies of interaction. This interaction flexibility renders it to be a high affinity ligand for all the DNA triplets.

Principal Component Analysis revealed three components accounting for 77.4% of the total variance in the target matrix: they are related to three different regions of selectivity in the minor groove space. On plotting the loadings of the PCA in the real 3D space of the molecules it is possible to evaluate the contributions of each zone around the molecules in determining the differences between triplets. It should be emphasized that these regions are not necessarily the locations with the highest energy of interaction between the probe and the triplets, but the regions where the probe can best distinguish between the triplets.

The score plot for the first two components is shown in Fig. 4. In this plot each point represents one DNA triplet. When the points are close to each order the corre-

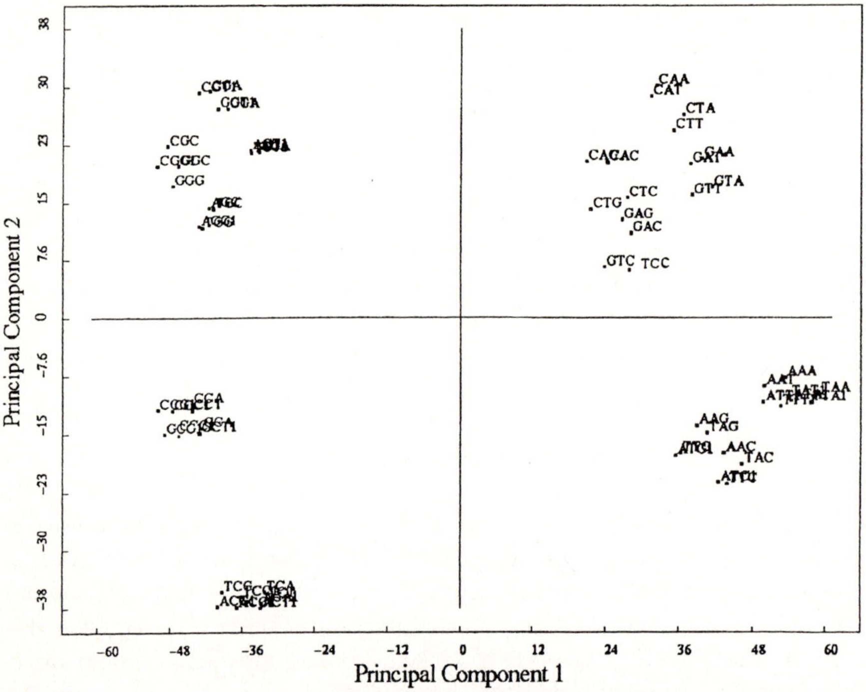

Figure 4. Score plot for the model describing the interaction of the $CONH_2$ probe. The first PC distinguishes two main groups of triplets while the second PC separates the two previous groups into three subgroups.

sponding triplets interact with the $CONH_2$ probe in a similar way, but where the points are apart their interaction energies are different, and the probe interacts in a more specific way with each individual triplet.

Fig. 4 shows that the triplets are clustered into two larger groups and five subgroups by means of this probe. The first principal component (PC) distinguishes between the two main groups, and these groups correspond to the two loading regions shown in Fig. 5. The probe can form multiple hydrogen bonds to triplets (right hand side of these score plot) from the larger region of Fig. 5, but there are fewer interactions from the smaller region (left hand side of the score plot). For example, the amide probe interacts with GTG (right) in the larger region, but with GCG (left) in the smaller region of Fig. 5, and, as mentioned above, this smaller region is less favourable for ligand design, simply because it is small. In fact, the size of a PC region is always an important factor to be considered, because it is not easy to exploit small regions for selective ligand design. Moreover, the two regions are not spatially distant from each other, and so it may not be easy to design a ligand which can place the amide group in the exact place for an exact required orientation.

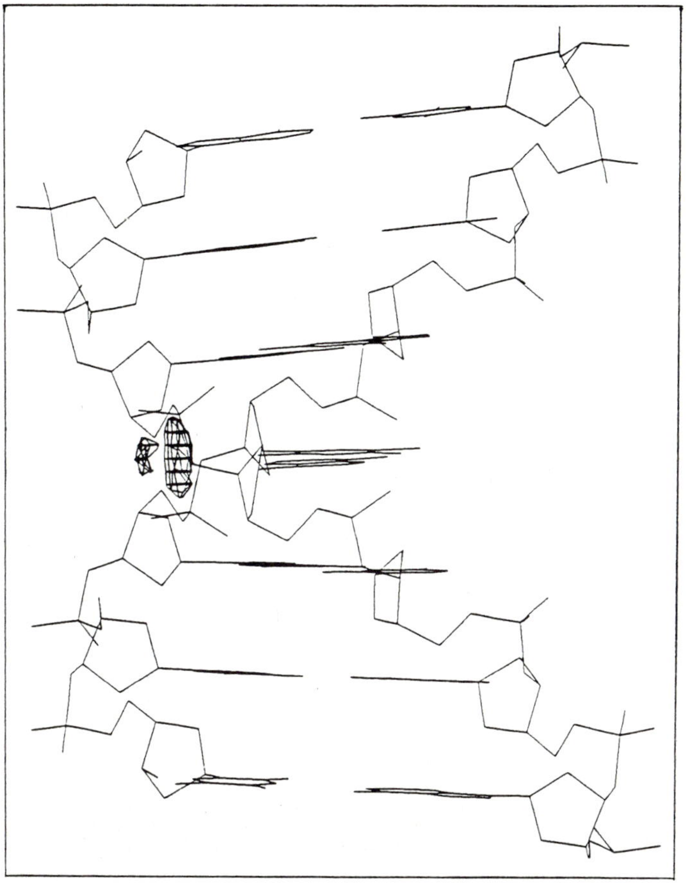

Figure 5. View of the minor groove space of TTCGGTT DNA heptuplet showing the selectivity regions defined by the first PC.

The situation is quite different with the second principal component which separates the three subgroups of triplets in Fig. 4. In this case, the selectivity regions are well separated from each other, as shown in the loading plot of Fig. 6, and could well be employed for ligand design.

2.3.3.2 PCA on the Probe Matrix

In this example, the binding of 31 different probes to the TTCGGTT double-stranded base-pair sequence is investigated by GRID. The whole data array is represented by a probe matrix containing 31 objects (the probes) and 9510 variables (the interaction

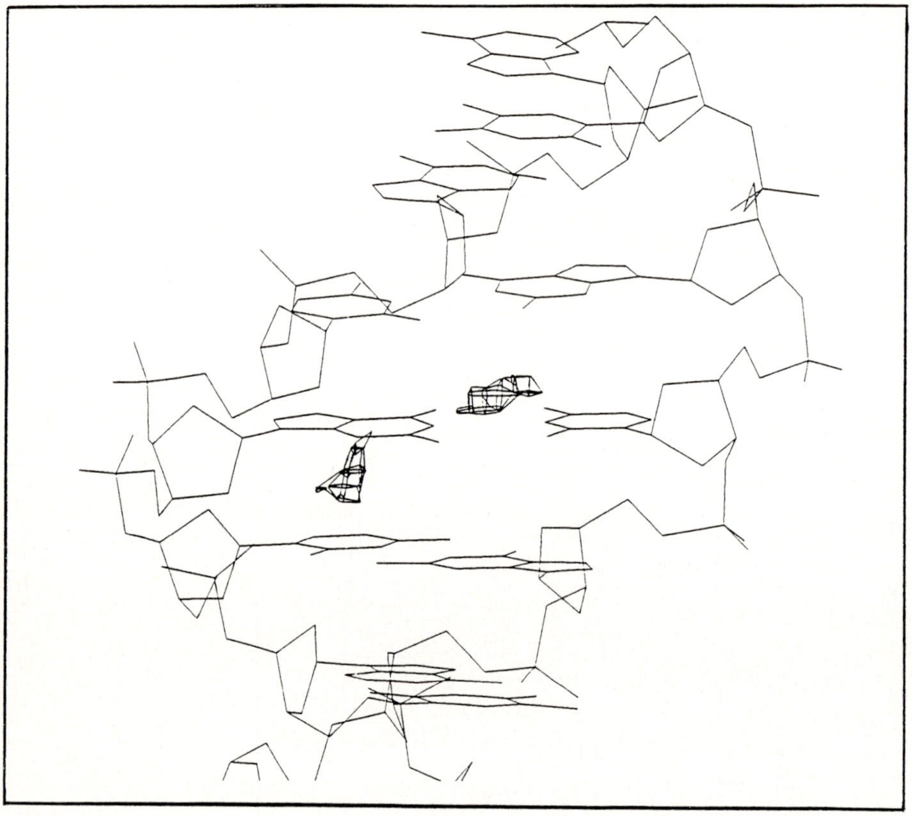

Figure 6. The minor groove of TTCGGTT DNA showing the selectivity regions defined by the second PC of the model.

energies calculated at each grid point in the minor groove of the TTCGGTT DNA heptuplet).

Two significant Principal Components are extracted from the matrix according to the cross-validation technique. These components account for about 87% of the total variance in the data. The score plot for the PC model is shown in Fig. 7, and from this plot it can be seen that the interaction of the different GRID probes with DNA may be classified into several distinct groups.

The first PC shows that probes carrying a partial negative charge (such as number 12; carboxy oxygen) are clustered to the right, while those with a partial positive charge (e.g. number 8, an NH_2 cation) are to the left of the plot. This shows that the charge on the probe influences its interaction energy with DNA. A numerical analysis of the interaction energies show that all the probes to the left of Fig. 6 tend to interact more favourably with the DNA heptuplet, since they have higher negative energy values.

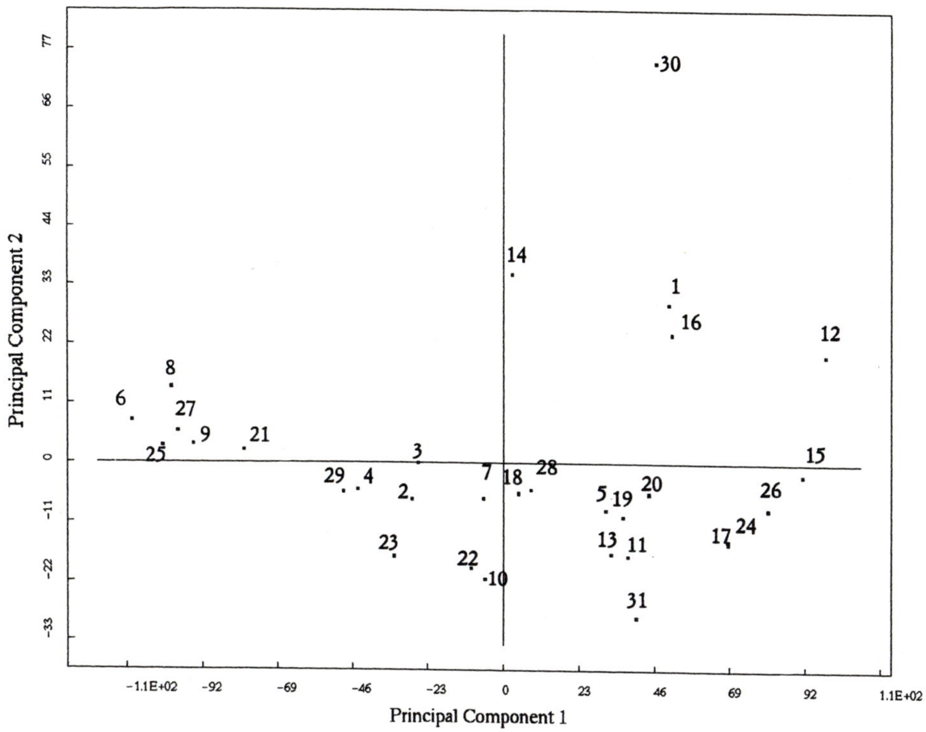

Figure 7. Score plot for the model describing the interactions of 31 probes with one target molecule (TTCGGTT DNA heptuplet).

The PCA Score Contribution (PCS) plots (Figs. 8 and 9) can improve the chemical interpretation of these findings. These plots are made by multiplying the PC loadings by the variable energy fields, and since the fields vary with the probe, the Score Contribution (PCS) plots are related to the type of probe employed. Fig. 8 shows the PCS plot for Probe 8 (sp^2 amine NH_2 cation) while Fig. 9 shows the PCS plot for Probe 15 (oxygen of sulphate or sulphonamide). The 3D regions in the minor groove are not the regions with highest interaction energies, nor the regions with the highest loadings: they are the regions in which the probes best differentiate their interaction with the same DNA target molecule.

The second PC shows that the carboxy (number 31) and the sulphone (number 14) probes are high leverage probes. PCS plots for the second PC shows that there are local regions in the minor groove space in which probes interact in different ways with DNA. The COO^- Probe interacts strongly with particular base-pairs of the heptuplet and this preference is determined both by the chemical properties of the probe and by the geometric properties of the hydrogen bonding pattern. Numerical

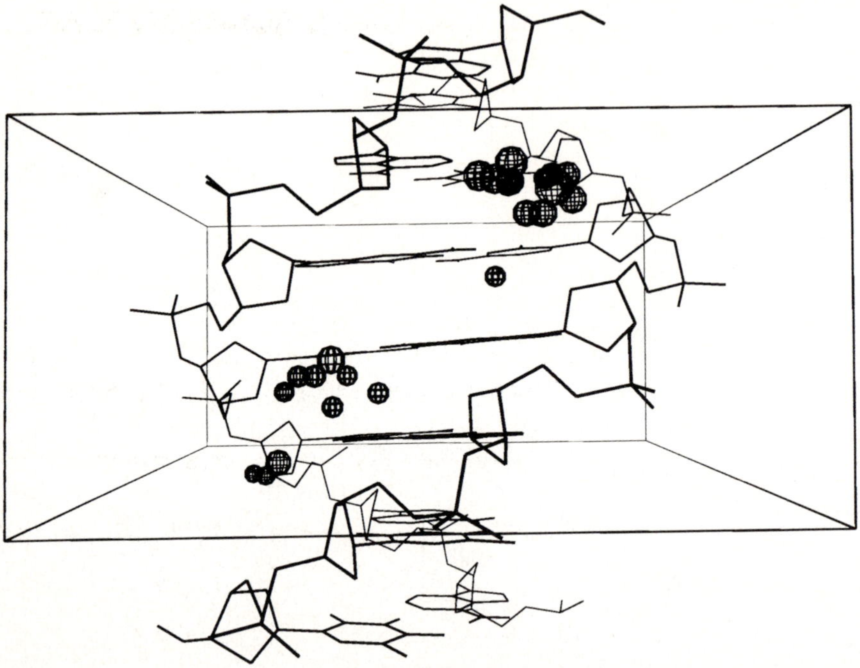

Figure 8. PCA Score Contribution plot describing the selectivity regions for interactions between the TTCGGTT molecule and the sp^2 amine NH_2 probe.

analysis shows that the energy binding differences between the carboxy and sulphone probes in these zones is about 4 Kcal/mol in favour of the carboxy probe.

In conclusion, the score plots of a probe matrix allows the chemical groups which selectivity interact with the target structure to be distinguished while the PCS plots are associated with the spatial regions around the target molecule in which probes interact preferentially in a selective way.

2.3.3.3 PLS Analysis on the Target Matrix

In this paragraph, the selection of variables from a target matrix is discussed using the PLS algorithm implemented in GOLPE [1]. The selection is validated for a series of 36 glucose analogue inhibitors, whose X-ray structures bound to the enzyme glycogen phosphorylase b (GPb) has been determined, either with the aim of overcoming the alignment problem, and to test if the variables (here 3D regions) predicted by GOLPE are as important for inhibition as those ascertained from the X-ray crystallographic studies [20].

It is worth mentioning that this series of compounds raises no doubts or problems as regards to alignment criteria or active conformations. In fact, the X-ray structure

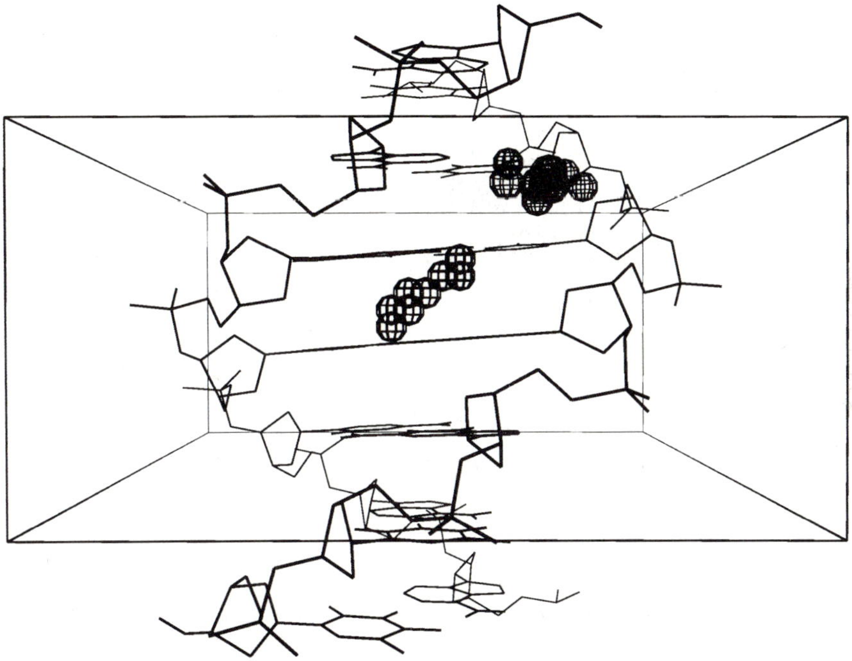

Figure 9. PCA score contribution plot describing the selectivity regions for interactions between the TTCGGTT molecule an the oxygen sulphate or sulphonamide probe.

of the 36 glucose analogue inhibitors in a crystal of GPb, shows that the glucose ring is always superimposed in all the compounds, while the substituents at the *alpha* or *beta* position to C1 exhibit different orientations throughout the protein active site cavities.

The data set consists of 36 compounds (objects) described by means of the interaction energies between the compounds and one specific interacting chemical group (the aromatic hydroxyl probe, OH). It is important to point out that in GRID force field the interaction energy is calculated as the overall sum of Lennard-Jones, electrostatic and hydrogen bond interactions between the probe and the target structure. The overall interaction energy may be negative (attractive) or positive (repulsive) and, in general, the positive interaction values are greater than the negative values. A GRID calculation for each inhibitor compound yields 8400 variables. Performing a PLS analysis on the whole data set, with a positive field cut-off value of 5 Kcal and with a minimum sigma cut-off value equal to one, shows that the resulting model contains all the important experimentally determined regions (Fig. 10), but that also a high number of regions which do not fit the information known from the crystallographic data (see Fig. 11). It is clear that the elimination of variables with small standard deviation does not eliminate all of the noise. Nevertheless, it is difficult to have

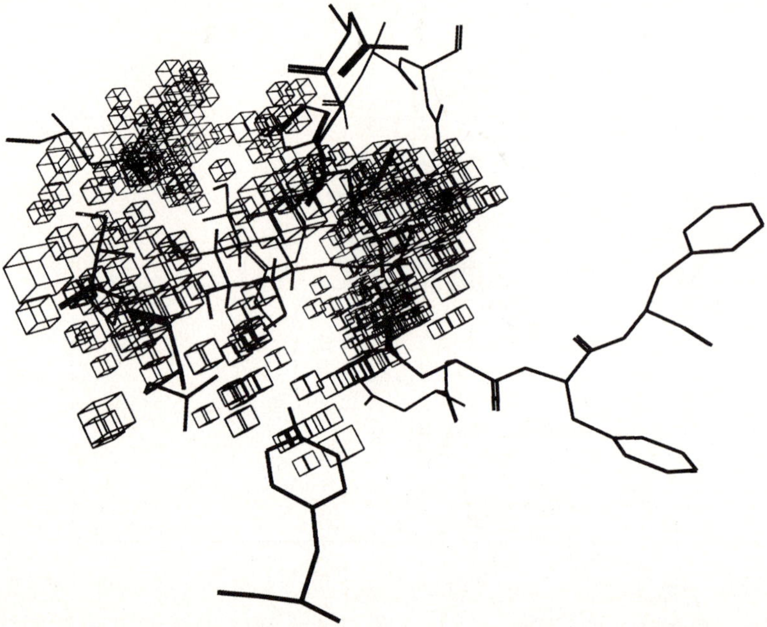

Figure 10. Contour map of the coefficients for the model of interactions between the OH probe and all the 36 target molecules. The regions which do not fit Fig. 11 are generated by noise.

a correct chemical interpretation of the model when so many variables with similar coefficients are present.

Performing the variable selection phase of GOLPE on the same data set leads to the elimination of 98% of variables. A more detailed study of the important three-dimensional regions involved in the final reduced model shows that there is a good agreement between the predicted regions and the experimentally determined important regions for inhibition (compare Figs. 11 and 12).

The reduced model (Fig. 12) clearly shows the effect of these regions on the inhibition capability of the compounds. Moreover, the predictive ability of the reduced model is notably increased, and the errors in the predictivity for the reduced model is in the order of magnitude of the average experimental errors in the K_i values [19–20]. This indicates that GOLPE, when variable selection is performed with a suitable pretreatment, minimizes the risk of overfitting and overpredicting.

2.3.3.4 PLS on Target Matrix as a Strategy to Ascertain the Active Conformation

In the previous example the active conformation of the inhibitor compound was well known from X-ray crystallographic studies. In order to simulate how to ascertain the

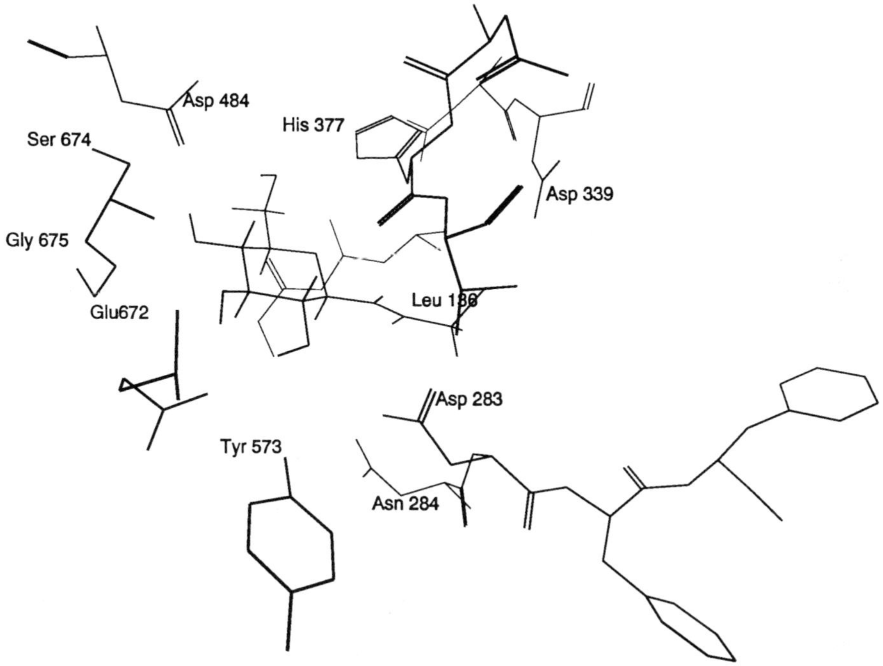

Figure 11. The active site of GPb showing the amino acid residues which interact with the glucose analogue ligands.

active conformation in a study, in which no previous knowledge of the compounds was available, three new inhibitor ligand molecules were added to the previous matrix. The inhibition constants for these new compounds are known, although no structural information from the crystallographic studies was used.

A conformational search procedure was carried out for each of the three new GP ligands using the Systematic Search option in the SYBYL package [21], and 10 energetically accessible conformers were selected as representative conformations for each molecule. Each conformer was then superimposed on to the molecules of the previous data set, maintaining the glucose ring in the same position. The substituted atoms at the *alpha* and *beta* positions to C-1 occupied different regions, which have been targeted in order to choose the active conformation.

Each conformer was then characterized by a vector of descriptor energies using the GRID program, thus, obtaining 10 vectors for each of the three new molecules. The biological response value of the individual compound was then assigned to all of these 10 vectors. It is obvious that a coherent choice of the active conformer should explain the observed variation of the inhibition data values. Finally, these 30 vectors were combined with the previous target matrix, obtaining a general matrix with 66 objects and 8400 GRID energy variables.

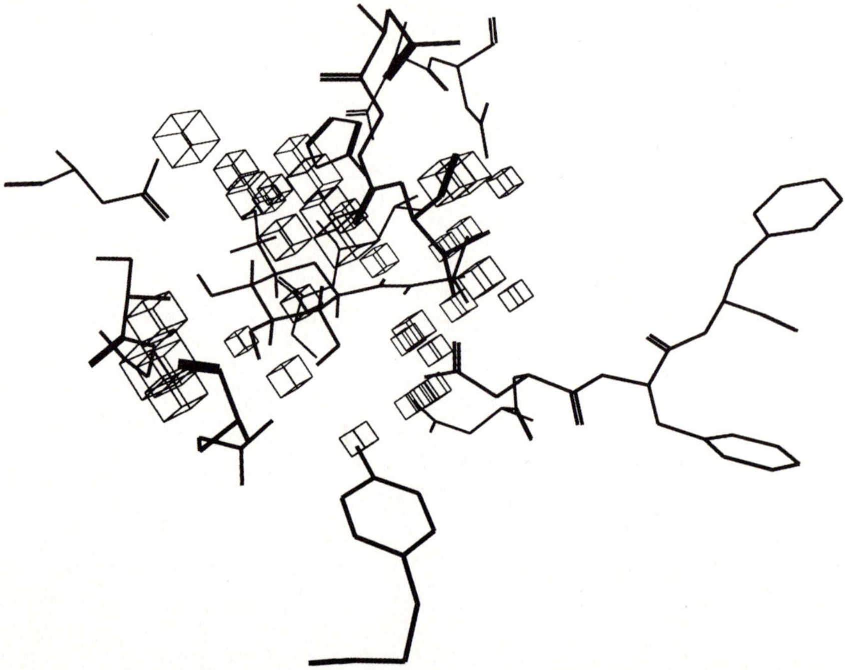

Figure 12. Contour map of the final model obtained by GOLPE showing the important interactions between the OH probe and all the 36 target molecules. The region shown are related to a change of the response value and are in really good agreement with the experimentally determined regions shown in Fig. 11.

The score plot of a PLS model drawn from this matrix (Fig. 13) demonstrates how it is relatively easy to select a few conformers which may represent the active conformation of the new compounds. All the conformers with the same *y* value lie at the same *y* level, but only a few conformers fit the model (continuous line). A chemical interpretation of this finding is that each of these conformer points represents different combinations of the important regions which define the PC model. All these combinations were forced to give the same experimental inhibition value, but only those combinations which explain the overall observed variation of the biological response fall within the model. Accordingly, the conformers n° 4 and 5 for each of the three compounds investigated may be chosen as candidates for the active conformations of the three new compounds. In a further step, GOLPE may be used to check the predictivity of all 2^3 combinations of conformations that can be obtained on adding the three new compounds to the 36 already available: the predictivities obtained will, hopefully, further reduce the number of candidate conformations.

It is important to point out that other authors use PCA, or cluster analysis, in order to classify similar conformations, or in order to ascertain the active conformers.

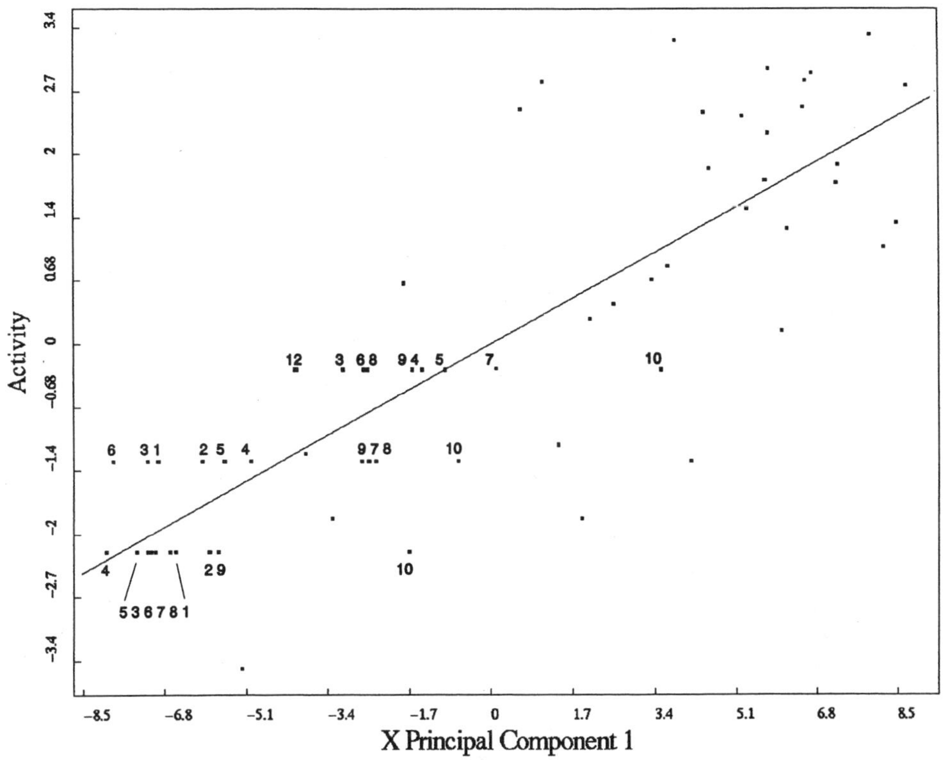

Figure 13. PLS score plot on the target matrix. The latent variable extracted from the *X* descriptors matrix is related to the activity. The numbers refer to the degenerated conformations of the three new compounds.

This is the correct strategy when the biological response in not known. However, if the response is known, we believe that PLS is the correct method for selecting the active conformers which would be consistent with the biological response model.

2.3.3.5 GOLPE with Different 3D Descriptors

The free energy of binding for a ligand-receptor complex may be partitioned into different spatial contributions for different regions of the complex, using empirical energy functions, following an approach similar to that of Williams et al. [22]. This type of treatment has been employed before, notably by Novotny et al. [23], and by Kollman et al. [24], with limited success. Several reasons can be given to explain the limited predictability of this approach. These are inaccuracies of the empirical potential energy functions and the difficulties of modeling the solvent effects in non-covalent interactions in solution. Nevertheless, the most important reason is probably related to the nature of the free energy surface of the ligand macromolecule complex, consisting in multiple minima with similar energy as a result of the compensation of the different contributing terms [25]. Therefore, as a result of the large number of energetic terms and their relative compensation, small errors usually accumulate producing large uncertainties in the calculation.

However, in a set of related compounds, probably not all the energetic terms contribute equally to the difference in the binding free energy, and only a subset of these energetic terms are perhaps responsible for most of the variance. This is the picture given by QSAR studies which show that enzyme inhibition is usually a function of the variance of certain physico-chemical properties at a specific site in the family of compounds under consideration [26].

The fact that QSAR studies are normally in good agreement with X-ray crystallographic studies of enzyme-ligand interactions [27] indicates that these site-specific effects are a consequence of the protein-ligand interaction. Then, if the above-mentioned errors are randomly distributed among the different energetic terms in all the complexes, it should be possible to separate the "energetic signals" from the "background noise". Statistical methods could then be used to obtain a correlation between the binding free energy (response) and a weighted subset of energetic contributions (independent variables).

Preliminary results [Ortiz, A. R., Pisabarro, M. T., Wade, R., *J. Am. Chem. Soc.*, submitted (1994)] obtained by GOLPE indicated that it is possible to determine the energetic terms responsible for the binding free energy, and to derive equations with good predictive properties which can be used for designing better inhibitors. GOLPE can, therefore, be used in combination with any type of descriptor variable by which a set of molecules is characterized in the 3D space.

2.3.4 Conclusions and Perspectives

The GOLPE procedure is implemented in a software package that renders it highly appropriate for QSAR studies. In fact, it works by a very fast algorithm, employing a suitable validation criterion and relies on a unique philosophy of variable selection which is based on the design criteria. This permits a sound evaluation of the predictivity of each individual variable, in this case grid location. This procedure is presumably more reliable than other methods which are based on a stepwise reduction of variables by means of their coefficients, or by means of neural networks, genetic algorithms or simulated annealing, which might necessitate some unwarranted pruning.

Some of the possibilities offered by GOLPE are illustrated in Fig. 14. Its speed is optimized by selecting the appropriate algorithms onyl in the presence of missing data. A data set containing 36 objects and 8400 variables with a single response takes 6 seconds on a 16 MB RAM R4000, and 14 seconds on an R3000 CPU for fitting

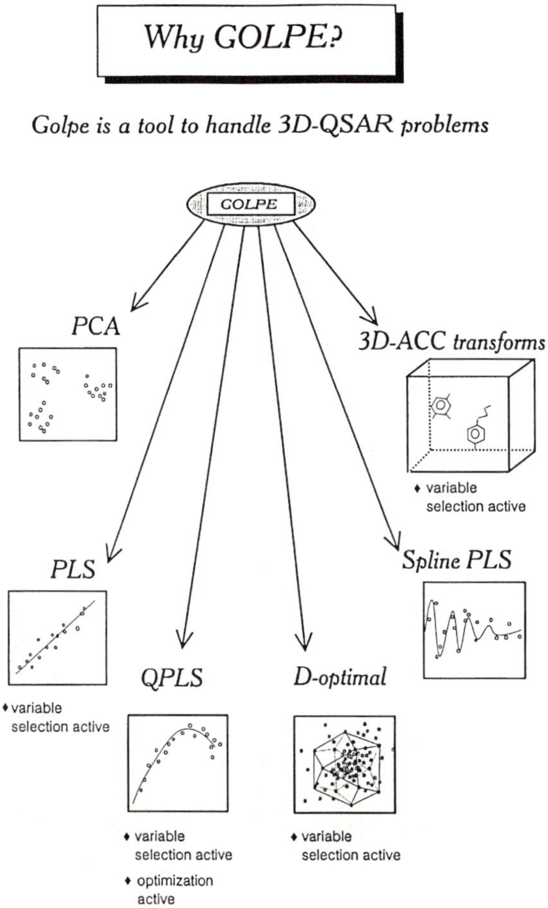

Figure 14. Some of the possibilities offered by the GOLPE procedure.

a PLS model with 5 latent variable. For a leave-one-out prediction run (36 PLS models), it takes 110 and 240 seconds, respectively. D-optimal variable selection with 8000 variables takes only a few seconds, while FFD variable selection on 600 variables takes about 10 hours.

The results of the 3D QSAR study are illustrated as isocontour plots, drawn up from the coefficients of the linear polynomial derived from the PLS model which still contains all the variables. However, we know that their relative importance depends upon model dimensionality, and the optimal dimensionality is not a objective result, as it depends, in turn, on the method of cross-validation [16]. On the other hand GOLPE gives apparently much worse contour plots because of the small number of variables remaining at the end of the analysis.

Accordingly, the presentation of the important variables for each compound is given as the "contribution to the activity" on multiplying each coefficient by the actual value of that variable for the compound, thus, measuring the contribution of the $\boldsymbol{b}_i \boldsymbol{x}_i$ term to y. This procedure gives simpler results on autoscaled data, although the contributions are, strictly speaking, not dependent upon scaling.

The GOLPE procedure was successfully applied by members of our research group to a dozen or so cases over the last two years. These include polychlorodibenzofurans, [1], benzodiazepine analogues [28], prazosine analogues [29], glucose analogues [19], kinurenic acids [Costantino, G., and Pellicciari, R., work in preparation], steroids [Clementi, S., Valigi, R., Cruciani, G., Riganelli, D., Cramer, R. D., and Patterson, D. E.; work in preparation], dipeptides [Riganelli, D., and Merz, A., work in preparation], triazines [Riganelli, D., Clementi, S., and Mabilia, M. A., work in preparation], monosubstituted benzenes [van de Waterbeemd, H., Carrupt, P. A., Clementi, S., Costantino, G., Cruciani, G., Valigi, R., work in preparation], dioxines [Clementi, S., Cruciani, G., Riganelli, D., and Valigi, R., work in preparation], xantines [Clementi, S., Riganelli, D., and Cruciani, G., work in preparation], ACE inhibitors [Davis, A., and Cruciani, G., work in preparation], PLA_2 inhibitors [Ortiz, A. R., Pisabarro, M. T., and Wade, R., *J. Am. Chem. Soc.*, submitted (1994)].

Meanwhile, however, we are well aware that there have been some rumours, casting doubts about the soundness of the GOLPE procedure, which is suspected of finding chance correlations and, therefore, of overpredicting because of the low number of finally selected variables. We can understand such doubts in the light of this problem, but we wish to point out in more detail why we believe that the procedure is foolproof.

We are well aware that, even with PLS, when there are so many variables compared to the number of objects, it is possible to find several combinations of variables which exhibit the same and apparent good predictivity, but this happens only when the data set does not meet the stated requirements and/or when cross-validation is carried out in an inappropriate way. We would like to emphasize once more that the LOO procedure is expected to give overpredictivity with grouped objects, such as they usually are in QSAR studies, although a data exploration by PCA is almost never done to test this. Moreover, change correlations are easily found only when variable selection is made stepwise, taking away those variables that do not predict well, i.e. that do not fit the reduced model(s). In such a case one "prunes" the variables according to the chemometric results.

It is because of this that we stated that GOLPE should only be performed if the whole model showed some predictivity in the first place, or with some designed data sets (see Sec. 2.3.2.1). Furthermore, GOLPE is derived from a predictivity parameter, *SDEP*, which is computed in such a way, from groups randomly formed several times to minimize the risk of obtaining false predictivity estimations (see Sec. 2.3.2). In this context, even the effort of evaluating the risk of chance correlations by extensive simulation studies on random numbers [30] does not appear to be appropriate. Since random numbers should exhibit no predictivity, but they do have "structure", however, as a result of being random, when the validation method is not sufficiently foolproof, they might appear to give predictivity in a few cases. Real QSAR data, however, do have considerable structure, as shown by the variance accounted for in PCA studies, and simulations with random numbers probably does not lead to a good approximation of the real problem.

To our knowledge the best way of checking the reliability of a regression model is to permutate the elements of the **y** vector several times, and in no instance should high Q^2 values be obtained unless one proceeds on to pruning which is forbidden because of the non-predictive nature of the model in the first place. So far, we have published only one illustrative example of such a procedure [1], which is also used in the simulation study [30], but we have used it successfully in several others of the data sets quoted above. Significantly, in one of these studies [Riganelli, D., Clementi, S., and Mabilia, M. A., work in preparation], we have found that GOLPE gave a positive Q^2 value in one case only out of the 32 possible combinations of aligning degenerate molecules as was expected.

Another criticism of the GOLPE procedure was its inability to handle symmetrical molecules. This is certainly true, and is a major drawback of the problem in question and not of the numerical analysis. In fact, we stated that the main problem in 3D QSAR is alignment, but this is, in turn, a consequence of the dependence of the 3D description upon the position of each molecule within the 3D grid. Only until a 3D description, which is independent of shifting or bending a molecule within the grid is derived, will a 3D QSAR table contain variables which are truly congruent and, therefore, appropriate for chemometric modeling.

Imaginative attempts at overcoming this problem have been reported, describing molecular structures in terms of similarity [31] or distance matrices [32]. However, in such cases, the symmetry of the matrices renders interpretation to be somewhat difficult: projection methods such as PLS describe objects in terms of variables and when rows and columns of a matrix are the same, the problem in question becomes intriguing. It seems to us that describing molecules in terms of similarity should be a good approach, provided that the similarity indices all refer to a target compound for each study, and that the similarity can be multivariately evaluated, dessecting the total similarity into a number of separate similarity concepts.

For the time being, we should restrict ourselves to using the alignment criteria developed so far: rigid fit, field fit, overlapping of hydrogen bonding points, fit of dummy atoms as in the APOLLO procedure [33], etc., although we are aware that all these methods are used depending on the problem under investigation and should not be considered as hard and fast rules. Furthermore, such a problem implicitly infers that all computed interactions simultaneously effect the biological response. It might be the case that we should dissect a ligand-receptor interaction into sequential steps, each depending upon specific properties; (a) crossing a membrane, presumably linked to some molecular hydrophobicity parameter, which produces the actual concentration in the cell; (b) molecular recognition, which is presumably an electrostatic interaction, across a large distance and driven, therefore, by molecular electrostatic potentials; (c) binding, which is, namely, due to hydrogen bonding and steric/lipophilic interactions. If this were the case, assigning the same importance to all fields in the PLS analysis might not be the best decision.

We have proposed [34] a promising approach to solving the problem of congruency in the 3D description: the ACC transforms. In fact, there are two major drawbacks with the present use of CoMFA or CoMFA-like approaches in 3D QSAR: the doubts concerning the congruency of the descriptor matrix and the absence of continuity constraints between the fields computed at neighboring grid nodes.

The auto- and cross-covariance (ACC) transforms, proposed by Wold [35] to describe biopolymers were extended to three dimensions to handle 3D descriptions [34]. These transforms are suitable tools for to recognizing the information contained in 3D fields in a way that is much more appropriate for 3D QSAR. In fact, this rearrangement of raw data provided new data which take into account neigboring effects, the required continuity between grid nodes, and which are independent of alignment within the grid lattice. The 3D ACC developed so far, allowed a unique description of degenerate numbering of molecules, but the method still requires considerable development.

ACC transforms cannot describe different conformers in a unique way, since different conformers give different ACC transforms due to the different relative positions of atoms in the 3D space. An overall strategy for handling ACC transforms should be developed, possibly similar to that presented elsewhere [36]. If 3D QSAR might be based on the PLS modeling of the activity vector against the ACC matrix by means of GOLPE, the results would allow the interactions between locations that affect the biological response to be determined, and this would constitute a new tool for mapping unknown receptors.

Acknowledgements

We would like to thank the Italian funding agencies MURST and CNR for granting funds to the research projects in chemometrics, the team of coworkers in Perugia (Massimo Baroni, Daniela Bonelli, Gabriele Costantino, Daniela Riganelli and Roberta Valigi), and M. I. A. (Multivariate Infometric Analysis, Perugia) for support in inplementing the programs.

References

[1] Baroni, M., Costantino, G., Cruciani, G., Riganelli, D., Valigi, R., and Clementi, S., *Quant. Struct. Act. Relat.* **12**,, 9–20 (1993)

[2] Franke, R., *Principal Component Analysis and Factor Analyis.* In: *Chemometric Methods in Molecular Design* (Methods and Principles in Medicinal Chemistry, Vol. **II**), Mannhold, R., Krogsgaard-Larsen, P. and Timmerman, H., eds., VCH, Weinheim, 1995

[3] Wold, S., and Dunn, W. J., *SIMCA.* In: *Chemometric Methods in Molecular Design* (Methods and Principles in Medicinal Chemistry, Vol. **II**), Mannhold, R., Krogsgaard-Larsen, P. and Timmerman, H., eds., VCH, Weinheim, 1995

[4] Wold, S., *PLS.* In: *Chemometric Methods in Molecular Design* (Methods of Principles in Medicinal Chemistry, Vol. **II**), Mannhold, R., Krogsgaard-Larsen, P. and Timmerman, H., eds., VCH, Weinheim, 1995

[5] Sjöström, M. and Eriksson, L., *Applications of Statistical Experimental Design and PLS modeling in QSAR.* In: *Chemometric Methods in Molecular Design* (Methods and Principles in Medicinal Chemistry, Vol. **II**), Mannhold, R., Krogsgaard-Larsen, P. and Timmerman, H., eds., VCH, Weinheim, 1995

[6] Davis, A.M., *3D QSAR Methods.* Sec. 2.2 of this volume

[7] Clementi, S. and Wold, S., *How to Select the Proper Statistical method?* In: *Chemometric Methods in Molecular Design* (Methods and Principles in Medicinal Chemistry, Vol. **II**), Mannhold, R., Krogsgaard-Larsen, P. and Timmerman, H., eds., VCH, Weinheim, 1995

[8] Cramer, R.D. III, Patterson, D.E. and Bunce, J.D., *J. Am. Chem. Soc.* **110**, 5959–5967 (1988)

[9] Goodford, P.J., *J. Med. Chem.* **28**, 849–857 (1985)

[10] Floersheim, P., Nozulak, J. and Weber, H.P., Experience with Comparative Molecular Field Analysis. In: *Trends in QSAR and Molecular Modeling '92,* Wermuth, C.G., ed., ESCOM, Leiden, 1993, p 227–232

[11] Kellog, G.E. and Abraham, D.J., *J. Mol. Graph.* **10**, 212–217 (1992)

[12] Wold, S., Albano, C., Dunn, W.J. III, Edlund, U., Esbensen, K., Geladi P., Hellberg, S., Johansson, E., Lindberg, W. and Sjöström, M., *Multivariate Data Analysis in Chemistry.* In: *Chemometrics.* Kowalski, B.R. ed., Reidel, Dordrecht, 1984, p 17–94

[13] Wold, S., Johansson, E. and Cocchi, M., *Partial Least Squares Projections to Latent Structures.* In: *3D QSAR in Drug Design: Theory, Methods and Applications,* Kubinyi, H., ed., ESCOM, Leiden, 1993

[14] Cruciani, G. and Goodford, P.J., *J. Mol. Graph.* **12**, 116–129 (1994)

[15] Cruciani, G., Baroni, M., Bonelli, D., Clementi, S., Ebert, C. and Skagerberg, B., *Quant. Struct.-Act. Relat.* **9**, 101–107 (1990)

[16] Cruciani, G., Clementi, S. and Baroni, M., Variable *Selection in PLS Analysis.* In: *3D QSAR in Drug Design: Theory, Methods and Applications,* Kubinyi, H., ed., ESCOM, Leiden, 1993

[17] Shao, J., *J. Amer. Stat. Assoc.* **88**, 486–494 (1993)

[18] Box, G.E.P., Hunter, W.G. and Hunter, J.S., *Statistics for Experimenters*, Wiley, New York, 1978

[19] Cruciani, G. and Watson, K.A., *J. Med. Chem.* **37**, 2589–2601 (1994)

[20] Martin, J.L., Veluraja, K., Ross, K., Johnson, L.N., Fleet, G.W.J., Ramsden, N.G., Bruce, I., Ochard, M.G., Oikonomakos, N.G., Papageorgiou, A.C., Leonidas, D.D. and Tsitoura, H.S., *Biochemistry,* **30**, 10101–10116 (1991)

[21] *SYBYL Molecular Modeling System,* Tripos Associates, St. Louis, MO, U.S.A.

[22] Williams, D.H., J.P.L., Doig, A.J., Garder, M., Gerhard, U., Kaye, P.T., Lal, A.R., Nicholls, I.A., Salter, C.J. and Mitchell, R.C., *J. Am. Chem. Soc.* **113**, 7020–7030 (1991)

[23] Novotny, J., Bruccoleri, R.E. and Saul, F.A., *Biochemistry* **28**, 4735 (1989)

[24] Kollman, P., Wipff, G. and Singh, U.C., *J. Am. Chem. Soc.* **107**, 2212–2219 (1985)

[25] Brooks III, C.L., Karplus, M., Pettitt, B.M., *Proteins, a Theoretical Perspective of Dynamics, Structure and Thermodynamics,* Advances in Chemical Physics. Vol. **LXXI**, 1988

[26] Gupta, S.P., *Chem. Rev.* **87**, 1193 (1987)

[27] Hansch, C. and Klein, T.E., *Methods in Enzymology* **202**, 512 (1991)

[28] Allen, M.S., La Loggia, A.J., Dorn, L.J., Martin, M.J., Costantino, G., Hagen, T.J., Koehler, K.K., Skolnick, P. and Cook, J.M., *J. Med. Chem.* **35**, 4001–4010 (1992)

[29] Cocchi, M., Cruciani, G., Menziani, M.C. and De Benedetti, P.G., *Use of Advanced Chemometric Tools and Comparison of Different 3D Descriptors in QSAR Analysis of Prazosin Analog α_1 Adrenergic Antagonists.* In: *Trends in QSAR and Molecular Modeling '92,* Wermuth, C.G., ed., ESCOM, Leiden, 1993, p 527–529

[30] Clark, M. and Cramer, D.R. III, *Quant. Struct.-Art. Relat.* **12**, 137–145 (1993)

[31] Good, A.C., So, S.S. and Richards, W.G., *J. Med. Chem.* **36**, 433–438 (1993)

[32] Bush, B.L. and Nachbar, R.B., Jr., *J. Comp. Aid. Mol. Des.* **7**, 587–619 (1993)

[33] Snyder, J.P., Rao, S., Koehler, K. and Pellicciari, R., *Pharmacochem. Lib.* **18**, 367–403 (1992)

[34] Clementi, S., Cruciani, G., Riganelli, D., Valigi, R., Costantino, G., Baroni, M. and Wold, S., *Pharm. Pharmacol. Lett.* **3**, 5–8 (1993)

[35] Wold, S., Jonsson, J., Sjöström, M., Sandberg, M. and Rännar, S., *DNA and Peptide Sequences and Chemical Processes Multivariately Modeled by PCA and PLS, Anal. Chim. Act.* in press (1994)

[36] Pitea, D., Cosentino, U., Moro, G., Bonati, L., Fraschini, E., Lasagni, M., Todeschini, R., Chemometrics and Molecular Modeling. In: *Chemometric Methods in Molecular Design* (Methods and Principles in Medicinal Chemistry, Vol. **II**, Mannhold, R., Krogsgaard-Larsen, P. and Timmerman, H., eds., VCH, Weinheim, 1995

3 Rational Use of Chemical and Sequence Databases

3.1 Molecular Similarity Analysis: Applications in Drug Discovery

Mark A. Johnson, Gerald M. Maggiora, Michael S. Lajiness, Joseph B. Moon, James D. Petke, and Douglas C. Rohrer

Abbreviations

2D	Two-Dimensional
3D	Three-Dimensional
SAM	Structure-Activity Map
SAR	Structure-Activity Relationship
MEP	Molecular Electrostatic Potential
MSA	Molecular Similarity Analysis
MSV	Molecular Steric Volume
x*MEP*	Extended Molecular Electrostatic Potential

3.1.1 Introduction

For some time, the atom-based superpositioning of 2D or 3D molecular structures has complemented traditional QSAR methods by highlighting structural commonalities and differences [1] and by generating useful chemical descriptors for QSAR models [2, 3]. The recent appearance of rapidly computable similarity measures, which are mainly 2D, has also opened up new opportunities in drug discovery. Notable amongst these are similarity searching [4–6]; substructure similarity searching [7]; similarity, cluster, and dissimilarity selection of compounds for screening [8–10]; and structure-activity mapping [11, 12]. Almost concurrently, atom-based superpositioning has evolved into the more general, but computationally extensive procedure of optimally superimposing or matching the fields which surround molecules.

The application of the concept of molecular similarity to problems in chemistry has grown significantly during the last decade, particularly within the last four to five years. This growth has coincided with an explosion in the number and variety of electronic databases which nowadays serve chemistry and related fields. In this context, the concept of molecular similarity has had an important and unifying effect on the development of powerful methods for "mining" and analyzing the information from these databases in new and novel ways (see [13]). The drug discovery process, which is our focus here, has been, and is continuing to be significantly impacted by these methods.

As the name implies, molecular similarity focuses on molecular features. The manner in which this information is represented is crucial in molecular similarity analysis (MSA). It is convenient to divide the types of representations into two general classes that utilize *atom-based* or *field-based descriptors,* respectively. Atom-based descriptors may include the atoms themselves, molecular fragments or substructures (e.g. functional groups), molecular indices derived by topological methods (e.g. path counts [14]), atomic properties (e.g. electrotopological indices [15] or atomic polarizability), and non-bonded 3D atomic configurations. Field-based descriptors, on the other hand, describe the "micro-environment" surrounding all or a portion of a molecule and may include one or more of the following descriptors: charge or electron density, molecular electrostatic potential (*MEP*), molecular steric volume (*MSV*), hydrophobic field [16], and shape. Additional descriptive features used in MSA can be found in several more recent books [5, 17, 18].

Development of similarity methods requires some means for evaluating which molecular features are shared and those which are not shared by the set of molecules under study. Methods for accomplishing this task of "measuring" the commonalities and differences amongst molecules depend upon some form of optimal *superpositioning* or *matching* of either atom-based or field-based molecular features [19]. Quantification of the degree of superpositioning or matching operationally gives rise to a *similarity measure.*

Most similarity measures that are in use today possess values that lie in the range 0 to 1 with 1 denoting maximum similarity and 0 indicating minimum similarity. In some cases, however, the range goes from -1 to 1 (see Sec. 3.1.4) and [20]). *Dissimilarity measures* complement similarity measures by emphasizing the number of unshared features. Although one can always construct a dissimilarity measure from a similarity measure by taking 1 $-$ (similarity value), dissimilarity measures are usually constructed without reference to similarity measures. Most dissimilarity measures possess values ranging from 0 upwards, with 0 reflecting minimum dissimilarity (i.e. maximum similarity) and large values indicating increasing dissimilarity. Both types of measures are often reflected in the generic terms "similarity measure" or "proximity measure".

Atom-based similarity measures can be applied to molecules portrayed as either 2D or 3D entities. Examples of how such measures can be employed in drug discovery will be presented and discussed in Secs. 3.1.2 and 3.1.3. When more detailed 3D comparisons of molecules are needed, as is the case for many molecules of pharmaceutical interest, field-based similarity measures are required. Field-based measures are much less restrictive than atom-based measures and, consequently, can provide a more flexible characterization of subtle commonalities and differences amongst molecules. An example of the application of field-based measures is given in Sec. 3.1.4.

Molecular similarity analysis can be viewed from both a *global* and a *local* perspective. Global similarity methods are needed in order to take advantage of the information stored in large chemical databases. These methods require that the "feature set" used to compute similarity is obtainable for essentially all of the compounds in a database. Local similarity methods, on the other hand, deal only with small subsets of molecules within a database and, thus, the feature sets need only be obtainable for the subset of molecules under study. Moreover, due to the size of most chemical databases, global methods must be relatively fast and must produce unambiguous results, while local methods can employ more elaborate and computationally demanding procedures that may, and in many cases do, exhibit a set of comparable similarity values for each pair of molecules considered. Hence, atom-based methods can generally be applied both globally and locally, while field-based methods are generally confined to local similarity studies.

3.1.2 Similarity-Based Compound Selection

3.1.2.1 Similarity Measures and Neighborhoods

As suggested in the introduction, molecular similarity concepts have been employed to address quite diverse problems. Underlying these diverse applications is the simple and intuitive concept or a *similarity neighborhood* that "surrounds" a compound.

In practice, construction of a similarity neighborhood requires three things in addition to a similarity measure. First, there must be a set of compounds. This is typically a database of structures of a collection of compounds. Second, there must be a reference compound from which the neighborhood is constructed. This is often called a "query compound" in database terminology. Finally, the similarity value at the neighborhood boundary or the neighborhood size must be specified. If a boundary value is specified, the neighborhood consists of all compounds whose similarity to the reference compound exceeds the boundary value. In this case, the number of compounds in the neighborhood (the neighborhood size) depends on the reference compound. If a neighborhood size, say N_{size}, is specified, then the neighborhood consists of the reference compound and the $N_{size} - 1$ compounds most similar to the reference compound. In this case, the similarity value at the neighborhood boundary depends on the reference compound.

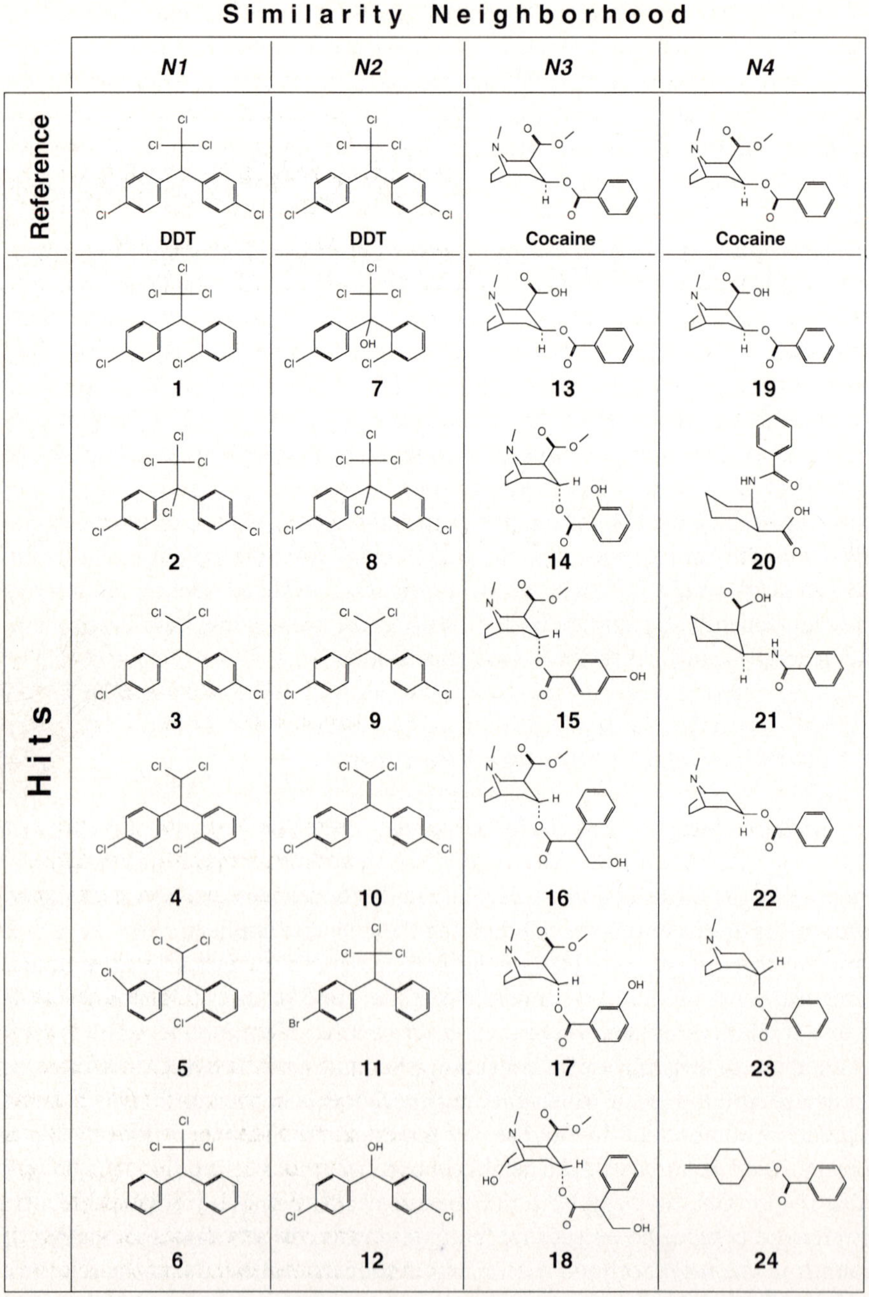

Figure 1. Four size 7 similarity neighborhoods in the Fine Chemicals Directory based on a rigid (DDT) and a flexible (cocaine) reference compound and on a fragment-based (*SM*1) and a 3D atom-based (*SM*2) similarity measure.

Four similarity neighborhoods, *N*1, *N*2, *N*3, and *N*4, with neighborhood sizes of seven (i.e. $N_{size} = 7$), are given in Fig. 1. The reference database is the Fine Chemicals Directory (available from Molecular Design Ltd). Each column of structures represents a separate neighborhood and is headed by the respective reference compound. The compounds of each neighborhood are ordered, so that the measured similarity values decrease as one proceeds down the list away from the reference compound.

The four neighborhoods arise from two reference compounds, *DDT* and cocaine, and two similarity measures, *SM*1 and *SM*2, which employ 2D and 3D atom-based similarity procedures, respectively. The relationship of the similarity measures to the neighboorhoods they induce is as follows: *SM*1 → (*N*1, *N*3) and *SM*2 → (*N*2, *N*4). This can also be written, using a notation that explicitly defines the similarity measure used to induce the given neighborhood, as $N1_{SM1}$, $N2_{SM2}$, $N3_{SM1}$, and $N4_{SM2}$. However, designation of the similarity measure will be omitted, except in cases where confusion may arise from its omission.

The nature of the four neighborhoods provides information about the two similarity measures: *N*1 and *N*2 employ DDT, which is a relatively rigid molecule, as the reference structure, while *N*3 and *N*4 employ cocaine, which is a more flexible molecule, as the reference structure. Clearly both similarity measures relate to our intuitive concept of molecular similarity. Only a very small percentage of the compounds in the Fine Chemicals Directory are "DDT-like" compounds. Yet both *N*1 and *N*2 are comprised entirely of such compounds. It is equally obvious that the two measures reflect different aspects of structure, for if they did not, then the two similarity measures would have necessarily generated identical neighborhoods.

On first encountering two similarity measures with differing neighborhoods, it is appropriate to question which one provides the best measure of molecular similarity. However, it is more practical, if not more realistic, to analyze the contexts in which each measure may be the most useful. In this regard, it is helpful to know how the similarity measures were constructed.

SM1 – Fragment-Based 2D Similarity Measure

*SM*1, giving rise to *N*1 and *N*3, is based upon a fixed set of structural fragments. Computationally, each structural fragment can be represented by a one-dimensional array, in which each element is represented by a single "bit". Each bit is set to 1 if that fragment is present in a particular molecular structure or is set to zero otherwise. The order of elements is arbitrary. A contrived example would be, if the fixed set of fragments consists of a OH, COOH, and phenyl group, then benzoic acid would be represented by the OH bit set to 0, the COOH bit set to 1 and the phenyl bit set to 1. Such bit representations, consisting of 300–2000 fragment bits, have traditionally been used to quickly screen out unwanted compounds in substructure searching.

Here the bit representation, which is implemented in COUSIN [21], a proprietary structural database and retrieval system, is used.

Fragment-based similarity measures relate the number of fragments common to two structures (*NC*) to the number of fragments (*ND*) that differ between these structures, i.e. are possessed by one of the structures, but not by the other. These numbers are combined into a single value by a similarity coefficient. The Tanimoto coefficient, recommended by Willett and Winterman [22], is one of the more common coefficients for fragment-based similarity measures and is used here. It is defined by $NC/(NC+ND)$.

SM2 – Atom-Based 3D Similarity Measure

*SM*2 giving rise to *N*2 and *N*4, is based upon the similarity of the spatial arrangement of the atoms of a reference molecule with comparable atoms of another molecule found in a given database [23]. In these examples, a 3D version of the Fine Chemicals Directory generated using the CONCORD program [24] was used. A clique-detection algorithm [25] was employed to determine any possible correspondences, based on interatomic distance between the atoms of a database molecule with those of the reference molecule. An optimal overlay is produced for each possible correspondence by least squares fitting. Several hundred different overlays may be generated for every database molecule: each overlay is scored according to the number of atoms that fall within some tolerance distance from similar atoms in the reference molecule. The best *k* overlays are retained as "hits" from the search, where *k* is a user-defined number. It is possible, but highly unlikely, that a single compound will have more than one of the top *k* overlays. Although the procedure actually generates a set of 3D overlays, only the 2D representations of the highest-scoring hits are shown in columns *N*2 and *N*4 of Fig. 1.

3.1.2.2 Application of 2D and 3D Similarity Measures

The resemblences of similarity neighborhoods generated by the two atom-based similarity measures, *SM*1 and *SM*2, can be understood in terms of the method in which the two measures were constructed. In the case of *SM*1, the fragment-based measure, similar fragment bit representations imply similar bonding structures. This also applies to *SM*2, the 3D atom-based measure, in the case of rigid molecules. In this case, 3D structure is essentially determined by bonding structure and, thus, similar 3D structures should possess similar fragment bit representations. This, however, does not obtain for *SM*2 in the case of flexible molecules, where the underlying bonding structures may differ appreciably due to conformational flexibility, hence, their fragment bit representations are also likely to be dissimilar.

The consequences of these differences in the two similarity measures lead to corresponding differences in the similarity neighborhoods that they generate. For example, the bonding structure of the relatively rigid DDT molecule largely determines its 3D shape, and since molecules with similar bonding structures possess similar fragment bit representations, both neighborhoods, *N*1 and *N*2, obtained from the two measures are rather similar (Fig. 1). A comparable result is not expected for the conformationally more flexible cocaine molecule. And this was found to be the case for *N*3 and *N*4 (Fig. 1). Moreover, the greater diversity of bonding structures found in *N*4 is not surprising, since the 3D atom-based method emphasizes the similarity of spatial configurations of comparable atoms among molecules and is not constrained by their underlying bonding structure.

The above example shows that different similarity measures may be useful for different purposes. However, this begs the question as to which method should be used for a given task. To clarify this issue, consider the following example. Suppose a compound with an interesting pharmacological activity is discovered. How can new, structurally-related molecules be found? Certainly, atom-based similarity methods ought to be of use here. But, the question remains as to which specific method should be used. If molecules that differ only slightly from the reference (i.e. active) molecule in bonding structure are desired, a fragment-based approach is recommended. This follows on from the fact that similar molecules have similar fragment bit representations and, hence, similar bonding structures. If, on the other hand, more general bonding patterns are desired, an atom-based 3D method is recommended. As discussed above, this approach is likely to produce a greater yield of diverse bonding structures. In both cases, however, the results obtained will depend not only upon the similarity measure used, but also upon the conformational flexibility of the molecules under study. The differences obtained by the two similarity methods discussed here, are accentuated significantly for conformationally flexible molecules.

If additional specific structural information, such as the nature of the *pharmacophore* is available, both 2D and 3D similarity methods may also be of use. Two-dimensional methods are, however, considerably more limited than 3D methods in cases where the subset of atoms in question are not bonded, a situation which arises in most instances involving pharmacophores.

3.1.2.3 Application of Dissimilarity-Based Compound Selection for Broad Screening

The concept of a similarity neighborhood is useful for many purposes besides browsing through a database and searching for similar compounds. In broad screening programs, it is usually desirable to screen as many diverse structures as possible. Lajiness, et al. [9] briefly described three different approaches for selecting a set of structurally diverse compounds from a large compound collection. One of these, cluster analysis, is discussed in Sec. 3 of this volume. Here a maximum dissimilarity approach is considered.

In principle, a maximum dissimilarity method can be based upon any proximity measure. However, as maximum dissimilarity-based compound selection procedures can be computationally demanding due to the size of typical compound collections, only fast computable and globally available measures, such as the fragment-based measures described above, are used in practice. The dissimilarity selection process is as follows. A compound, C_1, is selected at random from a given compound collection. A second compound, C_2, which is maximally dissimilar from the first one, is selected. A third compound, C_3, which is maximally dissimilar to the first two, is selected. And this process is continued until the desired number of dissimilar compounds is obtained, ensuring at each iteration that the compound selected is maximally dissimilar from all of the previously selected compounds.

The preceding examples illustrate the basic concept of a similarity neighborhood and the role it plays in defining similarity and dissimilarity searches. As noted earlier, such searches are used for browsing through a database, selecting compounds for screening, and examining the effects of small structural changes. Other uses are emerging, such as characterization of the diversity of structures in a database and development of diversity criteria for efficiently augmenting a compound collection. It is encouraging, and somewhat surprising, that such significant and diverse applications can arise from such a basic idea as a similarity neighborhood.

3.1.3 Structure-Activity Maps (SAMs)

3.1.3.1 A Visual Analogy

The diverse uses for the ordered sets of compounds defined by similarity neighborhoods is evidence of the amount of information contained in those sets. Yet more information is available in the similarity region surrounding the reference compound. For instance, the ordered list in *N*4 (Fig. 1) indicates correctly that compound **20** is less similar to cocaine than is compound **19**. It also implies, possibly incorrectly, that compound **19** lies between cocaine and compound **20** in this particular 3D similarity space. It would be similar to suggesting that New York is further away from Detroit than is Chicago, but incorrectly suggesting to the geographically naive, that Chicago lies in between Detroit and New York. Knowledge of the distance from Chicago to New York resolves the ambiguity. However, knowing all of the paired distances between a large number of cities is somewhat daunting and usually not necessary. A travel map efficiently prioritizes this paired distance information by arraying the cities on a plane and connecting neighboring cities by roads. An analogous concept of a map of structures will now be developed in which the cities are replaced by structures, and the lines representing links between neighboring cities are replaced by lines linking "neighboring" structures.

3.1.3.2 Representing Inter-Structure Distances

For the purpose of illustration, a region in similarity spacc consisting of the 20 compounds in the Derwent Standard Drug File (available from Molecular Design Ltd) most similar to carbamazepine – as defined by the preceding fragment-based similarity measure – is used. These compounds are given in Table 1 in the order of decreasing similarity to carbamazepine along with their principal pharmacological activity. The corresponding structures and activities can be found in Fig. 2.

To construct a structure map of the 20 compounds, first compute all of the 190 paired distances. In this example, each distance is calculated using the *bond-deletion dissimilarity measure* described in [12]. Basically, the bond-deletion distance between two molecules is the minimum number of bonds that must be deleted in order to obtain a substructure shared by both molecules. However, the details of this particular measure are unimportant here, as our construction procedure is valid for mostly any dissimilarity measure and is easily adapted to accomodate most similarity measures as well. The important feature is that every pair of structures has a value representing the distance between them.

These 190 distances can be reprsented by linking each pair of structures with a line whose length is related to the corresponding distance. However, as much of the distance information is redundant, 190 connecting lines gives rise to an informational overload which increases non-linearly with the number of structures under consider-

Table 1. Twenty compounds from the Derwent Standard Drug File which are the most similar to carbamazepine in comparison to a fragment-based similarity measure. The compounds are listed in decreasing order of similarity. Major pharmacological activities are indicated by C – anticonvulsant, D – antidepressant, and P – psychostimulant.

No.	Compound	No.	Compound
1	Carbamazepine (C)	**11**	CGP-16997 (C)
2	Dihydrocarbamazepine (C)	**12**	Carbadiol (C)
3	GP-37-375 (C)	**13**	19148-RP (P, D)
4	Hocarbam2 (C)	**14**	CGP-10000 (C)
5	CGP-9055 (C)	**15**	CGP-5924 (C)
6	Hocarbam3 (C)	**16**	Carbaepox (C)
7	CGP-10795 (C)	**17**	19749-RP (P, D)
8	CGP-077 (C)	**18**	Metapramine (P, D)
9	GP-47779 (C)	**19**	Didesipramine (P, D)
10	Oxcarbamzepine (C)	**20**	23669-RP (P, D)

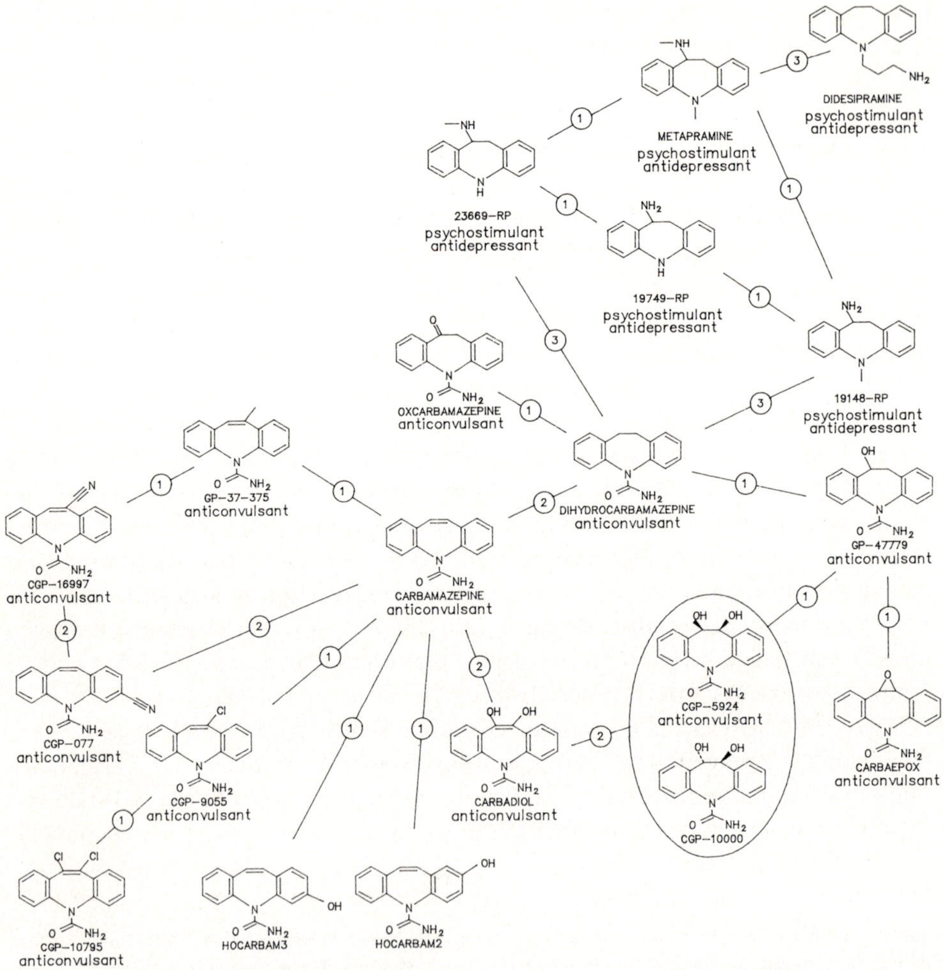

Figure 2. Structure-activity map of a size 20 similarity neighborhood of carbamazepine constructed from the Derwent Standard Drug File using a fragment-based similarity measure. Edge weights reflect bond-deletion distances. Edges were pruned using the Relative Neighbor Rule.

ation (for N structures there are $N(N-1)/2$ lines connecting them). To illustrate this, return to the analogy of a travel map.

Suppose the *shortest* route to Chicago from New York is via Detroit. Given that we indicated the distance from New York to Detroit and the distance from Detroit to Chicago by a connecting line, little additional information is gained by drawing a line directly from New York to Chicago. Such additional lines would only clutter the road map.

The key is to decide which pairwise distances need to be represented by connecting lines and, which do not. Different sets of rules have been developed for eliminating largely redundant lines [26]. Toussaint's simple relative neighbor rule [27] will be illustrated here. Consider any three structures and their connecting lines which are appropriately placed apart to form a triangle. If one of the lines is longer than either of the other two lines, then this line is tagged. When every possible combination of three structures has been examined in this manner, all tagged lines are deleted. The next section will illustrate the high information content in the remaining lines and will show how they focus attention on a number of simple comparisons available in the set of structures under consideration.

3.1.3.3 Structure Maps

The "net" of connected lines with structures at the nodes where the lines connect or end becomes a structure map when it is drawn out on a sheet of paper. However, just as many molecular structures cannot be drawn on a sheet of paper in such a way that the lengths of the drawn bonds are proportional to the actual bond lengths, most of these "nets" cannot be drawn on a sheet of paper without modifying the lengths of the connecting lines. However, the distance information can be preserved by labeling each line with its associated distance. Fig. 2 shows one such drawing amongst the many possibilities.

Before discussing the aesthetics of this particular mode of representation of the relationships, some of the logical considerations inherent in the "net" itself should be highlighted. Firstly, the "net" is an example of a weighted graph [28], which is called a *proximity graph.* As such, it consists of a set of vertices (the structures) and a set of edges (the lines) with assigned weights (the distances). What can be deduced from the weights of the edges of this particular proximity graph? Here the bond-deletion distance has a simple interpretation. The weight of an edge is the number of bonds that must be deleted in order to obtain a common substructure of the two structures linked via that edge. By scanning the structure map for edges with weights of 1, we quickly find all the pairs of structures that differ by only one bond.

What can be deduced from the distances or weights of the "tagged" edges that were deleted? Generally speaking, not much can be deduced with regard to any particular deleted edge, except that it was part of a triangle in which it formed the longest edge. However, much more could be deduced in the "topological" sense.

Consider the homologous series methane, ethane, propane, and butane. The complete structure net, based on the edge-deletion distance, is laid out in the top of Fig. 3 with one edge crossing. By applying the relative neighbor rule to the triangle consisting of methane, ethane, and propane, the methane-propane edge is tagged. After applying the rule to the metane, propane, and butane triangle, the methane-butane edge is tagged. Similarly, the ethane-butane edge is tagged. Removing all tagged edges and aesthetically drawing the resulting proximity graph, we obtain the structure map at

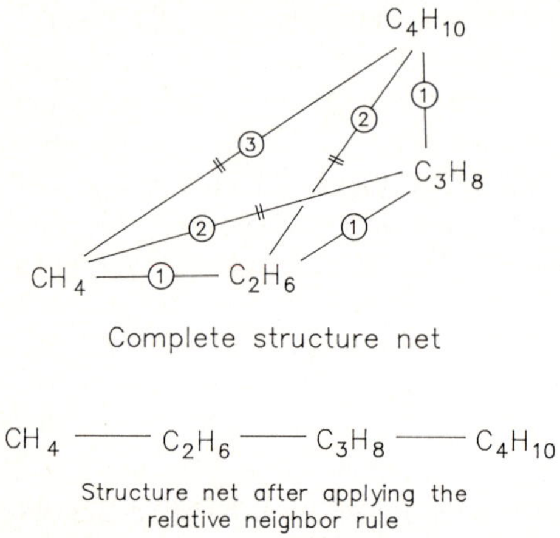

Figure 3. Structure maps of methane, ethane, propane and butane based on the bond-deletion distance. Hashed edges of the complete structure net are deleted according to the Relative Neighbor Rule to form a "path" of homologous structures.

the bottom of Fig. 3. In graph theory, the underlying proximity graph is called a path. Viewed topologically, the structure map in Fig. 2 is clearly not a path.

A proximity graph conveys topological information much better than actual distance information. In fact, not much is lost by erasing the weights in Fig. 2 and retaining only the edges. To illustrate this, consider the cycle of four compounds (i.e. a "4-cycle") appearing on the left side of the figure, namely, carbamazepine, GP-37-375, CGP-16997, and CGP-077. It is can be seen that GP-37-375 lies between carbamazepine and CGP-16997, and that no compound lies between CGP-077 and carbamazepine. This 4-cycle contains all of the structural changes related to the placement of the nitrile group. Such information is not easily conveyed in structural lists.

The topological information inherent in the proximity graph of a SAM is always preserved regardless of how the proximity graph is depicted on a page. However, different depictions focus attention on different features of the proximity graph and a choice arises. The proximity graph can be laid out in such a way that the physical distances between the structures on the page correlate with their distances in similarity space using various linear of non-linear mapping techniques dealt with elsewhere in this volume. Such mapping layouts may obscure similarity relationships amongst neighboring compounds if there are many edge crossings. In Fig. 2, the proximity graph is laid out so as to minimize edge crossings, thus, emphasizing the neighboring relationships. Both layouts can be useful.

3.1.3.4 Coloring a Structure Map

A structure map can be viewed as a layout of a proximity graph in which the vertices are "colored" (i.e. labeled) by the corresponding structures. This is analogous to viewing a chemical graph as a graph in which the vertices are colored by atom types and the edges are colored by bond types. The vertices of a proximity graph can also be colored by the values of one or more biological activities associated with the corresponding molecules. A structure map, additionally colored by activity, is called a *structure-activity map* or SAM. Fig. 2 is a SAM in which "activity coloring" gives the major pharmacological category of that compound as defined by the Derwent Standard Drug File.

A SAM can be viewed as a picture taken from a particular perspective within a "structure-activity space" generated by a particular proximity measure. While it is not the space itself, it can be very useful for acquiring an immediate insight into the nature of the space and into many of the structure-activity relationships (SARs) that exist on the space. For such relationships to exist, an important similarity principle must be satisfied, namely that *similar structures should generally possess similar properties or activities* [29].

Evidence to support this principle can be seen in Fig. 2: anticonvulsants lie in the lower half of the figure, while psychostimulants and antidepressants lie in the upper right corner. Clearly, neighboring structures in this small region of 20 compounds in SAR space are in general agreement as regards to their major pharmacological activity.

The qualifier "generally" is needed, however. A stricter version of the principle that similar structures always have similar properties would break down every time a small structural change led to a big effect on the property of interest. Such "breakdowns" provide critical scientific insights. The boundary in Fig. 2 where the anticonvulsants meet the psychostimulants and antidepressants is a good place to look for one such "breakdown". The SAM suggests that moving from the amines to the ureas greatly alters the dominant pharmacological activity in this region of structures.

SAMs, which deal with more complex situations, can be found in several references [11, 12, 30]. In Fig. 2, the activity is a simple "yes" or "no" coloring as regards to the pharmacological classification of the compound. In the work of Gifford et al. [30], activity coloring is again a "yes" or "no" coloring as regards to the occurrence of *N*-oxidation, but in this case the structures are replaced by metabolic sites where either *N*-demethylation or *N*-oxidation could occur. The resulting SAM suggested a steric descriptor upon which a quantitative prediction of the relative frequency of *N*-oxidation was based. Recently, Johnson [12] illustrated the use of SAMs in viewing structure-activity relationships involving quantitative potency estimates, and in studying how the effect of a particular structural change is a reflection of the structural environment of that change.

3.1.4 Field-Based Similarity Methods

The construction of structure maps usually requires considerably more computational effort per compound than is required with a similarity search, but these maps generaly provide more information on the similarity relationships amongst a set of compounds. Field-based similarity methods usually require even more extensive computations in molecular comparisons. However, this extra effort can reveal additional detailed and often less obvious information on the commonalities and differences among compounds. Thus, in many such cases, the resulting field-based superpositions of two compounds are as interesting, if not more so, than the actual value of the similarity measure upon which they are based.

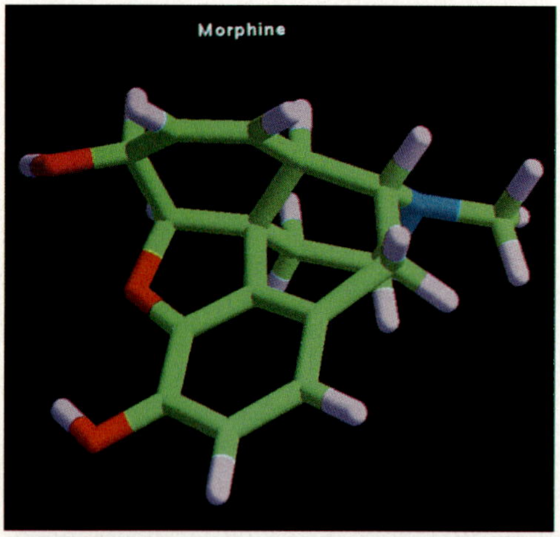

a

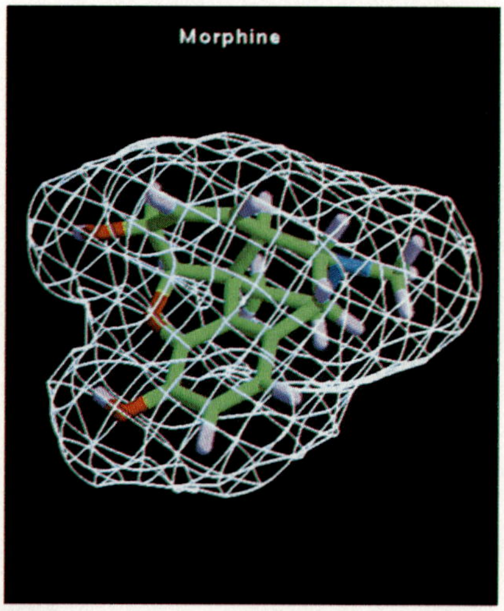

b

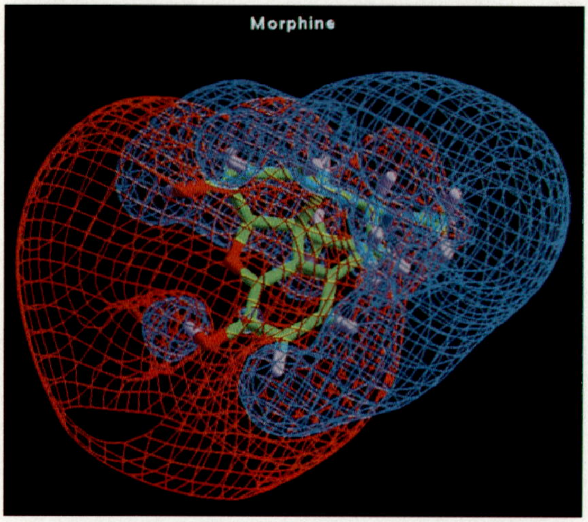

c

Figure 4a. Energy minimized structure of morphine; carbons are shown in green, hydrogens in magenta, oxygen in red and nitrogen in blue.

Figure 4b. The molecular electrostatic potential field of morphine with isopotentials contoured at +10, blue, and −10 kcal/mol, red.

Figure 4c. The molecular steric volume of morphine contoured at the van der Waal's distance from the atoms.

In field-based methods, the similarity of one or more molecular field properties evaluated at field points surrounding a molecule is used as the basis for quantitating molecular similarity. Field properties are determined primarily by the arrangement and electronic nature of the atoms constituting a molecule, and they furnish a much more detailed representation of the electronic and steric characteristics of a molecule than is provided by the spatial configuration of the atoms alone.

3.1.4.1 Field-Based Similarity Measures

The basic idea of field-based similarity is illustrated by the following example which uses the molecular electrostatic potential (*MEP*) to determine electrostatic molecular

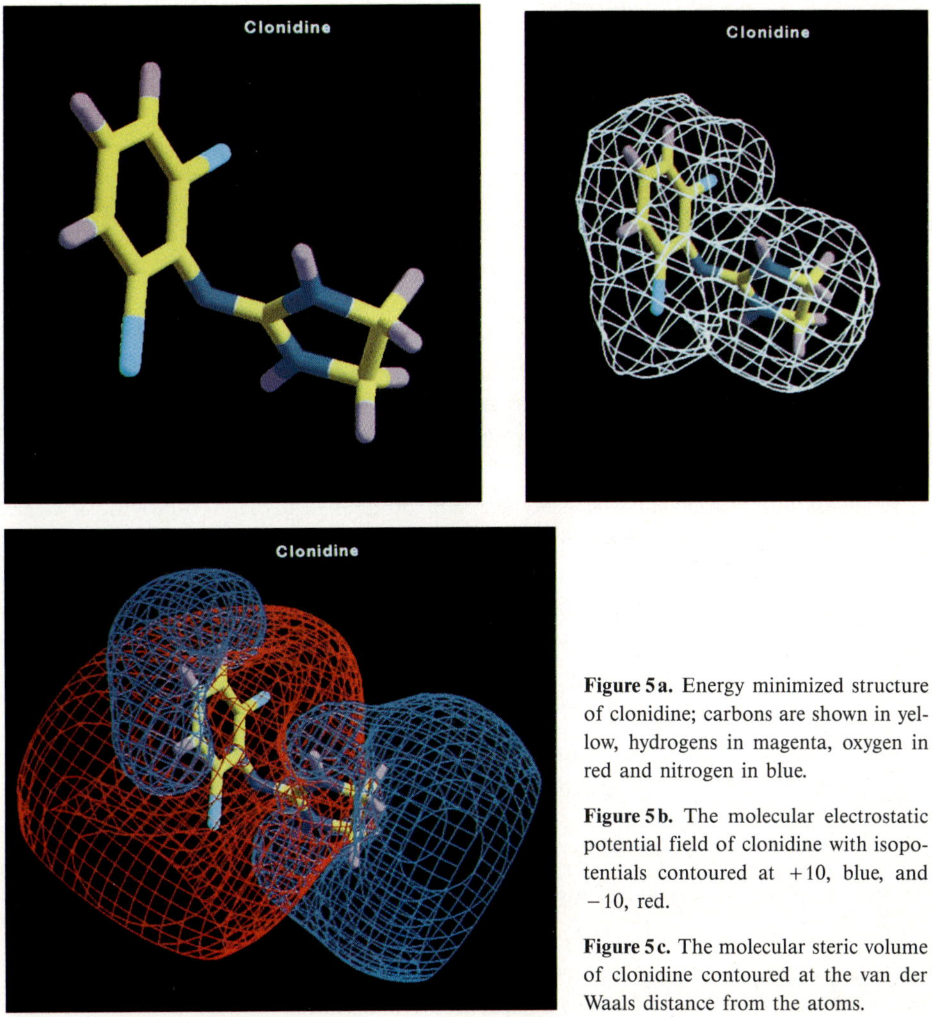

a

b

c

Figure 5a. Energy minimized structure of clonidine; carbons are shown in yellow, hydrogens in magenta, oxygen in red and nitrogen in blue.

Figure 5b. The molecular electrostatic potential field of clonidine with isopotentials contoured at +10, blue, and −10, red.

Figure 5c. The molecular steric volume of clonidine contoured at the van der Waals distance from the atoms.

similarity. Consider two molecules which have been structurally superimposed in a predetermined relative orientation. We may construct a grid of points surrounding the molecules, and evaluate the *MEP* of each molecule at every point. Let u_k and v_k be the value of the *MEP* of the respective molecules at the kth grid point. The molecular similarity measure may be determined using the Carbo Similarity Index [31].

Alternatively, the Hodgkin-Richards Similarity Index [32],

$$C = \frac{\Sigma_k u_k v_k}{\sqrt{\Sigma_k u_k^2}\sqrt{\Sigma_k v_k^2}} \tag{1}$$

$$H = \frac{2\Sigma_k u_k v_k}{\Sigma_k u_k^2 + \Sigma_k v_k^2} \tag{2}$$

or that developed by Petke may be used [20]:

$$J = \frac{\Sigma_k u_k v_k}{\max\,(\Sigma_k u_k^2, \Sigma_k v_k^2)} \tag{3}$$

These indices may be evaluated by summation of the grid points as shown in the following example, or alternatively, by integration of approximate expressions over all space [33]. Each of the above indices provides a numerical value for the similarity ranging from -1 (full dissimilarity) to $+1$ (full similarity). In addition to the *MEP*, other related electronic properties such as the electron density or electric field may be used.

Steric similarity may also be used and can be determined by employing a steric molecular "field" descriptor to represent the molecular steric volume (*MSV*). For example, such a descriptor may be constructed by representing the "size" of each atom by a spherical gaussian, and using the function

$$u_k = \Sigma_m c_m \exp\left[-a_m (r - R_m)^2\right] \tag{4}$$

where R_m specifies the position of atom m, c_m and a_m are parameters characteristic of atom m, and the sum is taken over all atoms in a given molecule.

3.1.4.2 Field-Based Molecular Superpositions

Field-based methods may also be used to provide robust procedures for determining structural superpositions. This approach involves a straightforward extension of the above, in which the similarity index is employed as an objective function in an optimization procedure, and the relative orientation of the two molecules is varied in order to maximize the value of the index. A number of options are possible in such "field-fitting" procedures, including optimizations based on combined electrostatic and steric similarity, and optimizations, in which selected torsional angles (i.e. flexible fitting") in one or both molecules are varied in addition to relative translation and rotation. The use of field-based similarity methods for evaluating and optimizing the match between molecules provides a means of overcoming a fundamental deficiency in atom-based matching, namely that of choosing which atoms to match in the fitting process.

3.1.4.3 An Example of Field-Based Fitting: Morphine and Clonidine

Two molecules with very different structures, morphine and clonidine, the latter being an effective α_2-adrenergic agent for treating morphine withdrawal symptoms [34], illustrate the use of field-based methods to explore the similarity between the two molecules. The 3D molecular structures were obtained by energy minimization using the MM2 molecular mechanics potential-energy function [35] starting from the X-ray crystal structure conformations [36, 37]. The MOPAC program [38, 39] was then used to evaluate the partial atomic charges of each molecule. Fig. 4a shows the structure of morphine which was used as the prototype molecule in the matching process. An isopotential drawing, Fig. 4b, contoured at +10 (blue) and −10 kcal/mole (red) shows the general location and strength of the electrostatic potential field surrounding the molecule. Similarly, Fig. 4c illustrates the steric field of the molecule contoured at the van der Waals radii. Fig. 5 shows the corresponding drawings for the clonidine molecule.

Three levels of field-based similarity matches have been performed; *MEP*, extended-*MEP* (x*MEP*), and x*MEP* plus *MSV*. The Petke similarity index was used in all of the calculations described here. The match, involving only the *MEP* fields, is the simplest of all the approaches and contains the least information about the actual molecular structure. Moreover, the nature of the function used to evaluate the "standard" *MEP* is comprised of terms containing "$1/r$", where r is the distance of the field point to an atomic center: as $r \rightarrow 0$, $1/r \rightarrow \infty$ and, thus, grid points that lie too close to an atomic center must be excluded. This gives rise to "holes" in the grid surrounding each atom position in the molecules being matched, a limitation that is removed when x*MEP* fields are used. In the latter case, the contribution to the *MEP* field in the region surrounding an atom is treated as a constant, removing the necessity to exclude grid points in this region.

Fig. 6 shows an optimized superposition obtained for morphine (green) and clonidine (yellow) with an *MEP*-based similarity value of 0.78. It is evident that while

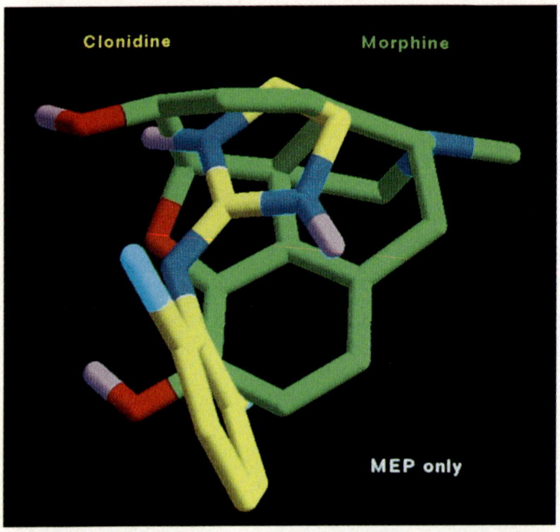

Figure 6. A superposition of morphine, green, and clonidine, yellow, obtained by matching the *MEP* field.

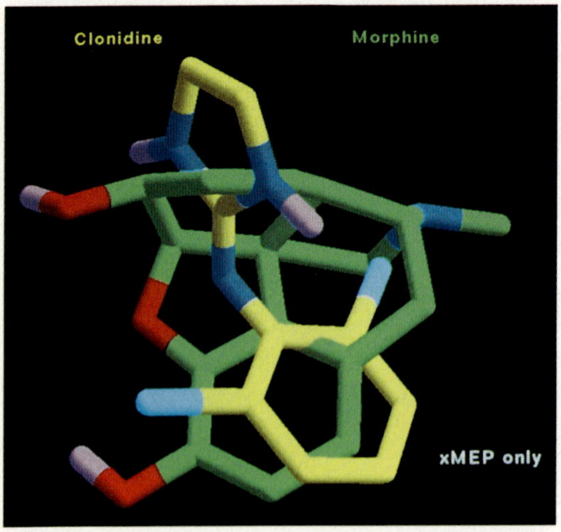

Figure 7. A superposition of morphine, green, and clonidine, yellow, obtained by matching the *MEP* field extended to include the region around the atoms.

the similarity match obtained is very good, structural features that might be expected to match, such as the planar six-membered rings, are far from being aligned. The overall structural similarity is improved, however, if x*MEP* rather than *MEP* is employed. Fig. 7 shows the optimized superposition of morphine onto clonidine obtained using the x*MEP*-based similarity measure. The relative positioning of the two molecules, especially the six-membered rings, is much closer to what might be expected from both an electronic and structural view point. The x*MEP*-based similarity measure had a value of 0.60. Although this value is lower than the *MEP*-based similarity value, the two values are not directly comparable as they are based on a different number of points.

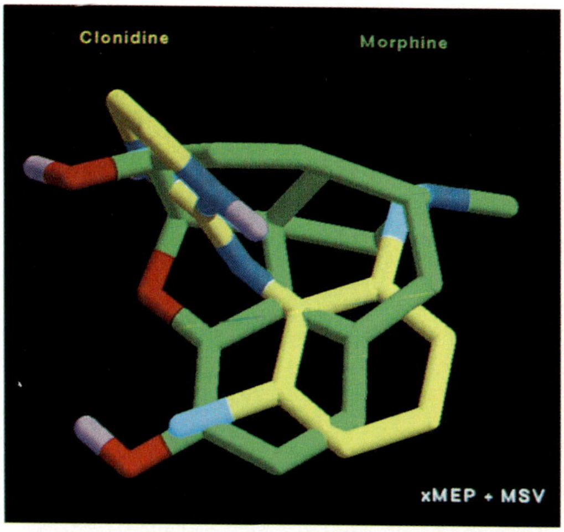

Figure 8. A superposition of morphine, green, and clonidine, yellow, obtained by matching bothe the extended *MEP* and *MSV* fields.

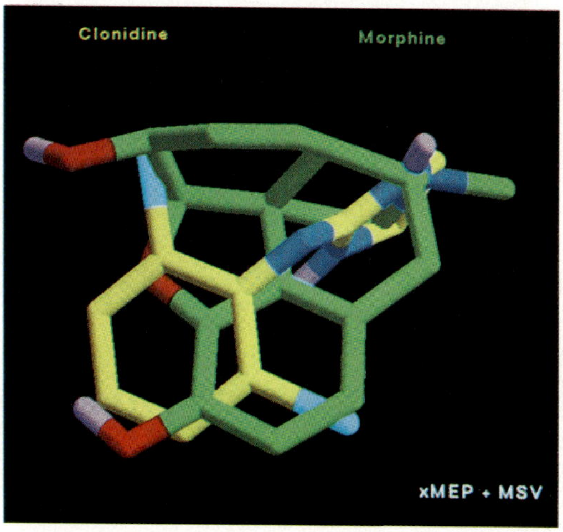

Figure 9. An alternate superposition of morphine, green, and clonidine, yellow, obtained by matching both the extended *MEP* and *MSV* fields.

Some measure of *MSV* should, however, be included in any similarity-based molecular superpositioning to properly account for the way in which both molecules interact with an opioid receptor. Thus, in the example shown in Fig. 8, x*MEP* and *MSV* have been combined to provide an explicit description of both the electronic and steric fields of morphine and clonidine. This figure clearly shows that the positioning of the two molecules is close to that obtained using the x*MEP* field alone, but that the match between the shapes of the molecules has been further improved to the point where the six-membered rings are now co-planar. This matching yielded a total similarity value of 0.61, which is a weighted average of the similarity values of 0.57 and 0.62 for the x*MEP* and *MSV* fields, respectively.

Interestingly, field-based similarity methods can exhibit multiple molecular superpositions with almost identical similarity values. Fig. 9 provides an example of an alternative superpositioning of morphine and clonidine, which has a similarity value identical to that of the example shown in Fig. 6, namely 0.61. The corresponding x*MEP* and *MSV* similarity values of 0.56 and 0.63, respectively, are also close to those in the previous example. However, the two superpositions differ from each other rather dramatically. While field-based similarity matching of these two molecules alone cannot resolve the actual mode of binding within an opioid receptor, consideration of a series of analogues could potentially lead to a unique solution.

3.1.5 Conclusions

The concept of molecular similarity is a powerful one that can impact the drug discovery process in many ways, some of which have been illustrated here. However, due to the vagueness of the concept, it is difficult to define unambiguously. (*"Similarity is like pornography; it is difficult to define, but you know it when you see it!"* G.M. Maggiora, 1993). In chemistry, for example, molecular features which are held to be important by one chemist may not be considered important by another, and such differences in opinion depend critically on their background and disciplinary orientation. Thus, which molecular features should be included in any definition of molecular similarity is somewhat dependent on the problem and this has led to a profusion of 2D and 3D methods, some of which are described above.

The role that conformational flexibility plays in similarity-based matching has not been discussed in this work. As dealt with in current similarity methods, conformational flexibility only considers rotations about single bonds. Hence, only 3D similarity methods are affected. As has been shown by a number of workers [40, 41], inclusion of conformational flexibility in atom-based similarity searches does lead to larger sets of "hits" for a given reference (sub)structure. This is also true of field-based methods, where additional superpositionings are found within a given dataset (J.D. Petke & D.C. Rohrer, unpublished results). However, for illustrative purposes only, rigid 3D similarity matchings have been considered here.

Although molecular similarity is a powerful concept, it must be applied in a flexible manner so as to achieve its maximal effectiveness. To obtain such adaptability a number of 2D and 3D similarity methods must be available, ideally within an integrated computer environment – an environment which can also enhance the potential for synergy. The methods described in this chapter provided examples of some of the molecular similarity techniques in use today, but the most important point to be gained from the above material is that many aspects of the drug discovery process can be impacted by the applications of MSA.

Acknowledgement

The authors wish to express their appreciation to J. D. Baker, B. V. Cheney, C. Cheng, T. R. Hagadone, W. J. Howe, and A. B. Miller for helpful discussions and assistance in the preparation of materials used in the manuscript.

References

[1] Marshall, G. R., Barry, C. D., Bosshard, H. E., Dammkoehler, R. A. and Dunn, D. A., *The Conformational Parameter in Drug Design: The Active Analog Approach.* In: *Computer-Assisted Drug Design,* Olson, E. C. and Christoffersen, R. E., eds. (ACS Symp. Ser. **112**), Amer. Chem. Soc. Washington D. C., 1979, p. 205–226

[2] Hopfinger, A. J., *J. Am. Chem. Soc.* **102**, 7196–7206 (1980)

[3] Cramer, R. D. III, Patterson, D. E. and Bunce, J. D., *J. Am. Chem. Soc.* **110**, 5959–5967 (1988)

[4] Carhart, R. E., Smith, D. H. and Vankataraghavan, R., *J. Chem. Inf. Comput. Sci* **25**, 64–73 (1985)

[5] Willett, P., *Similarity and Clustering in Chemical Information Systems,* Research Studies Press Ltd., Letchworth, 1987

[6] Pepperrell, C. A. and Willett, P., *J. Comput.-Aided Mol. Des.* **5**, 455–474 (1991)

[7] Hagadone, T. R., *J. Chem. Inf. Comput. Sci.* **32**, 515–521 (1992)

[8] Willett, P., Winterman, V. and Bawden, D., *J. Chem. Inf. Comput. Sci.* **26**, 109–118 (1986)

[9] Lajiness, M. S., Johnson, M. A., and Maggiora, G. M., *Implementing Drug Screening Programs Using Molecular Similarity Methods.* In: *QSAR: Quantitative Structure-Activity Relationships in Drug Design.* J. L. Fauchere ed., Alan R., Liss, Inc., New York, 1989, p. 173–176

[10] Bawden, D., *Applications of Two-Dimensional Chemical Similarity Measures to Database Analysis and Querying.* In: *Concepts and Applications of Molecular Similarity.* Johnson, M. A. and Maggiora, G. M., eds., Wiley Interscience, New York, 1990, p. 65–76

[11] Gifford, E. M., Johnson, M. A., Kaiser, D. G., and Tsai, C.-C., *J. Chem. Inf. Comput. Sci.***32**, 591–599 (1992)

[12] Johnson, M., *J. Biopharm. Statist.* **3**, 203–236 (1993)

[13] Maggiora, G. M., Johnson, M. A., Lajiness, M. S., Miller, A. B. and Hagadone, T. R., *Looking for Buried Treasure: The Search for New Drug Leads in Large Chemical Databases.* In: *Mathematical Modelling in Science and Technology.* Rodin, E. Y. and Avula, X. J. R. eds., Pergamon Press, New York, 1988, p 626–629

[14] Randić, M., Molecular Similarity Approaches to Property Optimization. In: *Concepts and Applictions of Molecular Similarity.* Johnson, M. A. and Maggiora, G. M., eds., Wiley Interscience, New York, 1990, p. 77–145

[15] Kier, L. B. and Hall, L. H., *Pharm. Res.* **7**, 801–807 (1990)

[16] Good, A. C., Hodgkin, E. E., and Richards, W. G., *J. Chem. Inf. Comput. Sci.* **32**, 188–191 (1992)

[17] Johnson, M. A. and Maggiora, G. M., eds., *Concepts and Applications of Molecular Similarity Analysis,* Wiley Interscience, New York, 1990

[18] Martin, Y. C. and Willett, P., eds., *Three-Dimensional Chemical Structure Handling,* Tetrahedron Comput. Meth., Vol. **III**, No. 6C, 1990

[19] Johnson, M. A., *J. Math. Chem.* **3**, 117–145 (1989)

[20] Petke, J. D., *J. Comput. Chem.* **14**, 928–933 (1993)

[21] Hagadone, T. R. and Howe, W. J., *J. Chem. Inf. Comput. Sci.* **22**, 182–186 (1982)

[22] Willett, P. and Winterman, V., *Quant. Struct.-Act. Relat.* **5**, 18–25 (1986)

[23] Moon, J. B. and Howe, W., *J. Tetrahedron Comput. Methodol.* **3**, 697–711 (1990)

[24] The CONCORD program was developed by R. Pearlman, University of Texas, Austin, and is available from Tripos Associates, St. Louis

[25] Willett, P., *Algorithms for the Calculation of Similarity in Chemical Structure Database.* In: *Concepts and Applications of Molecular Similarity.* Johnson, M.A. and Maggiora, G.M., eds., Wiley Interscience, New York, 1990, p. 43–63

[26] Dearholt, D.W. and Schvaneveldt, R.W., *Properties of Pathfinder Networks.* In: *Pathfinder Associative Networks: Studies in Knowledge Organization,* Schvaneveldt, R.W., ed., Ablex Pub. Corp. Norwood, NJ, 1990, p. 1–30

[27] Toussaint, G.T., *Pattern Recog.* **12**, 261–268 (1980)

[28] Harary, F., *Graph Theory,* Addison-Wesley, Redding, 1969

[29] Wilkins, C.L. and Randić, M., *Theoret. Chim. Acta* **58**, 45–68 (1980)

[30] Gifford, E.M., Johnson, M.A., Kaiser, D.G. and Tsai, C.-C., *SAR QSAR Environ. Res.* in press

[31] Carbo, R., Leyda, L. and Arnau, M., *Int. J. Quantum Chem.* **17**, 1185–1189 (1980)

[32] Hodgkin, E.E. and Richards, W.G., *Int. J. Quantum Chem. Quantum Biol. Symp.,* **14**, 105–110 (1987)

[33] Kellogg, G.E., Semus, S.F., and Abraham, D.J., *J. Comput. Mol. Des.* **5**, 545–552 (1991)

[34] Cheney, B.V. and Kalantar, J., *J. Mol. Graphics* **4**, 21–27, 35–36 (1986)

[35] Allinger, N.L., *J. Amer. Chem. Soc.* **99**, 8127–8134 (1977)

[36] Bye, E., *Acta Chem. Scand., Ser B* **30**, 549–554 (1976)

[37] Cody, V. and Detitta, G., *J. Cryst. Mol. Struct.* **9**, 33–43 (1979)

[38] Dewar, M.J.S. and Thiel, W., *J. Amer. Chem. Soc.* **99**, 4899–4907 (1977)

[39] Dewar, M.J.S., Zoebisch, E.G., Healy, E.F. and Stewart, J.J.P., *J. Amer. Chem. Soc.* **107**, 3902–3909 (1985)

[40] Haraki, K.S., Sheridan, R.P., Venkatraghvan, R., Dunn, D.A. and McCulloch, R., *Tetrahedron Comput. Meth.* **3**, 565–573 (1990)

[41] Clark, D.E., Willett, P. and Kenny, P.W., *J. Mol. Graphics* **11**, 146–156 (1993)

3.2 Clustering of Chemical Structure Databases for Compound Selection

Geoffrey M. Downs and Peter Willett

3.2.1 Introduction

Clustering is the process of subdividing a group of entities into more homogeneous subgroups on the basis of some measure of similarity between the entities. General books on the subject include those by Gordon [1], Everitt [2] and Sneath and Sokal [3]. The technique can be used to:

- identify the groups that are present in a dataset which is believed to contain many distinct populations;
- present a summary of the types of entity present in a dataset;
- construct a classification scheme for the entities in a dataset;
- test or define hypotheses about the relationships between the entities in a dataset;
- identify homogeneous subgroups within a dataset of entities of known behavior to assist in the prediction of the behavior of entities outside the dataset.

In the present context, the entities of principal interest are the compounds in a chemical database, and the attributes are the descriptors used to represent them. Clustering methods can be employed to identify and to summarize the classes of compounds or attributes that are present. Knowledge of these can then be used as the basis for an unbiased and systematic approach to the selection of a representative of each of the classes that have been identified. Many applications of clustering have been reported including the following:

- the clustering of small sets of compounds on the basis of their chemical and/or biological properties;
- the clustering of substituent properties as part of the process of experimental design of a new series of compounds;
- the clustering of hits resulting from a *substructure search* of a databases, i.e. the retrieval of all molecules that contain a user-defined partial structure, with compound selection being undertaken to provide an overview of the range of structural classes present in the search output;
- the clustering of an entire database, e.g. a corporate structure file, with compound selection being undertaken to identify candidates for biological screening.

The first two of these applications use clustering as one of the tools available for structure-activity relationship analysis, whereas the last two use clustering for larger-scale information analysis, visualization and sampling.

Clustering of small sets of compounds is frequently performed as a preliminary step in more extensive structure-activity analyses, as outlined by Downs and Willett [4], or as a more general analytical chemistry exercise, which is exemplified by the extensively employed MASLOC procedure [5]. More specific examples include the classification of 40 neuroleptics, using assays with rats, by Lewi [6], and of 29 antibiotics, with antibacterial data, by Takahashi et al. [7]. Similarly, Miyashita et al. [12] have clustered 62 cephalosporins, on the basis of antibacterial spectra, and 38 benzodiazepines, on the basis of 8 physico-chemical parameters, to obtain a representative subset of compounds. Clustering to ascertain the interdependence of variables is discussed by Chen et al. [8], who employed cluster analysis to eliminate redundant variables, so that the remaining variables could be used for more precise structure-activtiy analysis methods (in this case factor analysis). Lin et al. [9] gave a more recent account of clustering and principal component analysis to select descriptors and then compounds for input into a CoMFA analysis of 3D molecular shape [10].

Clustering of substituent properties was pioneered by Hansch et al. [13] as a means for selecting representative substituents from homogeneous subgroups formed on the basis of up to six physico-chemical parameters. This preliminary step in the data reduction was undertaken to facilitate the rapid formulation of a viable structure-activity relationship. The clusters formed were investigated further by Dunn et al. [14] to ascertain as to whether they shed light on the mechanisms of action by antitumor triazenes. There have been several subsequent discussions on the use of clustering methods with substituent data, which have been reviewed by Pleiss and Unger [15]. However, van de Waterbeemd et al. [16] have shown that cluster analysis employed in this way can give disappointing results in some cases. An alternative approach has been adopted by Wootton et al. [11], in which compounds are selected that are apart by at least a specified distance, as defined by the sum of their substituent physico-chemical parameters. The aim is to obtain a well distributed sample from the data set, and the approach is, thus, analogous to the maximum dissimilarity selection process, which is discussed later in Section 3.2.3.3.

Clustering the outputs of 2D substructure searches, where a user's query pattern consists of a set of atoms and bonds, has been reported by Willett et al. [17] and Barnard and Downs [18], but does not appear otherwise to have been discussed in the literature. However, the new generation of systems for 3D substructure searching, where a user's query pattern is a putative pharmacophoric pattern that (usually) consists of a set of atoms and the associated interatomic distances [19, 20], typically produce much larger hit lists than 2D substructure searches. This may encourage a re-evaluation of cluster analysis for the processing of search outputs, especially when flexible 3D searching becomes fully established, since this is known to produce still larger numbers of hits than the present generation of rigid 3D searching systems [21].

To-date, greater interest has been shown in the final application above, and this review hence focuses on the use of cluster analysis methods to select compounds for screening, a task that is normally accomplished by manual means. Cluster-based selection has the following advantages:

- a complex and time-consuming manual operation involving highly trained staff can be replaced by a cheaper automated procedure;
- an effective clustering procedure can help to ensure that no classes of compounds are overlooked when selecting structures for testing;
- the use of a parameterized clustering method permits the creation of different sets of structures to suit different screening requirements.

The general steps involved in the process of clustering a dataset are as follows:

1) Select an appropriate set of attributes (molecular properties).
2) Process the dataset of entities (compounds) to generate attribute lists for each entity. If the set of attributes is of reasonable size then use these attributes as descriptors. Otherwise, combine attributes or conduct a dimensional analysis to produce a reasonable number of descriptors. If appropriate, and where necessary, standardize the descriptors.
3) Use an appropriate similarity measure to calculate the similarity between each pair of entities.
4) Use an appropriate clustering method to group the entities.
5) Analyze the resultant clusters or classification hierarchy; the clustering can then be repeated, or the best set of clusters chosen as required.

The reader will note that we have used the term "similarity measure" in Step 3 of this algorithm; this should be understood to include not only similarity measures, but also dissimilarity and distance measures. Sneath and Sokal [3] have given a detailed account of the many ways in which the resemblance between pairs of entities can be calculated.

The attributes used to classify the entities are generally a subset of all of the possible attributes. The choice of attributes is critical to the success of the subsequent classification: if the set of chosen attributes is incomplete with respect to a particular application, then distinctive subgroups in the dataset will not be differentiated, since entities that are similar in one dimension may be very different in another. The chosen set of attributes is the basis for the *descriptors* employed in the similarity measure calculation (many attributes may be combined to form one descriptor). Examples of descriptors that can be used to measure chemical similarities include biological properties [7, 12]; topological indices, such as molecular connectivity χ (*chi*) and κ (*kappa*) shape indices [22] and electrotopological state indices [23]; structural fragments, such as augmented atoms [24], atom pairs [25], and three-dimensional (3D) information, such as interatomic distance ranges [26]; and property values, such as melting point, molar refractivity, volume and log *P* [27].

In addition to the chosen attributes, classification of a set of entities is also dependent on the similarity measure employed. The choice of similarity measure is normally left to the investigator; however, some of the available clustering methods demand, or are defined by, the use of a particular similarity measure. Attributes, descriptors and similarity measures have been discussed in greater detail earlier in Chap. 3.1 and by the present authors elsewhere [4, 28]. Once appropriate descriptors and similarity measures have been selected, a clustering method can be used to produce the required subgroups. The methods available for this purpose are discussed in the next section, with special reference to those methods that have already been used for clustering files of chemical compounds.

3.2.2 Review of Clustering Methods

Clustering methods can produce *overlapping* clusters, in which each object may be in more than one cluster, or *non-overlapping* clusters, in which each object occurs in only one cluster. The latter are far more widely used and, thus, most of the methods that are discussed below belong to this class. An example of the application of an overlapping method to compound selection has been provided by work carried out at the National Cancer Institute: this is discussed in Section 3.2.4.2.

The non-overlapping cluster methods can be classified as shown in Figure 1. There are two main classes of clustering method: *hierarchical* methods and *non-hierarchical* methods, which can be further subdivided as shown. This is not the only clas-

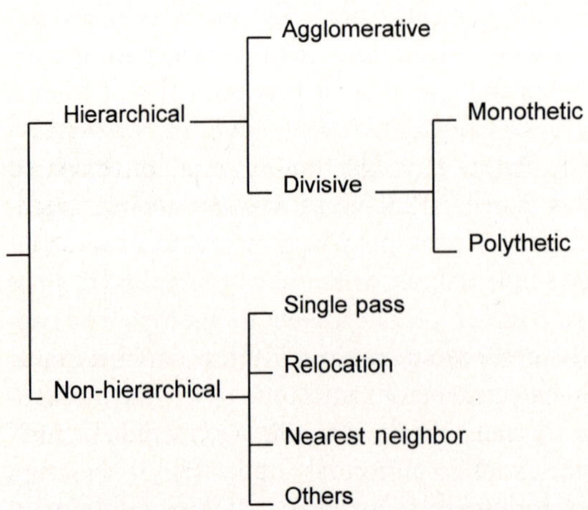

Figure 1. Simple classification of clustering methods.

sification for non-overlapping clustering methods, but is the most widely used and the most accepted.

Many of these methods are available in general-purpose statistical packages, such as SAS and SPSS, or in specific clustering packages, such as CLUSTAN. However, the routines in these packages are typically designed for processing small datasets, which contain at the most a few hundred entities, and they are thus not appropriate for processing chemical databases, which contain tens or hundreds of thousands of entities. Accordingly, users will generally need to encode their chosen clustering method to suit their own purposes, and so we have provided considerable algorithmic detail in the discussion below to facilitate such implementations.

3.2.2.1 Hierarchical Clustering Methods

An hierarchical clustering method produces a classification in which the smaller clusters of very similar molecules are nested within ever increasing larger clusters of less closely related molecules. This hierarchical arrangement of clusters is usually visualized as a *dendrogram,* which illustrates the ways in which molecules fuse into (or are subdivided into) clusters, and the similarity level at which this takes place. This is ideal for producing taxonomies, but it is difficult to interpret a dendrogram if it is used to describe a classification of a dataset containing more than a few tens of entities. Hierarchical *agglomerative* methods generate a classification in a "bottom-up" manner, by a series of agglomerations in which small clusters, initially containing individual molecules, are fused together to form progressively larger clusters. Conversely, hierarchical *divisive* methods generate a classification in a "top-down" manner, by progressively subdividing the single cluster which represents the entire dataset.

Hierarchical Agglomerative Methods

Many clustering applications use one of the *Sequential Agglomerative Hierarchical Non-overlapping* (SAHN) clustering methods. These methods can be implemented by means of the basic algorithm below, which is known as the *stored-matrix* algorithm since it involves random access to the interentity similarity matrix throughout the entire cluster generation process:

1) Calculate the similarity matrix, which contains the similarities between all pairs of entities in the dataset that is to be clustered.
2) Find the most similar pair of points (where a point denotes either a single entity or a cluster of entities) and merge them into a cluster to form a new single point.
3) Repeat Step 2 until only a single point remains, i.e. until all of the entities have been merged into one cluster.

Individual hierarchical agglomerative methods differ in the ways in which the most similar pair of points is defined, and in which the merged pair is represented as a single point. The original data is not required once the initial similarity matrix has been computed since the combinatorial solution to recompute the intercluster similarity is given by the Lance-Williams matrix-update formula [29].

$$d_{k(i,j)} = \alpha_i d_{ki} + \alpha_j d_{kj} + \beta d_{ij} + \gamma |d_{ki} - d_{kj}| \tag{1}$$

where $d_{k(i,j)}$ is the similarity measure between point k and a point (i, j) formed by fusing points i and j. Different SAHN methods have different values for the four parameters α_i, α_j, β and γ. Many other SAHN techniques exist, but do not fall within the scope of the matrix-update formula and, thus, are rarely used.

Murtagh [30] and others classified SAHN techniques into *graph-theoretic* or *linkage* methods and *geometric* or *cluster-centre* methods. Graph theoretic methods include the *single linkage, complete-linkage, weighted-average* and *group-average* methods, whilst geometric methods include the *centroid, median* and *Ward's*, or the *minimum variance*, methods. An important concept, which Murtagh [30] discusses, is that of the *reducibility property.* If a method satisfies the reducibility property then agglomerations can be undertaken in restricted areas of the similarity space and the results amalgamated to form the overall hierarchy of relationships. Satisfaction of the property also means that *reversals, or inversions,* of the hierarchy cannot occur. Reversals are a problem with geometric methods because a cluster may end up being more similar to its parent cluster than to any of its constituent entities. Both median and centroid geometric methods are subject to this problem, but Ward's method satisfies the reducibility property and, thus, this problem is not encountered. Graph-theoretic methods are also not subject to this problem, and the reducibility property is generally not applicable to them. However, under certain conditions, the group average method does satisfy the reducibility property and this point is discussed further below.

The advantage of isolating methods which satisfy the reducibility property is that the stored matrix algorithm can be replaced by the computationally more efficient *reciprocal nearest neighbor* (RNN) algorithm. In this algorithm, a path is traced through the similarity space until a pair of points is reached that are more similar to each other than to any other points, i.e. they are reciprocal nearest neighbors (RNN). These RNN points are fused to form a single new point, and the search continues until the last unfused (unused) point is reached. The basic RNN algorithm is, thus, as follows:

1) Mark all entities, i, as "unused".
2) Starting at an unused i, trace a path of unused nearest neighbors (NN) until a pair of reciprocal nearest neighbors is encountered; i.e. trace a path of the form $j := \mathrm{NN}(i)$, $k := \mathrm{NN}(j)$, $l := \mathrm{NN}(k)\ldots$ until a pair of points is reached for which $q = \mathrm{NN}(p)$ and $p = \mathrm{NN}(q)$.

3) Add the RNNs p and q to the list of RNNs along with the distance between them, mark q as "used", and replace the centroid of p with the combined centroid of p and q.
4) Continue the NN chain from the point in the path prior to p, or choose another unused starting point if p was a starting point.
5) Repeat Steps 2–4 until only one unused point remains.

The RNN method is applicable to geometric clustering methods in which the most similar pair at each stage is defined by a distance measure. In Ward's method, the intercluster variance is maximized as the intracluster variance is minimized. The Euclidean distance is used to determine distances between points and, hence, to define a cluster centroid. For two points i and j, the Euclidean distance, E, is given by:

$$E = \sqrt{\Sigma (x_{ik} - x_{jk})^2} \tag{2}$$

with the summation over all k.

Thus, if the Euclidean distance is used in the RNN algorithm, the clusters obtained are those that would be obtained from the stored-matrix algorithm using the update formula appropriate to Ward's method, i.e. the procedure results in a hierarchy of Ward clusters. To obtain the clusters, it is necessary to arrange the list of RNNs, produced in Step 3 of the algorithm above, in order of increasing distance between them. This list represents the Ward hierarchy; the first in the list is the first pair to be merged, and so on.

We have mentioned above that the reducibility property does not apply to the graph theoretic methods. However, the group-average method *can* be implemented using the RNN algorithm if, and only if, the Cosine coefficient is used instead of the Euclidean distance to calculate the interentity similarities. For two points, i and j, the Cosine coefficient, C, is given by:

$$C = \frac{\Sigma (x_{ik} \times x_{jk})}{\sqrt{(\Sigma x_{ik}^2 \times \Sigma x_{jk}^2)}} \tag{3}$$

with the summations over all k.

This changes the normally graph-theoretic group-average method into a geometric one, and the RNN approach is valid [31, 32]. Since a similarity coefficient is used instead of a distance, the closest points are those with the largest coefficients, and the RNN list needs to be arranged in order of decreasing coefficient (rather than in order of increasing distance as would be necessary for Ward's method).

Hierarchical Divisive Methods

Hierarchical divisive methods are generally much faster than the corresponding agglomerative methods, but often give poor levels of performance since they are *monothetic* in character, i.e. the divisions are based on just a single attribute. This is in marked contrast to all of the other clustering methods described in this review, which are *polythetic* in character, i.e. all of the attributes are considered simultaneously during the cluster-generation process [33].

One of the few successful polythetic divisive methods is the minimum diameter hierarchical divisive method of Guenoche et al. [34], a general outline of which is given below:

1) For the set of N entities to be clustered, produce an input list of all $N(N-1)/2$ dissimilarities, arranged in decreasing order of magnitude;
2) Take the top two entities in this dissimilarity list; these become the focus of the first bipartition of the dataset. Assign all other entities to the least dissimilar of these initial cluster centres;
3) Recursively select the cluster with the largest diameter and partition it into two clusters, such that the larger cluster has the smallest possible diameter;
4) Continue for a maximum of $N-1$ bipartitions.

The diameter of a cluster is defined as the largest dissimilarity between any two of its members, and does not necessarily reflect the size (number of members) of the cluster; the diameter of a singleton cluster is defined as zero.

One of the potential benefits of polythetic hierarchical divisive clustering is that users typically wish to have only a few clusters. Hierarchical agglomerative clustering requires the production of most the hierarchy, with the accompanying problem that erroneous assignments made early in the hierarchy are not corrected and, thus, become compounded. In the case of hierarchical divisive clustering, only the first part of the hierarchy need be produced, resulting in less risk of compounding erroneous assignments. However, this method requires random access to the full dissimilarity matrix (to calculate the diameters in Step 3 of the algorithm above) if the demands on time are not to become too great.

3.2.2.2 Non-Hierarchical Clustering Methods

A non-hierarchical method generates a classification by partitioning a dataset, giving a set of (generally) non-overlapping groups which have no hierarchical relationships between them. A systematic evaluation of all possible partitions is not feasible, and many different heuristics have, thus, been described which identify good, but possibly less than optimal, partitions; such methods are generally much less demanding of computational resources than the hierarchical methods. Three of the main categories of non-hierarchical methods are the *single-pass, relocation* and *nearest-neighbor* methods.

Single-Pass Methods

Single-pass methods are easy to implement and are very fast. As the name suggests, they require a single pass through the dataset to assign the entities to clusters. A threshold of similarity is used to decide as to whether to assign the next entity to an existing cluster or to use it as the start of a new cluster. As with other clustering methods, it is, thus, necessary to decide how to represent an existing cluster so that similarity between an entity and a cluster can be determined. This *cluster representative* is normally obtained by calculating the *centroid,* i.e. the arithmetic mean of the attribute vectors for each of the entities in a cluster. The basic algorithm, which is commonly referred to as the Leader Algorithm [35], is as follows:

1) Designate the first entity as the first cluster.
2) Assign the next entity to the most similar existing cluster, or, if the similarity does not equal or exceed the threshold similarity, designate it as the start of a new cluster.
3) Continue until all entities have been processed.

The major problem with single-pass clustering is that the resultant clusters are dependent on the order in which the structures are processed and upon the threshold used in Step 2. The former limitation has serious implications since it implies that an alternative, and possibly superior, classification could be obtained simply by permuting the dataset so that the entities are processed in a different order.

Relocation Methods

Relocation methods assign entities to a user-defined number of *seed* clusters and then iteratively reassign entities to see if improved clustering results. Such methods are prone to reaching local optima rather than a global optimum, and it is generally not possible to determine when, or whether, the global optimum solution has been reached. The two most common relocation methods are *k-means* (as first discussed by Forgy [36]) and *hill-climbing.* The basic algorithm is as follows:

1) Select a user-defined number of entities as the cluster seeds.
2) Assign all other entities to the most similar cluster seed.
3) Calculate the cluster representatives.

4) Assign each entity to the most similar cluster representative.
5) Repeat Steps 3 and 4 until fewer than some minimal threshold number of changes occur in the membership of the set of clusters, or until a user defined number of iterations has taken place.

The relocations, or repeated assignments, in Step 4 are an attempt to correct what is a potentially poor selection of seed entities. There are many ways of defining the cluster representative. The *k*-means method use the centroid, whereas the hill-climbing method uses a criterion function, such that an entity is relocated to another cluster if such a relocation results in an improvement of the chosen criterion function. The problem of stabilization around a local optimum can be lessened by reiteration of the algorithm either with different seed points, or with modified parameters to see if there is any improvement. However, consistent results, independent of the seed entities selected in Step 1, still cannot be guaranteed.

Nearest-Neighbor Methods

Nearest-neighbor methods assign structures to the same cluster as a defined number of their nearest neighbors. User-defined parameters determine how many nearest neighbors need to be considered and the necessary level of similarity between nearest-neighbor lists. There are several such methods, of which the Jarvis-Patrick method [37] has proved to be highly appropriate for the clustering of chemical structures represented by structural fragments [38].

The Jarvis-Patrick method involves the use of a list of the top K nearest neighbors for each of the N entities in a dataset. Nearest neighbors are typically identified using the Euclidean distance or the Tanimoto Coefficient as the similarity measure. The Tanimoto Coefficient, *TC*, is defined as:

$$TC = \frac{\Sigma(x_{ik} \times x_{jk})}{\Sigma(x_{ik}^2 + x_{jk}^2 - (x_{ik} \times x_{jk}))} \tag{4}$$

with the summations over all k.

Once the lists of top K nearest neighbors for each entity have been produced, a second stage is used to create the clusters. Two entities, i and j, are placed in the same cluster if all of the following conditions are satisfied.

- i is in the top K nearest-neighbor list of j,
- j is in the top K nearest-neighbor list of i,
- i and j have at least $K_{\min}$ of their top K nearest neighbors in common, where $K_{\min}$ is a user-defined parameter in the range $1 \leq K_{\min} \leq K$.

The second stage uses a label array of length N to store the cluster labels for each entity, and proceeds as follows:

1) Initialize the label array by setting each element to its position in the array; this sets each entity to its own initial cluster;
2) Compare the nearest-neighbor lists for all pairs of entities, i and j ($i<j$). If all of the clustering conditions above are satisfied then replace the label array entry for j, and all other occurrences of the label array entry for j in the label array, with the label array entry for i.
3) The label array contains the lowest array entries for the entities in each cluster, and members of the same cluster will have the same array entry. Scan the array to extract the members of each cluster.

As with the hierarchical methods, this process is not order-dependent. However, instead of choosing a required number of clusters, as is required for the relocation methods, the partition is governed by the choice of K_{min}; i.e. the emphasis is on partitioning by degree of similarity between structures rather than by a predefined number of clusters. It is necessary, therefore, to experiment with a range of K_{min} values until roughly the required number of clusters is obtained.

3.2.3 Choice of Clustering Method

Each clustering method has its own characteristics, both in terms of the resultant clusters and in terms of the computing resources required.

3.2.3.1 Computational Requirements

For a dataset of N entities, the standard stored matrix approach for hierarchical agglomerative clustering requires $O(N^2)$ time and $O(N^2)$ space to generate and store the $N \times N$ similarity matrix and $O(N^3)$ time for the clustering, whereas the Reciprocal Nearest Neighbor approach reduces the time required for clustering to $O(N^2)$ and the space requirement to $O(N)$. Other fast SAHN algorithms have also been described [39, 40]. The minimum diameter hierarchical divisive algorithm has an $O(N^2 \log N)$ clustering time requirement and an $O(N^2)$ space requirement (since it needs random access to the similarity matrix). The best of the non-hierarchical methods reduce the clustering time requirement to only $O(MN)$, where M is the number of clusters generated, with a storage requirement of $O(M+N)$; however, the worst non-hierarchical methods are slower than hierarchical ones. The Jarvis-Patrick method requires $O(N^2)$ time and $O(N)$ space for the generation of the nearest-neighbor lists.

Methods requiring $O(N^3)$ time for the clustering, i.e. SAHN methods using the stored-matrix algorithm, can only be used for datasets containing a few tens or a few hundreds of entities, whereas the best of the non-hierarchical methods can be used for datasets containing tens or hundreds of thousands of entities given an appropriate implementation. Methods requiring $O(N^2)$ time for the calculation of the similarities are feasible for this size of dataset only if a subset of the total number of similarities are required in the clustering process.

3.2.3.2 Cluster Shapes

For the SAHN methods the two extremes of cluster characteristics are represented by single linkage and complete linkage. Single-linkage clusters are based on connectedness, in that a single edge between two clusters is sufficient to merge them. These clusters are maximally connected subgraphs and so are characterized by the minimum path length among all pairs of entities in the cluster. The result is a tendency for clusters to be chained together to form long, straggly clusters. At the other extreme, complete-linkage clusters are based on the diameter of maximally complete subgraphs, where the diameter of a complete subgraph is the smallest similarity for all pairs of entities in the cluster. Complete linkage does not generate the clusters with minimum diameter (unlike the Guenoche Algorithm) and has a tendency to produce compact, but not very well separated clusters [41]. Ward's method and the group-average method lie in between these two extremes and tend to produce globular clusters. The minimum-diameter divisive method produces clusters which tend to be well separated and of variable shape. Of the non-hierarchical methods, the Jarvis-Patrick method tends to produce clusters similar to single linkage, though this depends on the relative values of K and K_{min}, with k-means clusters being more similar to complete linkage.

As to which of these methods is the most appropriate depends largely on the nature of the datasets. Clustering methods will form clusters from any dataset, even those containing evenly distributed entities. It is necessary, therefore, to ensure, first of all, that the dataset does contain distinct clusters, and there is an extensive literature devoted to this problem of cluster validation [42]. In the case of heterogeneous datasets of chemical structures, this first condition can be assumed. The next decision is whether the natural clusters are best perceived as, for instance, straggly or globular, so that the most appropriate cluster method can be applied. There is no automatic way of arriving at this decision. The most common approach is to use a range of different clustering methods and then to analyze visually the resultant clusters to determine which appear to be the most realistic or appropriate for a given application. However, visualization of the results of a cluster analysis is always a problem if large datasets are to be processed [2].

For hierarchical clustering the standard output is a dendrogram, but these become incomprehensible for more than a few hundred entities. For larger datasets, it is necessary to take just one or two partitions from the hierarchy (which requires some manual or automated technique for deciding which partition(s) should be chosen), and then either display the structures in each cluster (or a representative structure from each cluster), or perform some form of data reduction to enable the entities to be plotted in a few dimensions. The former approach is the more common for chemical applications of non-hierarchical clustering methods. Structures from each chosen partition are transferred to a structure display routine, with the user being able to view each member of a cluster in 2D or 3D.

3.2.3.3 Comparative Studies

The review given in Section 3.2.2 emphasized the wide range of clustering methods that are currently available, and new methods are being described all the time. Hence there is a need for a set of guidelines to determine which of the clustering methods is the most appropriate for a particular application. Three main approaches to the selection of a method have been described: theoretical analyses to identify those methods with characteristics that closely match a set of predefined criteria of effectiveness (see, e.g. [43, 44]); simulation studies, which use artifical datasets, for which the groupings are already known, and investigate the extent to which different methods are able to recover this structure (see, e.g. [45]); and purely empirical comparisons, which use evaluation criteria specific to the problem being studied. This last approach has been advocated by Everitt [46] and was adopted by Willett [28] and co-workers in a long series of experiments that evaluated over 30 different hierarchical and non-hierarchical methods. These methods were used for clustering sets of compounds which were represented by their constituent atom-centred or bond-centred fragment substructures.

If we wish to carry out an empirical comparison of clustering methods, some quantitative measure of effectiveness is required in order to compare one method with another. In the context of selecting compounds for biological screening, the most important characteristic of a method is its ability to group together molecules with similar properties (or activities), while separating them from clusters that contain molecules with different properties. Adamson and Bush [24] have described a "leave-one-out" approach for the evaluation of chemical classifications that has been extensively used. The property value of a molecule, I, within a dataset is assumed to be unknown, and the classification resulting from the use of some particular clustering method is scanned to identify the cluster that contains the molecule, I. The predicted property value for I, $P(I)$, is then set equal to the arithmetic mean of the observed property values of the other compounds in that cluster. This procedure results in the calculation of a $P(I)$ value for each of the N structures in a dataset, and an overall figure of merit for the classification is then obtained by calculating the product moment correlation coefficient between the sets of N observed and predicted values. The most generally useful clustering methods will be those that give high correlation coefficients in as wide a range of datasets as possible. This approach of comparing clustering methods was used by Willett [28], who found that the best results were given by Ward's hierarchical-agglomerative method and by the Jarvis-Patrick non-hierarchical nearest-neighbor method. As mentioned in Section 3.2.3.1 these have comparable time and storage complexities. However, the Jarvis-Patrick method is noticeably more efficient in practice for at least three reasons:

- The intermolecular and intercluster similarity calculations are simpler and, hence, less time-consuming;
- the calculation of the intermolecular similarities needed for the generation of the nearest-neighbor lists can be performed in a large number of small computer runs as computer resources allow, since the lists are calculated in isolation from each other;
- it is relatively easy to update these lists when new compounds are added to a dataset, whereas most of the hierarchical agglomerative methods (including Ward's method) would require re-clustering of the complete dataset if new compounds became available.

Accordingly, it was concluded that the Jarvis-Patrick method was the most appropriate method for the clustering of files of compounds characterized by fragment substructures [38]. This method now forms the basis for the clustering package of the chemical database software produced by Daylight Chemical Information Systems, and for the CLASS routine in the CAVEAT package for molecular design that is distributed by the University of California.

Lajiness and his co-workers [47–49] at the Upjohn Company have described a rather different approach to compound selection that has also been shown to be highly effective in operation. The basic algorithm, which is referred to as *Maximum Dissimilarity Selection,* attempts to select a set of compounds that are as dissimilar to each other as possible in a single pass of the dataset to be processed. The identification of the maximally dissimilar set of structures is computationally infeasible; instead, the initial compound is selected at random. Then, at each stage in the processing, the next compound is selected which exhibits the minimum similarities to all compounds previously selected. While this procedure cannot be expected to identify an optimal set of compounds, it has been found to work reasonably well in practice. Indeed, a simulaton study involving a test dataset of 2,000 structures with the addition of five groups of known active compounds suggested that this approach identified a greater number of active compounds than an approach which was based on clustering followed by the selection of a representative compound from each cluster.

3.2.4 Examples of the Selection of Compounds from Databases by Clustering Techniques

The need to process large files of compounds means that only the simplest and most efficient similarity measures and clustering methods can be used to select compounds for screening. To date, most of the studies published on cluster-based compound selection have used the fragment-based structure representations discussed in Section 3.2.3.3. This approach to the calculation of intermolecular similarity is not new, having originally been studied by Harrison [50] and by Adamson and Bush [24] over two decades ago. Both of the selection systems discussed below use this approach, as does the system at the Upjohn Company developed by Lajiness et al. [47–49]. However, it must be emphasized that other similarity and clustering procedures could be used for large-scale compound selection if the processing requirements were computationally feasible (see, e.g. [14, 27, 51, 52]). The popularity of fragment-based approaches stems from the proven effectiveness of this approach [18, 28, 47] and the fact that the requisite fragment data may already be available in a chemical organization in the form of the fragment bit-screen records that are a central component of 2D chemical substructure searching systems [53].

3.2.4.1 The Jarvis-Patrick Method

The first report of large-scale clustering for compound selection described a system that was implemented at Pfizer Central Research (U.K.) and which was based on the Jarvis-Patrick method [38]. The company had maintained a Structural Representatives File (SRF) for some years, which contained approximately 5% of the total corporate structure file. The structures in the SRF had been chosen on the basis of their availability in sufficient quantities for testing purposes and of their being representative of the structural variation in the entire file. The method of selection had been intellectual and somewhat *ad hoc*, generally amounting to the inclusion of compounds containing ring systems or functionalities that had been unrepresented previously.

The structures in the Pfizer database, which contained about 240,000 structures at the time of this work, were represented for the search by a bit-map B in which the bit $B(I, J)$ is set to true if the J-th screening fragment was present in the I-th compound. A total of 1315 screens were employed, those being atom-centred or bond-centred fragment substructures that had already been selected for inclusion in the screen set on the basis of their frequencies of occurrence in the database [53]. The bit-map is stored on disk so as to allow quick access to its columns, i.e. to the lists of compounds that contain specific fragment screens. This mode of access is needed for the implementation of an efficient nearest-neighbor searching algorithm that minimizes the time required for the generation of the Jarvis-Patrick nearest-neighbor lists, calculated using the Tanimoto Coefficient as the intermolecular similarity measure. This nearest-neighbor algorithm, and others, are discussed in detail by Willett [28].

The nearest-neighbor lists for each of the compounds in the database are generated in an overnight batch run, and the resulting sets of nearest neighbors – 20 for each compound – are written to disk for use in the subsequent interactive clustering stage of the Jarvis-Patrick method. A classification is generated by the user specifying the particular clustering parameters that are to be used, and the resulting clusters are then output together with any *singletons* that have been identified, where a singleton is a cluster that contains only one molecule. The user can then, if he or she so wishes, allocate each singleton to the non-singleton cluster with which it is most similar. Once the final clusters have been obtained, a representative molecule is chosen from each cluster for biological screening; this is acccomplished by calculating the centroid of the bit-strings describing each of the compounds in a particular cluster, and then choosing the compound whose bit-string is the most similar to the centroid.

3.2.4.2 The Leader Method

For many years, the National Cancer Institute (NCI) Division of Cancer Treatment has run a large-scale program for the computer-assisted testing of molecules for antitumor activity (see, e.g. [54]). NCI has recently adopted a cluster-based mechanism for the selection of compounds for biological screening that is rather different in approach to that adopted by the Pfizer system [55, 56].

The compounds in the NCI database are again represented by lists of fragment substructures, but an open-ended fragment description is employed here, which results in large numbers of detailed and highly specific fragments being available for the characterization of each structure. Moreover, each of these fragments has an associated weight that describes its multiplicity, its size (in terms of the number of atoms and bonds) and its frequency throughout the entire database. In the Pfizer work, conversely, each molecule was characterized by an unweighted bit-string describing merely the presence or absence of a limited number of fairly generalized fragments.

The clustering method used is a modification of the basic Leader method that has been described in Section 3.2.2.2. The NCI version assigns a compound, *I*, to all existing clusters, *J*, for which the similarity between *I* and the first compound to have joined cluster *J* is greater than a user-defined threshold, thus resulting in an overlapping classification (since an individual compound can belong to more than one cluster). The similarity between a pair of structures is calculated using a weighted form of the Asymmetric Coefficient [57], since this was found to give better results with these weighted fragment descriptions than the Tanimoto Coefficient used at Pfizer.

A file of compounds to be clustered is first arranged in order of increasing sum of fragment weights. Not only does this ensure that a unique classification is obtained from a given set of compounds, which is often not the case with single pass methods, but it also allows for the use of an optimization procedure that minimizes the number of compound-to-cluster similarities that need to be calculated [55].

The threshold similarity above which a compound joins a cluster is deliberately set at a high value, to ensure the identification of a large number of small clusters, which is in marked contrast to the clusters produced in the Pfizer work. Moreover, no attempt has been made to reallocate singletons to other non-singleton clusters. Instead, the view is taken that a compound that does not cluster is clearly dissimilar to the rest of the dataset and is, thus, of especial interest in a program which maximizes the diversity of the compounds that are to be submitted for biological testing. Thus, while the NCI work uses a similarity-based approach to clustering, it also takes into account at least some of the work carried out on dissimilarity selection.

3.2.5 Conclusions

In this review, we have summarized the algorithms and methods that have been suggested for clustering files of chemical compounds as a precursor to the selection of compounds for inclusion in biological screening programs. The basic computational techniques necessary for this were described some years ago [24, 38], but only recently have they started to become more widely used [18] as high-performance UNIX workstations became available. The availability of such hardware has facilitated the application of time-consuming clustering procedures to corporate databases that may contain in excess of a quarter of a million chemical compounds. The fact that most work on compound selection takes place within the fine-chemicals industries using proprietary databases means that it is difficult to obtain an accurate picture of the extent to which clustering procedures are routinely used. However, the widespread recognition of the need to maximize structural diversity in the sets of compounds that undergo biological screening means that cluster-based selection procedures are likely to become increasingly widely used in the near future.

The current status of cluster-based selection is exemplified by the system that is being developed at the European Communities Joint Research Centre [4, 18]. The aim is to provide analytical tools for processing the 100,000 EINECS substances in the ECDIN database [58]. Physico-chemical property values are available for only a few of these substances, and Jarvis-Patrick clustering is being used to reveal natural groupings in the dataset which are likely to have similar properties. To assist this process, the latest enhancements include the ability to produce descriptor sets tailored to specific properties by analysis of the variety of fragment descriptors present in those substances with known property values. The resultant clusters provide homogeneous subsets of the input data that are suitable for more exact, but more computationally demanding, QSAR techniques for the prediction of properties. Clustering is also being used to form a basic classification of the EINECS substances. Visual inspection of the structure diagrams of the substances in each cluster has been facilitated by a cluster display routine, in which any cluster can be selected and the constituent structure diagrams displayed as required.

The greatest change that we are likely to see in the next few years is in the type of structure representation that is used to calculate measures of intermolecular similarity. The systems for compound selection described in this review are based on the substructural fragments that are used for conventional, 2D chemical substructure searching [52]. Although efficient, and surprisingly effective, in operation, such a fragment-based representation provides only a relatively crude decription of a chemical structure; much more precise measures of intermolecular structural resemblance are possible using 3D structural information, since the 3D shape of a molecule is known to play a significant role in the recognition of a molecule at a biological receptor site. Techniques for processing databases of 3D molecules are at present undergoing rapid development [19, 20], and this has already resulted in several 3D similarity measures that use interatomic distance information [59]. These measures are sufficiently fast for similarity searching, where a single target molecule is matched against an entire database to find its nearest neighbors, but at present they are far too slow to facilitate the clustering of large databases. Once appropriate algorithms have been developed, it is likely that clustering based on 3D similarity measures will become more widely used since they should provide a classification of a database that reflects the biological properties of the constituent molecules more accurately than the present generation of clustering procedures that employ 2D similarity measures.

References

[1] Gordon, A.D., *Classification,* Chapman and Hall, London, 1981
[2] Everitt, B.S., *Cluster Analysis*, 3rd edn, Edward Arnold, London, 1993
[3] Sneath, P.H.A. and Sokal, R.R., *Numerical Taxonomy,* WH Freeman, San Francisco, 1973
[4] Downs, G.M. and Willett, P., *The Use of Similarity and Clustering Techniques for the Prediction of Molecular Properties.* In: *Applied Multivariate Analysis in SAR and Environmental Studies.* Devillers, J. and Karcher, W., eds., European Communities, Brussels, 1991, p. 247–279
[5] Massart, D.L. and Kaufman, L., *The Interpretation of Analytical Chemical Data by the Use of Cluster Analysis,* Wiley, New York, 1983
[6] Lewi, P.J., *Drug Res.* **26**, 1295–1300 (1976)
[7] Takahashi, Y, Miyashita, Y., Abe, H. and Sasaki, S.I., *Anal. Chim. Acta* **122**, 615–620 (1981)
[8] Chen, B.-K., Horvath, C. and Bertino, J.R., *J. Med. Chem.* **22**, 483–491 (1979)
[9] Lin, C.T., Pavlick, P.A. and Martin, Y.C., *Tetr. Comput. Methodol.* **3**, 723–738 (1990)
[10] Cramer, R.D. III, Patterson D.E. and Bruce, J.D., *J. Amer. Chem. Soc.* **110**, 5959–5967 (1988)
[11] Wootton, R., Cranfield, R., Sheppey, G.C. and Goodford, P.J., *J. Med. Chem.* **18**, 607–613 (1975)
[12] Miyashita, Y., Takahashi, Y., Yotsui, Y. and Abe, H., *Anal. Chim. Acta* **133**, 615–620 (1981)

[13] Hansch, C., Unger, S.H. and Forsythe, A.B., *J. Med. Chem.* **16**, 1217–1222 (1973)
[14] Dunn, W.J. III, Greenberg, M.J. and Callejas, S.S., *J. Med. Chem.* **19**, 1299–1301 (1976)
[15] Pleiss, M.A. and Unger, S.H., *The Design of Test Series and the Significance of QSAR Relationships.* In: *Comprehensive Medicinal Chemistry,* **4**, Ramsden, C.A., ed., Pergamon, New York, 1990, p. 561–587
[16] van de Waterbeemd, H., El Tayar, N., Carrupt, P.-A. and Testa, B., *J. Comput.-Aided Mol. Des.* **3**, 111–132 (1989)
[17] Willett, P. and Winterman, V., *Quant. Struct.-Act. Relat.* **5**, 18–25 (1986)
[18] Barnard, J.M. and Downs, G.M., *J. Chem. Inf. Comput. Sci.* **32**, 644–649 (1992)
[19] Willett, P., *Three-Dimensional Chemical Structure Handling,* Research Studies Press, Taunton, 1991
[20] Martin, Y.C., *J. Med. Chem.* **35**, 2145–2154 (1992)
[21] Clark, D.E., Willett, P. and Kenny, P.W., *J. Mol. Graph.* **10**, 194–204 (1992)
[22] Hall, L.H. and Kier, L.B., *The Molecular Connectivity Chi Indexes and Kappa Shape Indexes in Structure-Property Modelling.* In: *Reviews of Computational Chemistry* **25**, Boyd, D.B. and Lipkowitz, K., eds., VCH, New York, 1991, p. 367–422
[23] Hall, L.H., Mohney, B.K. and Kier, L.B., *J. Chem. Inf. Comp. Sci.* **31**, 76–82 (1991)
[24] Adamson, G.W. and Bush, J.A., *Inf. Stor. Retr.* **9**, 561–568 (1973)
[25] Carhart, R.E., Smith, D.H. and Venkataraghavan, R., *J. Chem. Inf. Com. Sci.* **2**, 64–73 (1985)
[26] Pepperrell, C.A. and Willett, P., *J. Comput.-Aided Mol. Des.* **5**, 455–474 (1991)
[27] Lawson, R.G. and Jurs, P.C., *J. Chem. Inf. Comp. Sci.* **30**, 137–144 (1990)
[28] Willett, P., *Similarity and Clustering in Chemical Information Systems,* Research Studies Press, Letchworth, 1987
[29] Lance, G.N. and Williams, W.T., *Comp. J.* **9**, 373–380 (1967)
[30] Murtagh, F., *Multidimensional Clustering Algorithms. COMPSTAT Lectures,* **4**, Physica-Verlag, Vienna, 1985
[31] Voorhees, E.M., *Inf. Proc. Man.* **22**, 465–476 (1986)
[32] El-Hamdouchi, A. and Willett, P., *Comp. J.* **32**, 220–227 (1989)
[33] Rubin, V. and Willett, P., *Anal. Chim. Acta* **151**, 161–166 (1983)
[34] Guenoche, A., Hansen, P. and Jaumard, B., *J. Class.* **8**, 5–30 (1991)
[35] Hartigan, J.A., *Clustering Algorithms,* John Wiley, New York, 1975
[36] Forgy, E., *Biometrics* **21**, 768 (1965)
[37] Jarvis, R.A. and Patrick, E.A., *IEEE Trans. Comput.* **C-22**, 1025–1034 (1973)
[38] Willett, P., Winterman, V. and Bawden, D., *J. Chem. Inf. Comp. Sci.* **26**, 109–118 (1986)
[39] Sibson, R., *Comp. J.* **16**, 30–34 (1973)
[40] Day, W.H.E., *J. Class.* **1**, 7–24 (1984)
[41] Podani, J., *Vegetatio,* **81**, 61–77 (1989)
[42] Dubes, R. and Jain, A.K., *Patt. Recog.* **11**, 235–254 (1976)
[43] Fisher, L. and van Ness, J.W., *Biometrika* **58**, 91–104 (1971)
[44] Jardine, N. and Sibson, R., *Mathematical Taxonomy,* John Wiley, New York, 1971
[45] Milligan, G.W., *Multi. Behav. Res.* **16**, 379–407 (1981)
[46] Everitt, B.S., *Biometrics* **35**, 169–181 (1979)
[47] Johnson, M.A., Lajiness, M.S. and Maggiora, G., *Molecular Similarity: a Basis for Designing Drug Screening Programs.* In: *QSAR: Quantitative Structure-Activity Relationships in Drug Design,* Fauchere, J.L., ed., Alan R. Liss Inc., New York, 1989, p. 167–171
[48] Lajiness, M.S., Johnson, M.A. and Maggiora, G., *Implementing Drug Screening Programs Using Molecular Similarity Methods.* In: *QSAR: Quantitative Structure-Activity Relationships in Drug Design,* Fauchere, J.L., ed., Alan R. Liss Inc., New York, 1989, p. 173–176
[49] Lajiness, M.S., *An Evaluation of the Performance of Dissimilarity Selection.* In: *QSAR: Rational Approaches to the Design of Bioactive Compounds.* Silipo, C. and Vittoria, A., eds., Elsevier Science Publishers, Amsterdam, 1991, p. 201–204

[50] Harrison, P.J., *Appl. Stat.* **17**, 226–236 (1968)

[51] Basak, S.C., Magnuson, V.R., Niemi, G.J. and Regal, R.R., *Discr. Appl. Math.* **19**, 17–44 (1988)

[52] Jerman-Blazic, B. and Fabic-Petrac, I., *Chemom. Intell. Lab. Syst.* **6**, 49–63 (1989)

[53] Ash, J.E., Warr, W.A. and Willett, P., eds., *Chemical Structure Systems*, Ellis Horwood, Chichester, 1991

[54] Hodes, L., Hazard, G.F., Geran, R.I. and Richman, S., *J. Med. Chem.* **20**, 469–475 (1977)

[55] Hodes, L., *J. Chem. Inf. Comp. Sci.* **29**, 66–71 (1989)

[56] Whaley, R. and Hodes, L., *J. Chem. Inf. Comp. Sci.* **31**, 345–347 (1991)

[57] Salton, G. and McGill, M.J., *Introduction to Modern Information Retrieval,* McGraw-Hill, New York, 1983

[58] Norager, O., ECDIN, *Environmental Chemicals Data and Information Network.* In: *Chemical Structures. The International Language of Chemistry.* Warr, W.A., ed., Springer-Verlag, Heidelberg, 1988

[59] Willett, P., *Similarity Searching in Databases of Three-Dimensional Chemical Structures.* In: *Information Systems and Data Analysis.* (Studies in Classification, Data Analysis and Knowledge Organization Vol. **IV**) Bock, H.H., Lenski, W. and Richter, M.M. (Eds.). Springer Verlag, Heidelberg; in press

3.3 Receptor Mapping and Phylogenetic Clustering

Paul J. Lewi and Henri Moereels

Abbreviations and Symbols

G	Guanosine
GDP, GTP	Guanosine Diphosphate, -Triphosphate
DNA	DeoxyriboNucleic Acid
$\boldsymbol{D}$	distance matrix
$\boldsymbol{L}$	similarity matrix
n	number of objects
i, i'	indices for objects
0	origin of space
$\boldsymbol{\theta}$	angular distance matrix
$\boldsymbol{C}$	variance-covariance matrix
EVD	EigenValue Decomposition
$\boldsymbol{U}$	eigenvector (or factor) matrix
$\boldsymbol{\Lambda}^2$	eigenvalue matrix
$\boldsymbol{I}$	identity matrix
r	number of factors (in the context of receptor mapping)
c	global dispersion
$\boldsymbol{S}$	scores matrix
P	number of measurements
$\boldsymbol{X}$	measurement table
M1, M2, ...	muscarinic cholinergic receptors
D1, D2, ...	dopaminergic receptors
A1, A2, ...	α-adrenergic receptors
B1, B2, ...	β-adrenergic receptors
5HT1, 5HT2	serotonergic receptors
H1, H2	histaminergic receptors
$\boldsymbol{W}$	weight matrix
UPGMA	Unweighted Pair-Group Mean Arithmetic
FM	Fitch and Margoliash
a, b, c, ...	tips of tree
r	root of tree (in the context of phylogenetic clustering)
x	node of tree
d	root mean square deviation of tips from root

$\bar{D}_r$	average distance of tips from root
s	root mean square relative deviation between observed and computed distances
D^*	reconstructed distance matrix

3.3.1 G-protein Coupled Receptors

G-protein coupled receptors are proteins that play an important role in the chemical transduction of signals within cells. The length of these proteins varies from 350 to 850 amino acid residues, with an average length of about 450 residues [1]. The secondary structure of a typical G-protein coupled receptor is represented in Fig. 1. The N-terminal of the protein is located in the extracellular space, while the C-terminal is within the cytoplasm. Seven domains, each with an average length of 24 residues, span the lipid bilayer of the cell plasma membrane. These seven transmembrane domains are linked together by means of outer (o) and inner (i) segments of variable lengths. A characteristic S-S bond forms a bridge between two cysteines which are located in the second and third outer segment (o_2 and o_3 in Fig. 1). The generally hydrophobic amino acids of the transmembrane segments form alpha helices which are arranged more or less in the form of a cylindrical shaft as indicated in Fig. 2. The amino acids in the extracellular and intracellular domains are generally hydrophilic, the former being generally acidic and the latter tending to be basic. Small molecules, such as neurotransmitters, are the first messengers in the signal transduc-

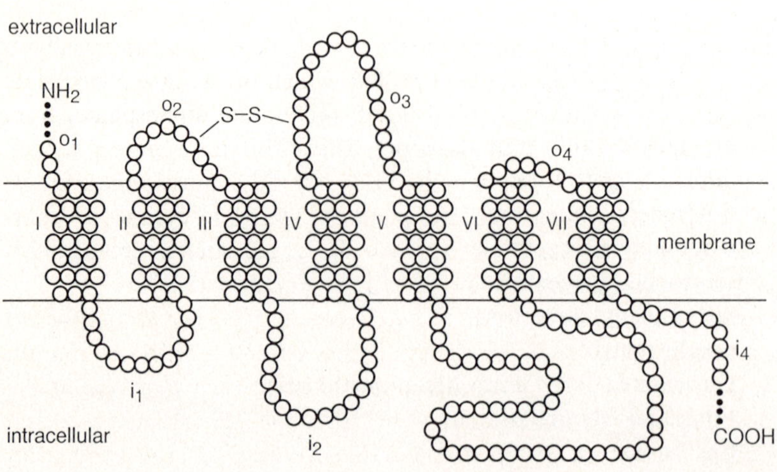

Figure 1. Secondary structure of a G-protein coupled receptor protein, showing the transmembrane, extracellular (o) and intracellular (i) segments.

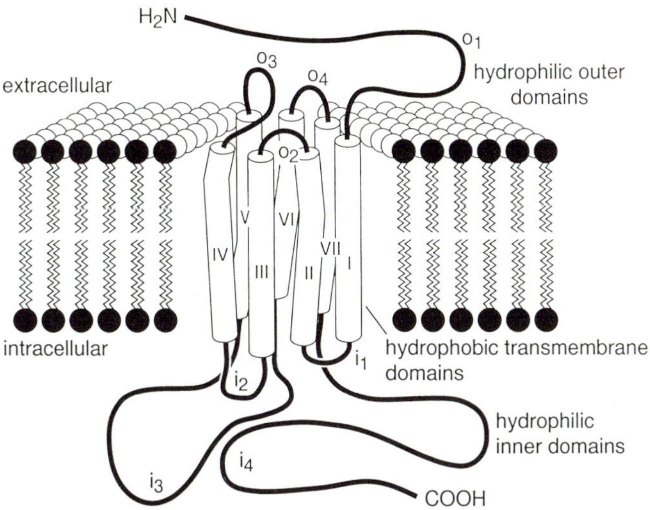

Figure 2. Tertiary structure of a G-protein coupled receptor with the shaft-like arrangement of the seven transmembrane segments.

tion pathway of a cell [2] and are released by other cells in the synaptic clefts. They are able to bind reversibly to specific amino acids which are located on the inside of the transmembrane shaft of a receptor. Typical neurotransmitters are acetylcholine, dopamine, noradrenaline, serotonin and histamine. Some of the G-protein coupled receptors also interact with peptides and odorants. A special class is formed by the photoreceptors in the retina of the eye, so called opsines, which are activated by photons rather than by chemical substances.

Signal transduction, as it is presently understood, is represented schematically in Fig. 3. Each receptor is associated with a G-protein which in the inactivated state consists of three parts (α, β and γ). It binds GDP (guanosine diphosphate) from which this class of receptors derives its name. Upon activation, the G-protein dissociates into two parts (α and $\beta\gamma$) by exchanging GDP for GTP (guanosine triphosphate). This in turn stimulates an enzyme (such as adenylate cyclase or phospholipase C) and causes the release of a secondary messenger (such as cyclic adenosine monophosphate or inositol phosphate) into the cytoplasm [3].

G-protein coupled receptors are thought to have evolved from an ancestral protein by random mutations of the DNA in the corresponding gene. It is estimated that the first divergence took place between 600 million and 1 billion years ago from an ancestral bacterial rhodopsin [4]. At the time of writing some 247 different receptors have been identified, which can be broadly classified into eight families according to the primary messengers that bind them specifically and which have been mentioned previously. The ancestral relationship between the various proteins is called a phylogeny. When a particular protein is studied in the different species, one can

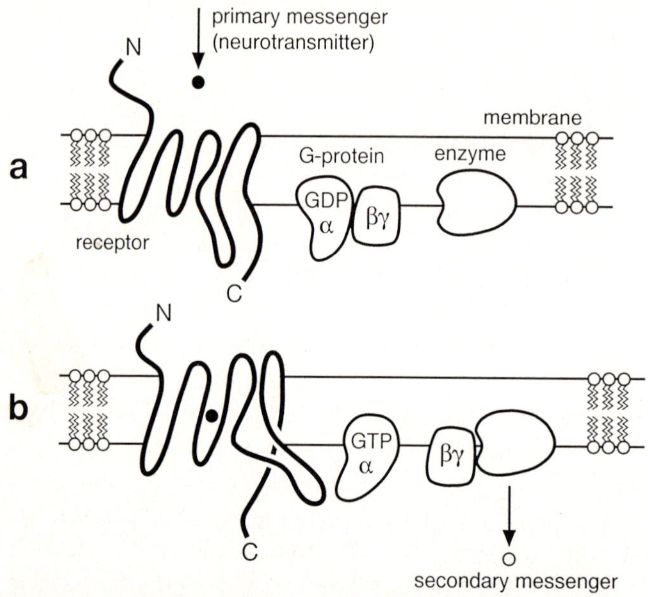

Figure 3. Schematic representation of a signal transduction pathway within a cell by means of a G-protein coupled receptor. a) The pathway starts with the binding of a neurotransmitter that is released in the synaptic cleft on the extracellular side, followed by b) activation of the receptor, GDP-GTP mediated dissociation of the G-protein, activation of an enzyme (phospholipase C or adenylate cyclase) and release of a secondary messenger (inositol phosphate or cyclic AMP) into the cytoplasm.

often trace the path of evolution of the species going from primitive organisms such as fungi and bacteria to fishes, amphibians, reptiles, birds, mammals and then up to primates and man. This is also the case with, for example, hemoglobin and cytochrome C, both of which have evolved over a period of 1 billion years [5].

G-protein coupled receptors play an important role in the central and peripheral nervous system. They are also implicated in various hereditary or congenital diseases, such as retinitis pigmentosa and diabetes mellitus, which are attributed to mutant receptor proteins encoded by impaired genes. In this study we only considered the muscarinic cholinergic receptors that bind acetylcholine, the α- and β-adrenergic receptors that bind noradrenaline, and those that bind serotonin and histamine. These receptors can be isolated in vitro for the purpose of screening new drugs. When a synthetic compound binds to a particular receptor, the compound may either mimick the action of the natural transmitter, in which case, it is called an agonist, or it may inhibit the receptor, in which case, it is referred to as an antagonist. The antagonist binds with a greater affinity to the receptor than the natural transmitter, but will not lead to the production of a second messenger down the transduction pathway. Agonists find therapeutic use either when there is a deficiency in the natural transmitter compounds in the neighbourhood of the receptor, or when the sensitivity of the receptor has decreased, or when the number of receptors has diminished. Conversely, antagonists play a therapeutic role either when there is an excess of natural transmitter molecules, or when the sensitivity of the receptor has increased, or when the receptors have proliferated abnormally.

Screening for synthetic agonists or antagonists is still largely an empirical undertaking. Molecular biology is only beginning to reveal the structure and the function of the active site of receptors. Although there are reasonable grounds for speculation on the binding and effect of naturally occurring transmitters, such as serotonin, we still have very little understanding about the way chemically unrelated synthetic agonists and antagonists exert their activity at the molecular level [6]. New sequences of G-protein coupled receptors are added continuously to the list of 247 that are known today. These may be duplicates of previously discovered sequences, but which have been expressed in different species, or they may be novel types or subtypes of a known receptor that exhibit distinct pharmacological properties. In order to organize and structure this large mass of information which is stored in our databases of amino acid sequences, we employed an approach which we call receptor mapping. Basically, our receptor mapping produces a two-dimensional representation of a group of receptors which reveals their similarity (or dissimilarity) together with their ancestral relationship. Similarities and dissimilarities can be mapped by means of Principal Coordinates Analysis as described by Gower [7]. Ancestral relationships are determined by phylogenetic clustering according to the method of Fitch and Margoliash [8].

3.3.2 Principal Coordinates Analysis of 71 Receptor Sequences

A preliminary analysis was first carried out on 71 sequences of G-protein coupled receptors. These receptors can be classified into 26 pharmacological classes, according to distinct biological functions and to the specific binding of various biogenic amines. The 71 sequences have been isolated from ten different vertebrate species, including human (25), rat (24), mouse (5), porcine (4), bovine (3), dog (3), hamster (3), chicken (2), turkey (1) and xenopus, a type of toad (1). The amino acid sequences have been compared pairwise, which resulted in a similarity score, ranging from 0 to 100 for each of the 2,485 (i.e. $71 \times 70/2$) different pairs. Similarity is defined here as the relative number of identical amino acid residues in the corresponding positions of two aligned sequences, with respect to the number of compared positions. Sometimes, similarity is defined as the relative number of "like" amino acid residues in the corresponding positions of the alignment according to the amino acid replacements that are acceptable in natural selection [9]. The former are referred to as identity scores, while the latter are called similarity scores. In this study, our measure of similarity is based on identity scores of amino acids. The computer program VGAP has been devised for the optimal alignment of the sequences using a variable gap penalty and for the derivation of identity scores [10]. The result of this calculation is a 71×71 symmetrical matrix of similarities (more precisely, identity scores) in which 0 means complete difference and where 100 indicates identity. This square matrix is identified by the symbol, ***L*** (for likeness).

The first step in the analysis is the transformation of the matrix of similarities, ***L***, into a matrix of dissimilarities, ***D***, by subtracting from 100:

$$\boldsymbol{D}_{ii'} = 100 - \boldsymbol{L}_{ii'} \qquad \text{with } i, i' = 1, \ldots, n \tag{1}$$

where i, i' are indices for the rows and columns of the matrices ***D*** and ***L***, and where n refers to the number of sequences. In Fig. 4 we have represented the distribution of the 2,485 dissimilarities in ***D***. A strong mode appears at about 70 percent, with minor modes around 50 and 10 percent. This diagram gives an impression of the rather large differences that exist between the 71 individual sequences (individual data are not shown).

Our objective is to derive a spatial configuration of the 71 sequences such that the distances between representative points correspond with the observed dissimilarities in ***D***. The problem can be equated to the reconstruction of a road map from a table of distances between cities, such as can be found on the reverse of a traveler's map [11]. The reconstruction can be performed with a ruler and compass, provided that the numbers represent point-to-point distances (as the crow flies), and provided that the configuration of points is truly two-dimensional. These assumptions may be largely satisfied in the case of a small geographical area for which the spherical shape of the earth can be neglected. In the case of n locations, one, thus, reduces $n \times (n-1)/2$ distances to $n \times 2$ coordinates. The reduction of data, thus, amounts to a factor $(n-1)/4$, which increases linearly with the number of points, n. Further-

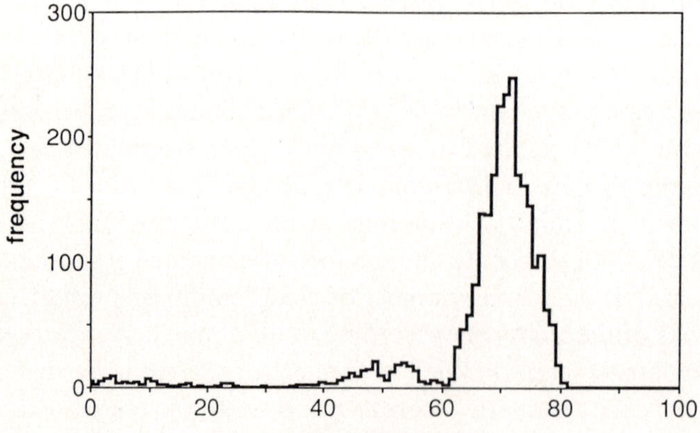

Figure 4. Distribution of the 2,485 dissimilarities among 71 G-protein coupled receptor sequences.

more, information represented by a map can be more readily comprehended, remembered and reproduced than from a table of distances [12].

In the case of protein sequences, we may consider each sequence as a point in space. Sequences that are very similar will be found close to one another, while those that are highly dissimilar will be a large distance apart. Here too, we may wish to convert the table of distances (more precisely, dissimilarities) into a kind of map which visualizes similarities and dissimilarities. The above assumptions, however, are not met in the case of protein sequences. Observed dissimilarities between sequences are not expressed as Euclidean distances, but rather as scores between 0 and 100. The latter also exhibit a degree of uncertainty, which arises from the occurrence of variable gaps in the alignment of the sequences. Furthermore, the dimensionality of the spatial configuration is unknown a-priori. In this case, the procedure with ruler and compass is inadequate and methods derived from factor analysis should be employed. Principal Coordinates Analysis is one such factor-analysis method developed by Gower [7]. The method is applied to a square table of distances (or dissimilarities) and produces a two-dimensional display which accounts for a maximum amount of the information in the table. It is closely related to Principal Components Analysis [13] which pursues the same objective, but with a rectangular table of measurements taken from a collection of objects, such as a table of biological and physico-chemical properties of proteins for example.

A crucial step in Principal Coordinates Analysis is the transformation of the $n \times n$ dissimilarity matrix $\boldsymbol{D}$ defined above into an $n \times n$ dispersion (or variance-covariance) matrix $\boldsymbol{C}$ by means of the formula:

$$C_{ii'} = -\frac{1}{2}(D^2_{ii'} - D^2_{i.} - D^2_{.i'} + D^2_{..}) \qquad \text{with } i, i' = 1, \ldots, n \tag{2}$$

where $D^2_{i.}$, $D^2_{.i'}$ and $D^2_{..}$ refer to row means, column means and global means of the squared dissimilarity matrix, $\boldsymbol{D}^2$:

$$D^2_{i.} = \frac{1}{n} \sum_{i'=1}^{n} D^2_{ii'} \qquad \text{with } i = 1, \ldots, n \tag{3}$$

$$D^2_{.i'} = \frac{1}{n} \sum_{i=1}^{n} D^2_{ii'} \qquad \text{with } i' = 1, \ldots, n \tag{4}$$

$$D^2_{..} = \frac{1}{n^2} \sum_{i=1}^{n} \sum_{i'=1}^{n} D^2_{ii'} \tag{5}$$

The expression between brackets in Eq. (2) represents the operation of centering of $\boldsymbol{D}^2$ simultaneously by rows and by columns, which is also called double-centering.

(The last term in Eq. (2) is required in order that the expression possesses zero global mean.) The formulae in Eq. (2) can be derived by considering the triangle relationship which involves two points, i, i' and the origin 0 of space:

$$D_{ii'}^2 = D_{oi}^2 + D_{oi'}^2 - 2D_{oi}D_{oi'}\cos\vartheta_{ii'} \tag{6}$$

or

$$D_{ii'}^2 = D_{oi}^2 + D_{oi'}^2 - 2C_{ii'} \tag{7}$$

where D_{oi} and $D_{oi'}$ denote the distances of points i and i' from the origin of space 0, where $D_{ii'}$ is the Euclidean distance between points i and i', and where $\vartheta_{ii'}$ is their angular distance as seen from the origin. Double centering of $\boldsymbol{D}^2$ removes the first two terms in the above expression (Eq. (7)) and leaves the term $-2\boldsymbol{C}$.

The resulting dispersion matrix $\boldsymbol{C}$ can now be factor-analyzed by means of a suitable computer program for eigenvalue decomposition (EVD), such as the QR-algorithm by Householder [14]. This results in a $n \times r$ orthonormal matrix of eigenvectors $\boldsymbol{U}$ (also called factor matrix) and an $r \times r$ diagonal matrix of eigenvalues Λ^2 such that:

$$\frac{1}{n}\boldsymbol{C} = \boldsymbol{U} \cdot \Lambda^2 \cdot \boldsymbol{U}^{\mathrm{T}} \tag{8}$$

with the orthogonality condition imposed on the columns of $\boldsymbol{U}$:

$$\boldsymbol{I} = \boldsymbol{U}^{\mathrm{T}} \cdot \boldsymbol{U} \tag{9}$$

where $\boldsymbol{I}$ represents the $r \times r$ identity matrix of dimension r. (Note that our definition of eigenvalue decomposition deviates from the usual notation by the introduction of the constant weight, $1/n$. The weight is introduced here in order to make the notation of this section compatible with the following section in which variable weights are introduced.) The superscript T indicates transposition of rows and columns of a matrix, and the dot indicates matrix multiplication. The number r is the rank of the dispersion matrix, $\boldsymbol{C}$, i.e. the number of independent dimensions of the space that is required to fully represent the geometrical properties of the n sequences. In general, r is at most equal to $n-1$. (Double centering removes one dimension from the original number of dimensions, which is at the most equal to n.) Hence, in the case of 71 sequences, we may require up to 70 dimensions to fully represent the distances between receptors sequences. Fortunately, as we shall see, not all dimensions contribute in the same way to the geometrical representation, and a large number of these may be artificial, or represent noise. Artifacts and noise may be introduced by our choice of distance metric, whose scale at the higher end is limited to 100 percent similarity. The distance metric is obtained by comparing sequences of variable gap lengths. The independent dimensions extracted by the algorithm are called factors of the dispersion table, hence the term factor analysis which is given as the generic term for the procedure described above by Eq. (8). In this context, the terms factor, eigenvector and principal component are used synonymously. We prefer the former, however, for the sake of brevity. Because of the orthogonality of U, we can also define the eigenvalue decomposition of the dispersion matrix in the form:

$$\Lambda^2 = \frac{1}{n} U^{\mathrm{T}} \cdot C \cdot U \tag{10}$$

The rows of the factor matrix U refer to n proteins. The columns of U define the orientations of the r factors in the original space that contains the pattern of points as defined by D. By definition, these factors are orthogonal, which means mutually uncorrelated or independent. The diagonal elements of the matrix Λ^2 represent the contributions of the corresponding factors to the global dispersion, c, in the matrix C. (Contribution to global dispersion and eigenvalue are used synonymously here, although we prefer the former.) The global dispersion, c, is equal to the mean of the diagonal elements of C, which is also equal to the sum of the diagonal elements of Λ^2:

$$\operatorname{trace}(\Lambda^2) = \frac{1}{n} \operatorname{trace}(C) = c \tag{11}$$

where trace defines the sum of elements on the main diagonal of a square matrix.

The global dispersion, c, is invariant when the original dimensions of space are rotated towards the computed orientations of the factors. These factors can also be regarded as the principal axes of inertia for the pattern of points that are defined by the distances in D. In the case of an ellipsoidal structure, the factors represent the main axes of symmetry of the pattern. The global dispersion is a measure of the distribution of the points around the centroid of the pattern, which is the center of mass, assuming that all points have unit mass. By convention, factors are arranged in decreasing order of their contributions to the global dispersion. Note that these contributions have positive values and that their number is at most equal to $n-1$ in the present application to distance matrices. The relative magnitude of the contributions indicates roughly whether a factor represents structural information or noise. We did not include trivial factors, i.e. those that do not contribute to the global dispersion. Such trivial factors arise when points within the pattern are coincidental, collinear or coplanar. Any such occurrence decreases the number of nontrivial factors, r, that can be extracted from the $n \times n$ distance matrix, which explains why the number of factors (also called rank) can be less than $n-1$.

The principal coordinates S of the n points can be reconstructed in r-dimensional factor space by means of:

$$S = \sqrt{n}\, U \cdot \Lambda \tag{12}$$

where S represents the $n \times r$ matrix of principal coordinates (also called factor scores), and where the $r \times r$ diagonal matrix of factor contributions has been defined above. Each row of S contains the r coordinates of one of the n points.

It can be shown that the original dispersion matrix C can be reconstructed from the principal coordinates in S (Eq. (13)) by virtue of the orthonormality of U and the eigenvalue decomposition of the dispersion matrix C (see Eq. (8)).

$$S \cdot S^{T} = n\, U \cdot \Lambda^{2} \cdot U^{T} = C \tag{13}$$

The main difference between Principal Coordinates Analysis and Principal Components Analysis is that the former is applied to an $n \times n$ distance matrix, D, while the latter requires an $n \times p$ measurement table, X, which describes the same n objects by means of p measurements. The table, X, can thus, be regarded as a table of coordinates which defines the n objects in p-dimensional measurement space. (For the sake of simplicity, we assumed that the pattern of points is centered about the origin of space.) In Principal Components Analysis one first computes the dispersion matrix, C, from the $n \times p$ measurement table, X:

$$C = \frac{1}{p} X \cdot X^{T} \tag{14}$$

Subsequently, one then applies the factor analysis described above. The results are identical to those obtained by Principal Coordinates Analysis. With both methods a substantial reduction in the apparent dimensionality of the data is achieved. In the case of an $n \times n$ distance table, D, there is a large degree of redundancy whenever the number of structural dimensions is smaller than n. In the case of a measurement table, redundancy is derived from the possible correlations between the p measurements. Both methods can be executed with the SPECTRAMAP program for exploratory multivariate data analysis which places more emphasis on the visual display of the results [15].

In Fig. 5 we have reproduced the 71 sequences of G-protein coupled receptors by means of their principal coordinates in the plane of the first two dominant factors. Dominant factors are the structural dimensions that account for a large part of the global dispersion, c, (Eq. (11)) of the pattern of points in multidimensional space. The horizontal and vertical axes of the diagram (Fig. 5) represent the first and second dominant factors which contribute 18 and 12 percent to the global dispersion, respectively. Together they account for 30 percent of the global dispersion, which is rather low. The third dominant factor is oriented perpendicular to the plane of the plot and contributes another 9 percent of the dispersion. This factor is indicated on the diagram by means of the variable thickness of the contours of the circular symbols. A thick outline signals that the corresponding point is represented above the plane of the map. A thin outline indicates that the point lies below the plane. By

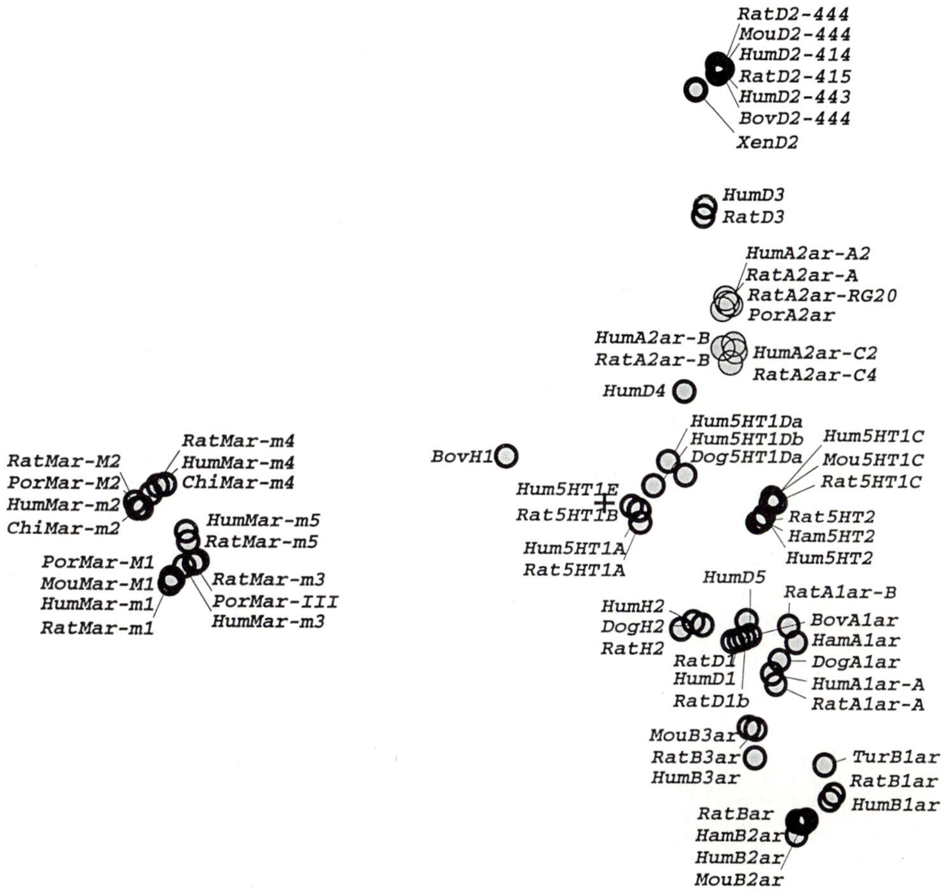

Figure 5. Principal coordinates plot of the 71 sequences, as obtained by the SPECTRAMAP program [15]. The two most dominant factors are represented along the horizontal and vertical axes of the plot. A third factor which is perpendicular to the plane of the plot is indicated by means of the variable thickness of the contours of the symbols. The three factors contribute 18, 12 and 9 percent to the total dispersion of the points around the centroid, respectively. The latter is represented by a small cross (+) at the center of the plot. All 71 points are assigned equal mass (or weight). Distances between points on the map are related to differences in amino acid sequences.

means of the horizontal and vertical dimensions of the map and the visual depth guide, we are able to reproduce 39 percent of the global dispersion, which leaves 61 percent unaccounted for. The latter is to be attributed to higher-order structural factors and to factors that represent artifacts and noise in the data. Nevertheless, we will show below that the two- (or three-) dimensional map which is shown in Fig. 5 provides useful information about the similarities and dissimilarities between protein sequences, their classification and their evolutionary descendence. We have arranged the labels of the 71 sequences on the map in order to emphasize these points. First, we shall explain briefly the structure of the labels. The initial three letters indicate the type of species from which the receptors have been isolated: Hum (human), Mou (mouse), Chi (chicken), Tur (turkey), Ham (hamster), Bov (bovine), Por (porcine), and Xen (xenopus). The next two or three letters identify the six types of receptors studied: Mar (muscarinic cholinergic), D (dopaminergic), A and B (alpha- and beta-adrenergic), 5HT (serotonergic) and H (histaminergic). These are broad categories of receptors that can be distinguished in pharmacological assays by their differential effects on the main neurotransmitters (acetylcholine, dopamine, noradrenaline, serotonin and histamine). A finer subdivision of the 71 receptors can be made according to their pharmacological subtype. In total, we can classify the 71 sequences into 26 subtypes of G-protein coupled receptors. These include 5 muscarinic cholinergic (M1, M2, M3, M4, M5), 5 dopaminergic (D1, D2, D3, D4, D5), 5 alpha-adrenergic (A1A, -B and A2A, -B, -C), 3 beta-adrenergic (B1, B2, B3), 6 serotonergic (5HT1A, -B, -C, -D, -E and 5HT2) and, finally, 2 histaminergic (H1 and H2) subtypes. Many of these subtypes are replicated several times in the map shown in Fig. 5 due to the same gene being expressed in the different species outlined above. Table 1 shows how the 71 sequences are distributed over the 26 subtypes. Conventionally, some subtypes are identified idiosyncratically by means of Greek letters, lower case letters, Roman numerals and subscripts, such as m_3, M_3, III, etc. Here, we adopted a notation of our own, which is compatible with the limitations of current computer representations. It must also be mentioned that the naming of subtypes is historical rather than systematic and is always subject to change; as new receptors are discovered, new classifications emerge. This accounts for some complicated labels, such as Rat A2ar-RG20, which is most likely to be an A2A receptor expressed in rats, judging from its proximity to other A2A sequences in Fig. 5.

Interpretation of the map in Fig. 5 is rather straightforward. Proteins that are represented close together possess highly similar sequences, while those that are further away are dissimilar. In other words, distances between points on the map can be interpreted in terms of the original dissimilarities in the original data table, $\boldsymbol{D}$. The small cross (+) in the center of the plot indicates the centroid (or center of mass) of the pattern of 71 points, where it is assumed that each point has the same mass. Along the horizontal axis of the map, which represents the first dominant factor, there is a considerable contrast between the 5 muscarinic cholinergic subtypes on the left and the 21 other subtypes on the right. The vertical axis, which accounts for the second dominant factor, showed a considerable contrast between D2, D3 subtypes at the top and the B1, B2, B3, A1, D1 and H2 subtypes at the bottom on the right side of the map. A third contrast, due to the effect of the third dominant factor, is shown along the depth axis which is encoded by the variable thickness of the contours of the circular symbols. This contrast distinguishes the three A2 subtypes (below the plane) and D2 (above the plane).

Table 1. Compilation of the 26 receptor subtypes, their number of sequences expressed in different species, evolutionary distance from the computed root (percent dissimilarity) and length of sequence (number of amino acid residues).

Type of receptor	Subtype of receptor	Number of sequences	Evolutionary distance	Length of sequence
Muscarinic acetylcholine	M1	4	37.4	460
	M2	4	39.0	466
	M3	3	37.4	590
	M4	3	38.2	479
	M5	2	38.2	532
Dopamine	D1	3	39.3	447
	D2	7	36.6	443
	D3	2	36.2	423
	D4	1	35.9	387
	D5	1	39.6	477
α-adrenergic	A1A	5	39.7	494
	A1B	1	39.1	515
	A2A	4	37.0	450
	A2B	2	37.8	457
	A2C	2	38.0	454
β-adrenergic	B1	3	38.4	477
	B2	4	39.1	418
	B3	3	36.3	400
Serotonin	5HT1A	2	34.9	422
	5HT1B	1	34.0	386
	5HT1C	3	39.7	459
	5HT1D	3	33.9	377
	5HT1E	1	34.1	365
	5HT2	3	40.2	471
Histamine	H1	1	38.0	491
	H2	3	37.1	359
All	26	71	37.5	450

Many of the 26 pharmacologically distinct classes can be distinguished as separate groups on the map of Fig. 5. An overlap, however, is observed between A2B and -C, between 5HT1A, -B and -E, and between D1 and D5 receptor subtypes. Two sequences also appear to be separate from the main group which they were thought to belong to for pharmacological reasons. This is the case for Turkey B1 and Bovine A1A. These discrepancies will be discussed in Sec. 3.3.5. Despite these shortcomings, one obtains a visual display of the main relationships between G-protein coupled receptors, in as far as they are known at present. A notable feature of the landscape is the isolated position of the histaminergic H1 receptor which appears as being quite unrelated to all the others, especially H2. Another feature of the pattern is the separation of the three dopaminergic receptors D2, D3 at the top, D4 in the middle and D1, D5 near the bottom of the map. The separation between the A1 and A2 receptors is also remarkable. Other salient features are, on the one hand, the division of the serotonergic subtypes into 5HT2, 5HT1C and on the other hand, the group composed of 5HT1A, -B, -D, -E. In the following section we shall reduce the 71 sequences to 26 receptor subtypes and illustrate how a more succinct mapping can be produced which still exhibits the same features discussed above.

3.3.3 Principal Coordinates Analysis of 26 Receptor Subtypes

In the previous section we have seen how the 71 sequences of G-protein coupled receptors can be grouped into 26 pharmacological subtypes. Principal Coordinates Analysis has also shown that these 26 subtypes can be distinguished on the basis of the similarities between their amino acid sequences with the exception of a few overlaps and discrepancies which will be discussed later on. In this section we shall attempt to simplify the analysis by reducing the 71×71 matrix of sequence similarities to a 26×26 matrix of subtype similarities. To this end, we have computed average similarities by averaging the rows and columns of the 71×71 matrix whose corresponding sequences belong to the same subtype as shown in Fig. 5 and in Table 1. (For example, the four rows in the 71×71 table, corresponding to the M1 subtype, are averaged into a single row. The four columns in the same 71×71 table corresponding to the M1 subtype are also averaged into a single column. By convention, the similarity between an averaged subtype and itself is set to 100 percent.) The result which has been subjected to Principal Coordinates Analysis once again is presented in Table 2. In contrast to the previous analysis, we now assign variable masses (or weights) to the 26 points in the pattern of subtypes according to the number of sequences present in each. This requires the construction of a 26×26 diagonal weighting matrix, W, in which the diagonal elements are proportional to the number of sequences in each of the 26 subgroups (Table 1). These masses (or weights) are normalized to a unit sum and are shown in the last column of Table 2. Variable weighting, rather than constant weighting, is applied here in order to account for the fact that subtypes with many replicates in different species are defined more precisely in space than those with fewer or no replicates. This type of variable weighting also ensures that the new result obtained from the 26 subtypes most resembles the previous one (Fig. 5) derived from the 71 original sequences.

Generalized Principal Coordinates Analysis involves the eigenvalue decomposition (or factorization) of the weighted dispersion matrix $W^{1/2} \cdot C \cdot W^{1/2}$, which results in an orthogonal matrix of eigenvectors (or factors) U and a diagonal matrix of eigenvalues (or contributions to the global weighted dispersion) Λ^2:

$$W^{1/2} \cdot C \cdot W^{1/2} = U \cdot \Lambda^2 \cdot U^{\mathrm{T}} \tag{15}$$

Table 2. Similarities (percent) between 26 receptor subtypes as computed by the VGAP computer algorithm [10]. The last column contains the weights which are proportional to the number of sequences in each subtype (Table 1).

	Subtype	1	2	3	4	5	6	7	8	9	10	11	12	13	14	15	16	17	18	19	20	21	22	23	24	25	26	Weight
1	M1	100	44	54	47	55	24	26	28	30	24	24	23	27	27	26	23	24	32	29	29	22	29	30	23	30	28	0.056
2	M2	44	100	46	60	45	23	27	27	27	22	23	24	25	24	23	21	22	25	28	30	20	28	30	20	28	28	0.056
3	M3	54	46	100	46	52	25	27	27	25	25	22	22	25	24	25	24	27	27	29	31	25	31	32	25	30	29	0.042
4	M4	47	60	46	100	47	23	29	28	29	22	22	22	26	24	24	21	24	27	31	31	21	29	29	21	30	26	0.042
5	M5	55	45	52	47	100	25	28	27	27	22	20	21	25	24	24	23	26	26	27	31	23	30	31	21	29	29	0.028
6	D1	24	23	25	23	25	100	29	29	31	66	30	30	29	27	27	32	32	31	31	32	25	32	31	26	22	36	0.042
7	D2	26	27	27	29	28	29	100	53	40	28	28	30	34	32	31	27	28	30	30	33	30	35	32	29	30	31	0.099
8	D3	28	27	27	28	27	29	53	100	41	30	28	30	35	33	33	29	29	34	32	33	28	34	33	28	28	29	0.028
9	D4	30	27	25	29	27	31	40	41	100	33	30	29	34	35	36	34	28	35	33	28	29	29	30	30	26	30	0.014
10	D5	24	22	25	22	22	66	28	30	33	100	29	28	29	27	27	32	32	34	33	32	25	32	32	24	21	36	0.014
11	A1A	24	23	22	22	20	30	28	28	30	29	100	67	27	28	28	31	32	34	29	31	27	31	29	24	22	33	0.07
12	A1B	23	24	22	22	21	30	30	30	29	28	67	100	27	28	28	31	30	35	32	31	28	32	30	27	21	35	0.014
13	A2A	27	25	25	26	25	29	34	35	34	29	27	27	100	52	52	29	26	32	34	33	28	36	34	26	25	30	0.056
14	A2B	27	24	24	24	24	27	32	33	35	27	28	28	52	100	69	28	26	31	34	32	28	33	33	27	25	30	0.028
15	A2C	26	23	25	24	24	27	31	33	36	27	28	28	52	69	100	27	26	31	33	32	28	33	32	26	24	30	0.028
16	B1	23	21	24	21	23	32	27	29	34	32	31	31	29	28	27	100	51	51	32	35	27	35	32	28	22	36	0.042
17	B2	24	22	27	24	26	32	28	29	28	32	32	30	26	26	26	51	100	44	29	30	29	31	32	29	25	33	0.056
18	B3	32	25	27	27	26	31	30	34	35	34	34	35	32	31	31	51	44	100	32	33	29	31	31	29	28	34	0.042
19	5HT1A	29	28	29	31	27	31	30	32	33	33	29	32	34	34	33	32	29	32	100	43	26	42	39	27	30	31	0.028
20	5HT1B	29	30	31	31	31	32	33	33	28	32	31	31	33	32	32	35	30	33	43	100	29	72	47	29	32	31	0.014
21	5HT1C	22	20	25	21	23	25	30	28	29	25	27	28	28	28	28	27	29	29	26	29	100	30	32	50	23	28	0.042
22	5HT1D	29	28	31	29	30	32	35	34	29	32	31	32	36	33	33	35	31	31	42	72	30	100	47	29	31	31	0.042
23	5HT1E	30	30	32	29	31	31	32	33	30	32	29	30	34	33	32	32	32	31	39	47	32	47	100	33	33	29	0.014
24	5HT2	23	20	25	21	21	26	29	28	30	24	24	27	26	27	26	28	29	29	27	29	50	29	33	100	21	28	0.042
25	H1	30	28	30	30	29	22	30	28	26	21	22	21	25	25	24	22	25	28	30	32	23	31	33	21	100	27	0.014
26	H2	28	28	29	26	29	36	31	29	30	36	33	35	30	30	30	36	33	34	31	31	28	31	29	28	27	100	0.042

with the usual condition for orthonormality of the columns of U (Eq. (9)):

$$I = U^{T} \cdot U \tag{16}$$

where I represents the identity matrix of r-dimensional factor space. The number of non-trivial factors, r, that can be extracted is called the rank of the dispersion matrix C. The coordinates (or scores) S of the 26 subtypes in r-dimensional factor space are now defined by the following:

$$S = W^{-1/2} \cdot U \cdot \Lambda \tag{17}$$

In the previous case of the 71 sequences, we assumed that all 71 points in the pattern had a constant mass (or weight). In the present case of 26 subtypes, we assign to each of the 26 points a mass (or weight) which is proportional to the number of sequences in the corresponding subtype. This explains the difference between Eqs. (8) and (12) and Eqs. (15) and (17). The latter is reduced to the former, however, when the variable weight matrix W is replaced by the constant weight $1/n$.

Similarly, as in the previous section, we can show that the sum of the diagonal elements (or trace) of Λ^2 is equal to the global dispersion of C:

$$\text{trace}\,(\Lambda^2) = \text{trace}\,(W^{1/2} \cdot C \cdot W^{1/2}) = c \tag{18}$$

In other words, the sum of the eigenvalues is equal to the global dispersion, c. The latter is invariant when the original coordinate axes are rotated toward the computed orthogonal factors.

It can also be shown that the dispersion, C, of the 26 receptor subtypes around their weighted centroid can be reconstructed from the matrix of their coordinates, S:

$$S \cdot S^{T} = W^{-1/2} \cdot U \cdot \Lambda^2 \cdot U^{T} \cdot W^{-1/2} = C \tag{19}$$

by virtue of the orthogonality of the columns of U and from the definition of eigenvalue decomposition (Eqs. (15) and (16)).

The result of the variable weighted analysis of the 26 receptor subtypes is shown in Fig. 6. The rules for interpretation are the same as those for Fig. 5. In the present case, we find that the three most dominant factors account for 19, 13 and 9 percent of the global dispersion. (These three factors are displayed along the horizontal, vertical and depth axes, respectively.) Together they account for 41 percent of the global dispersion, which is comparable to the result of the previous analysis, in which 39 percent was accounted for by the three dominant factors. The points on the map in Fig. 6 are virtually identical to the centroids of the corresponding clusters in the map of Fig. 5. In fact, one could also have obtained the map of the 26 receptor subtypes by computing averages of the factor coordinates of the corresponding sequences

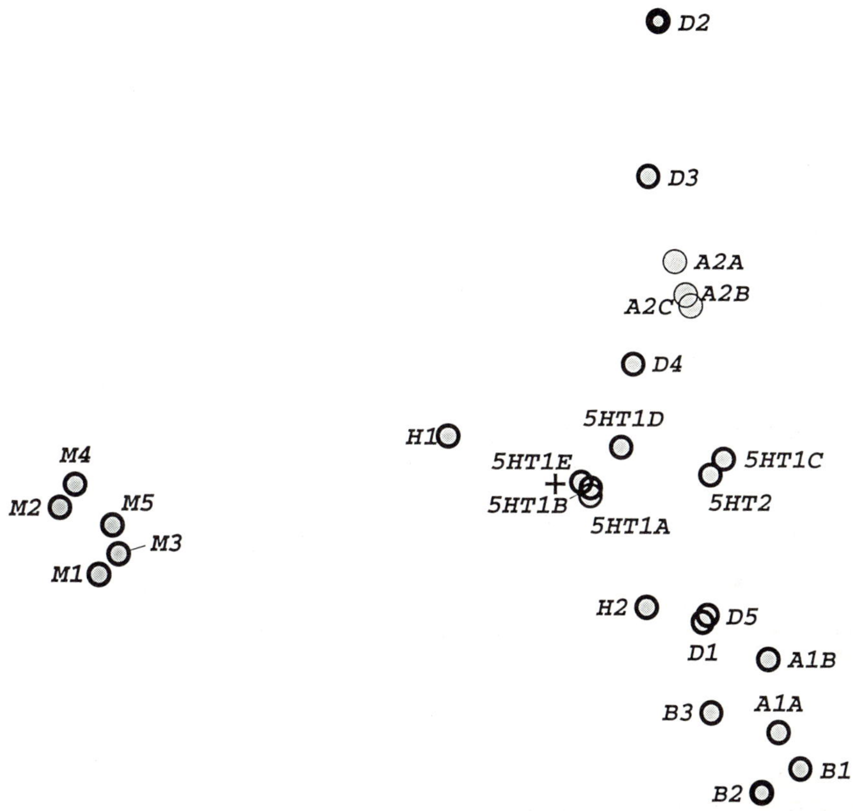

Figure 6. Weighted principal coordinates plot of the 26 receptor subtypes, as obtained by the SPECTRAMAP program [15]. The three factors contribute 19, 13 and 9 percent to the total dispersion of the points around the centroid, respectively. The individual 26 points are assigned a mass (or weight) which is proportional to the number of sequences in each subgroup, as listed in Table 1. The analysis has been produced from the data in Table 2. Distances between points on the map are related to differences in the average amino acid sequences of receptor subtypes.

from the results of the analysis on the complete set of 71 sequences. In the case of a large collection of sequences (several hundreds) it is more convenient, however, to compute averaged distances and to apply a weighted analysis as described above. The reason being, that the time required for the extraction of factors from distance data is proportional to the dimension of the table raised to a power of about 2.5. The generalized Principal Coordinates Analysis and the resulting map in Fig. 6 were obtained with the SPECTRAMAP program [15].

Our previous findings have been confirmed in the work described here. The overlaps between A2B and -C, between 5HT1A, and -B, between 5HT1A and -E, and between D1 and D5 receptor subtypes are also not clearly resolved in this map.

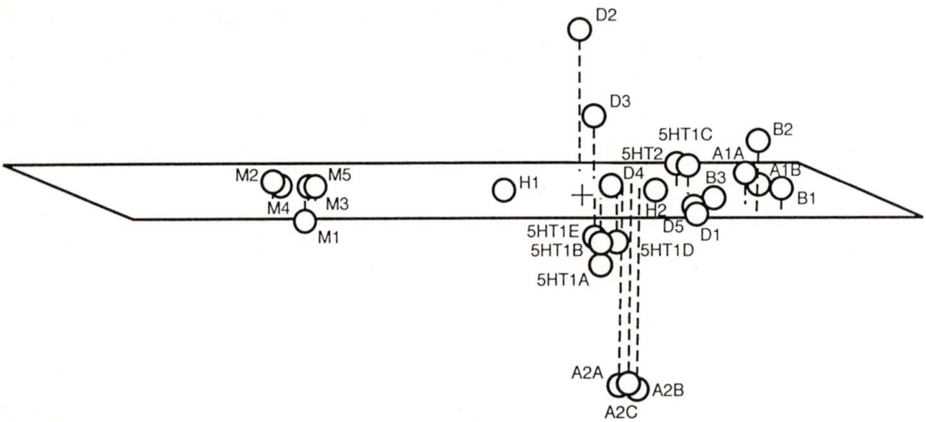

Figure 7. Three-dimensional perspective drawing of the arrangement of 26 receptor subtypes in the space defined by the three dominant factors obtained by weighted Principal Components Analysis, the first two factors of which are shown in Fig. 6.

In order to enhance the three-dimensional structure of the 26 receptors we also have presented the three dominant factors in perspective in Fig. 7 with the point of view being on the left lower side and slightly above the plane of the map in Fig. 6. In this representation one can see more clearly how the third factor accounts for a contrast between the dopaminergic receptors D2, D3 (top) and the cluster of alpha-adrenergic receptors A2A, -B, -C (bottom). The latter cannot be resolved in three-factor space.

3.3.4 Phylogenetic Clustering

As we have mentioned in the introduction, dissimilarities between amino acid sequences of proteins are the result of mutations in the corresponding DNA sequences over periods of several hundreds of millions of years. Estimates for the mutation rate indicate that it takes between 2.5 to 10 million years to produce a one percent change in the amino acid sequenes of functional proteins, such as hemoglobins, cytochrome C and G-protein coupled receptors [4, 16, 17]. It is, therefore, reasonable to assume that these proteins have descended from common ancestors and, hence, they are related to each other by means of a tree-like structure. Such an ancestral tree, if it exists, is called a phylogeny, and the process of its formation is referred to as phylogenetic clustering. There are two approaches to obtaining a phylogeny, depending whether clustering is performed upon sequences of DNA bases or upon sequences of amino acid residues. The former is called *parsimony clustering* and considers the minimal number of DNA mutations that are needed to transform one DNA sequence into another. The latter is referred to as *distance clustering* and requires a matrix of dissimilarities between sequences. Here we describe a method of distance clustering which has been developed by Fitch and Margoliash [8].

In order to illustrate the method of phylogenetic distance clustering, we have taken an excerpt from Table 2, which is shown in Fig. 8. The reduced table shows the dissimilarities between four selected receptor subtypes (B1, B2, D2 and M1). Note that the dissimilarities were derived from the similarities by Eq. (1).

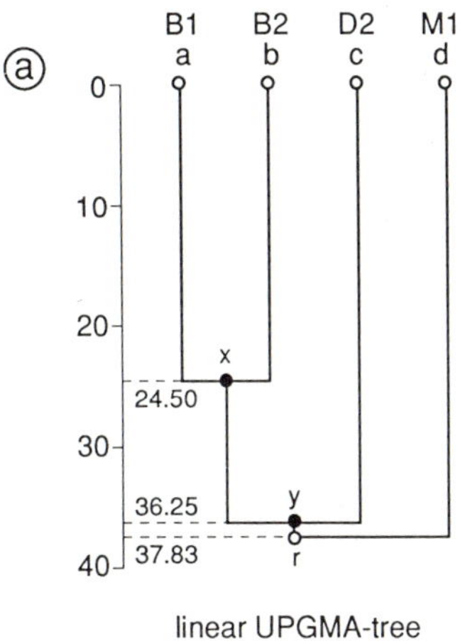

		B1	B2	D2	M1
		a	b	c	d
B1	a	0	49	73	77
B2	b	49	0	72	76
D2	c	73	72	0	74
M1	d	77	76	74	0

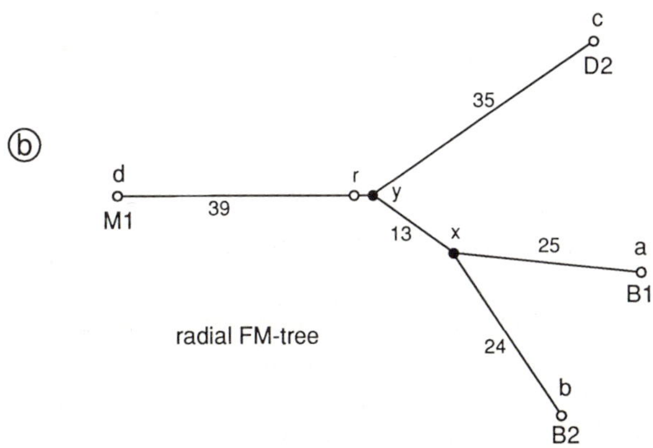

Figure 8. Illustration of phylogenetic clustering using an excerpt from Table 2 (insert). The table lists the dissimilarities (100 minus similarities) between four selected receptor subtypes. a) The corresponding linear UPGMA-tree according to Sneath and Sokal [18]. The tips of the tree (a, b, c, d) are arranged along a horizontal line. The two nodes are labeled x and y. The root of the tree is identified as r. The numbers along the vertical scale indicate mean clustering distance from the nodes and root to the tips of the tree. b) The corresponding radial tree, according to Fitch and Margoliash [8]. The numbers along the branches indicate the branch lengths from the nodes x and y to the tips of the tree. (See text for explanation).

As a starting point, we first built a hierarchical tree by means of the unweighted pair-group mean arithmetic method (UPGMA) by Sneath and Sokal [18].

The UPGMA-tree from the reduced distance table in Fig. 8 has been obtained as follows. First, we selected the two closest points. In this case we found that the smallest distance, $D_{a,b}$, of 49 was between points a and b. The ancestor of points a and b is labeled x, and is placed half-way between $D_{a,b}$, i.e. at 24.5 units from both a and b. Next, we computed the distances $D_{c,ab}$ and $D_{d,ab}$ between the remaining points c and d to the mean position of the previously identified group of two points ab:

$$D_{c,ab} = \frac{1}{2}(D_{c,a} + D_{c,b}) = \frac{1}{2}(73+72) = 72.5 \tag{20}$$

$$D_{d,ab} = \frac{1}{2}(D_{d,a} + D_{d,b}) = \frac{1}{2}(77+76) = 76.5 \tag{21}$$

We observed that $D_{c,ab}$ was the shortest of the two distances. Consequently, point c is joined to the group ab and their common ancestor, which is labeled y, is placed half-way between $D_{c,ab}$, i.e. at 36.25 units from the group abc. The remaining point d has a mean distance $D_{d,abc}$ from the three points a, b, c that are already clustered together:

$$D_{d,abc} = \frac{1}{3}(D_{d,a} + D_{d,b} + D_{d,c}) = \frac{1}{3}(77+76+74) = 75.67 \tag{22}$$

The terminal node or root of this artificial group of four points is labeled r, and is placed half-way between $D_{d,abc}$, i.e. at 37.83 units from the group abcd. The complete tree is shown in the form of a linear tree as illustrated in Fig. 8. In this linear tree, the tips of the tree are arranged along the horizontal line. The vertical scale represents the arithmetic mean distances of the nodes to the tips of the tree. In the UPGMA-method, mean values are not weighted due to the variable number of points in each group. For example, in the case when a group abc is joined with another group of two points de, the unweighted pair-group mean arithmetic (UPGMA) distance is obtained from:

$$D_{abc,de} = \frac{1}{3\times 2}(D_{a,d} + D_{a,e} + D_{b,d} + D_{b,e} + D_{c,d} + D_{c,e}) \tag{23}$$

Once the initial UPGMA-tree was obtained, we proceeded to calculate the phylogenetic branch lengths. This calculation is based upon a general formula which relates the distance, $D_{a,x}$ between an arbitrary point a and its ancestor x to the distance $D_{a,b}$ between the two points that descend from x and to the distances $D_{a,c}$ and $D_{b,c}$ between each of these points, a, b and a third point c that does not descend from x:

$$D_{a,x} = \frac{1}{2}(D_{a,b} + D_{a,c} - D_{b,c}) \tag{24}$$

This formula can be understood by regarding the branch lengths in Fig. 8a as evolutionary distances, i.e. as being proportional to the number of years required to mutate one amino acid sequence.

In Eq. (24), we have chosen point c as the pivotal point that does not descend from the node x. We could have chosen, however, point d instead. In the general case, we will obtain n_x points c_x that do not not descend from the node x. In order to accommodate for multiple points c_x we must rewrite Eq. (24) in the following way:

$$D_{a,x} = \frac{1}{2}(D_{a,b} + \frac{1}{n_x} \sum_{c_x} (D_{a,c_x} - D_{b,c_x})) \tag{25}$$

which involes the mean of the differences $D_{a,x} - D_{b,x}$ over all n_x points c_x that do not descend from x (cf. Fitch and Margoliash [8], abbreviated FM).

The formula in Eq. (25) must be applied recursively to all the intermediate nodes of the FM-tree. This is a tedious calculation which should be left to a computer algorithm, such as the PHYLIP program developed by Felsenstein [19]. In our simplified example from the insert of Fig. 8, however, the calculation can still be done on paper. First, we compute the distance $D_{a,x}$ according to Eq. (25):

$$\begin{aligned} D_{a,x} &= \frac{1}{2}\{D_{a,b} + \frac{1}{2}[(D_{a,c} - D_{b,c}) + (D_{a,d} - D_{b,d})]\} \\ &= \frac{1}{2}\{49 + \frac{1}{2}[(73-72) + (77-76)]\} = 25 \end{aligned} \tag{26}$$

From this result we can easily derive the branch length, $D_{b,x}$:

$$D_{b,x} = D_{a,b} - D_{a,x} = 49 - 25 = 24 \tag{27}$$

Next, we obtain the distance $D_{c,y}$ from the basic Eq. (24):

$$D_{c,y} = \frac{1}{2}(D_{c,ab} + D_{c,d} - D_{d,ab}) = \frac{1}{2}(72.5 + 74 - 76.5) = 35 \tag{28}$$

since:

$$D_{c,ab} = \frac{1}{2}(D_{c,a} + D_{c,b}) = \frac{1}{2}(73 + 72) = 72.5 \tag{29}$$

$$D_{d,ab} = \frac{1}{2}(D_{d,a} + D_{d,b}) = \frac{1}{2}(77 + 76) = 76.5 \tag{30}$$

At this stage, we can compute the distance $D_{x,y}$ between the two nodes x and y from the results obtained previously and using the clustering tree of Fig. 8a:

$$D_{x,y} = D_{c,ab} - D_{c,y} - D_{x,ab} = 72.5 - 35 - 24.5 = 13 \tag{31}$$

since:

$$D_{x,ab} = \frac{1}{2}(D_{x,a} + D_{x,b}) = \frac{1}{2} D_{a,b} = \frac{1}{2} 49 = 24.5 \tag{32}$$

Finally, we must compute the distance $D_{d,y}$ from Eq. (24):

$$D_{d,y} = \frac{1}{2}(D_{d,ab} + D_{d,c} - D_{c,ab}) = \frac{1}{2}(76.5 + 74 - 72.5) = 39 \tag{33}$$

since $D_{c,ab}$ and $D_{d,ab}$ have already been computed in Eqs. (29) and (30).

Using the branch lengths which we have calculated above, we can now construct the radial FM-tree which is shown in Fig. 8b.

Additionally, we may wish to estimate the position of the root of the radial tree. This root can be placed on the branch which joins the last two nodes in the UPGMA-tree. In our example shown in Fig. 8, we except the root to be close to node y on the branch that joins points d and y. (In this respect, we consider each tip of the tree as a primary node). The position of the root is assumed from a condition which is imposed on the branches of the tree. A feasible condition requires that the distances from all the tips to the assumed root of the tree must have minimal variance. Zero variance would result if the rate of mutation from one primordial ancestor had always been the same at any time along all branches of the tree. We can express that the root mean square deviation, d, of the distances from the tips to the root (labeled r) of the tree must be minimal:

$$d = \left(\frac{1}{n}\sum_{l}^{n}(D_{r,l} - \bar{D}_r)^2\right)^{1/2} \tag{34}$$

where $D_{r,l}$ represents the distance from tip l to the root r of the tree, and where $\bar{D}_r$ is the mean distance from the n tips to the root r:

$$\bar{D}_r = \frac{1}{n}\sum_{l}^{n} D_{r,l} \tag{35}$$

From the numerical results obtained from the FM-tree in Fig. 8 we found that:

$$D_{r,a} = 38 + D_{r,y} \tag{36}$$

$$D_{r,b} = 37 + D_{r,y} \tag{37}$$

$$D_{r,c} = 35 + D_{r,y} \tag{38}$$

$$D_{r,d} = 39 - D_{r,y} \tag{39}$$

and, hence, we compute the mean distance $\bar{D}_r$ from Eq. (35) as follows:

$$\bar{D}_r = \frac{1}{4}(149 + 2D_{r,y}) = 37.25 + 0.50D_{r,y} \tag{40}$$

When this result is substituted into the expression for the root mean square distance, d, in Eq. (34) we obtain:

$$\sum_{l}^{n} (D_{r,l} - \bar{D}_r)^2 = (0.75 + 0.50D_{r,y})^2 + (-0.25 + 0.50D_{r,y})^2 + (-2.25 + 0.50D_{r,y})^2$$
$$+ (1.75 - 1.50D_{r,y})^2 = 3.00D_{r,y}^2 - 7.00D_{r,y} + 8.75 \tag{41}$$

which is minimal when:

$$6.00D_{r,y} - 7.00 = 0 \tag{42}$$

which follows that $D_{r,y} = 1.17$.

Substitution of this result into Eq. (40) produces a mean distance from the tips to the root $\bar{D}_r$ or 37.84. Substitution of $\bar{D}_r$ into Eq. (34) leads to the root mean square deviation, d, being equal to 1.08, which constitutes 2.85 percent of the mean distance, $\bar{D}_r$.

By means of the branch lengths of the radial FM-tree, we can reconstruct the distances between the tips. The distance $D_{a,d}$ in the UPGMA-tree in Fig. 8 is equal to the sum of the branch lengths $25 + 13 + 39 = 77$, which is exactly equal to the tabulated dissimilarity. In this illustrative example we are able to reconstruct all the original dissimilarities from the computed branch lengths of the FM-tree. In real applications, however, this is not generally the case. In the analysis of amino acid sequences, deviations arise due to the nature of the dissimilarities, which are only estimates of the true values because of unequal sequence lengths and variable gaps in the sequences. With real data it is even possible that some of the computed branch lengths turn out to be negative. In the following example we shall show how an improved alternative tree may be produced, starting from the UPGMA-tree as an initial estimate.

In order to measure the goodness of fit between the computed phylogenetic tree and the measured dissimilarities, we adopted the root mean square relative deviation, s, which is expressed as a percentage:

$$s = 100 \left(\frac{1}{\frac{n(n-1)}{2}} \sum_{i}^{n} \sum_{i'>i}^{n} \left(\frac{D^*_{i,i'} - D_{i,i'}}{D_{i,i'}} \right)^2 \right)^{1/2} \tag{43}$$

where $D^*_{i,i'}$ and $D_{i,i'}$ represent the reconstructed and tabulated dissimilarities between sequences i and i', respectively, and where the sums extend over all n sequences. Note that the expression assumes that all sequences are different from each other, since $D_{i,i'}$ must be different from zero. In the example of Fig. 8 we find that the root mean square relative deviation is exactly equal to zero, since the tree fits the data exactly.

In general, there is no guarantee that the UPGMA-tree is the one that produces the smallest possible root mean square deviation in Eq. (43). One should, therefore, examine alternative trees with branching patterns that differ from that of the UPGMA-tree. In the method developed by Fitch and Margoliash [8] one produces alternative trees by systematic variation of the branching of the initial UPGMA-tree. It the radial tree of Fig. 8b, the branch connecting x to b may be removed from the tree and reinserted between y and c. This produces an alternative tree which is analyzed by the method outlined above, and the resulting relative deviation, s, (Eq. (43)) is compared with the value obtained previously. If the new tree leads to an improvement, then it will be retained as the best solution temporarily, and the old value of the relative deviation is replaced by the new one. The search then continues, until no further improvement is obtained. In practice, the number of possible alternative trees may be too large to be examined within a reasonable time, even with the use of a powerful computer. Searching for good alternative trees can be made more efficient by means of a so-called branch-and-bound algorithm. This reduces the number of possible alternatives to a smaller number with a higher probability of producing an improved tree. Even then, the number of alternatives may still be too large. It is usual, therefore, to limit the scope of the search process by allowing only one or two branches to be removed and replaced at the same time. Some of these alternative trees are represented in Fig. 9, in a similar way to the UPGMA-tree described in Fig. 8. In this case, none of these can be retained as an improvement, since the initial radial tree configuration reproduced exactly the tabulated dissimilarities.

It may be worth mentioning that phylogenetic clustering according to Fitch and Margoliash (FM) is a special case of hierarchical and agglomerative clustering, in which nodes are combined such that the sum of the branches that connect any two primary nodes (tips) is in optimal agreement with the observed distance between them. In this respect, the result is different to that of a minimal spanning tree, in which points are joined so that the sum of all branch lengths is minimal.

We now return to Table 2 which contains the 26×26 similarities between G-protein coupled receptor subtypes. The result of phylogenetic clustering is shown in the form of a radial FM-tree in Fig. 10. This tree is derived from an initial UPGMA-tree. The

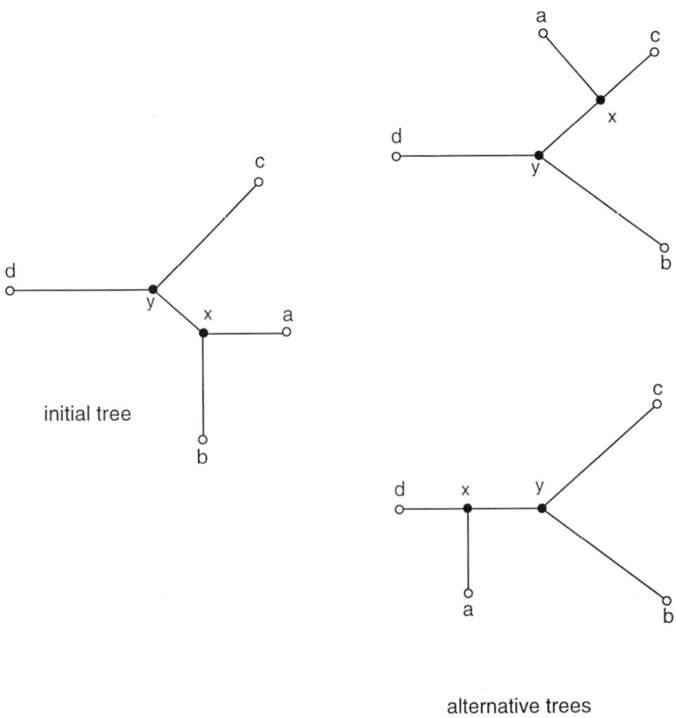

Figure 9. Construction of alternative trees (shown right) starting from an initial tree (shown left) by systematic removal and reconnection of the branches.

average distance, $\bar{D}_r$, (Eq. (35)) from the 26 tips to the computed root, r, is 37.2 units of percentage dissimilarity (see also Table 2). An open circle near the center of the radial tree marks the position of the root. The root mean square deviation, *d*, (Eq. (34)) of the 26 tips from the root equals 5.9 percent dissimilarity. The root mean square relative deviation, *s*, (Eq. (43)) between the 325 reconstructed and measured dissimilarities is 2.2 percent dissimilarity.

Three branches in the neighborhood of the root resulted in small negative branch lengths of the order of –.5 percent on the dissimilarity scale. These have been set to equal zero and are disregarded, as they cannot be represented conveniently on the radial tree of Fig. 10. A search for a better alternative tree yielded only marginal improvements of the root mean square deviations, *s*, and, *d*, from their initial values of 2.2 and 5.9 to the final values of 2.1 and 4.0 percent dissimilarity, after 52 trials. This occurred, however, at the expense of an increase of the total negative branch length from -1.4 to -2.8 percent dissimilarity. It was concluded, therefore, that the initial UPGMA configuration is an adequate representation of the phylogenetic relationship between the 26 subtypes of G-protein coupled receptors.

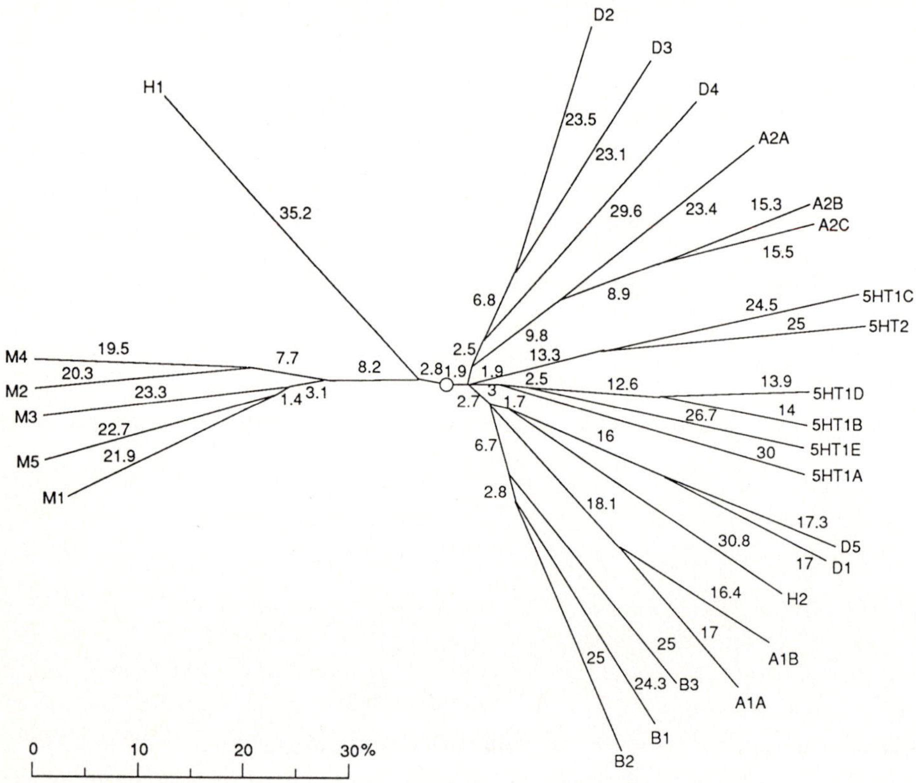

Figure 10. Radial tree according to Fitch and Margoliash [8] as derived from an initial UPGMA-tree according to Sneath and Sokal [18] from the 26 receptor subtypes. The radial branches have been arranged around the computed root (open circle) such as to agree maximally with the disposition of the receptors in the map of Fig. 6 around the center of the plot (small cross). The agreement is fair. Only the D4 dopaminergic receptor does not fit well.

From Fig. 11 it is evident that any two branches emerging from a node may be switched without affecting the topological properties of the tree. In the radial tree of the 26 receptors in Fig. 10, the pair consisting of M1 and M5 may be interchanged by switching the branches emerging from their common node. Likewise, the pair consisting of the group M1, M5 and M3 may be interchanged, and so on. It can be understood that the number of possible equivalent changes amounts to 2^{n-1}, where n is the number of receptors that have been clustered. In our case of 26 receptors, this corresponds with 2^{25} or about 34 million topologically equivalent trees. We have proposed elsewhere [20] to arrange the branches of the radial tree such as to match as closely as possible the disposition of the receptors in the two-dimensional map obtained by generalized Principal Coordinates Analysis. The arrangement of the 26 tips around the computed root of the tree in Fig. 10 is in such a way that it corre-

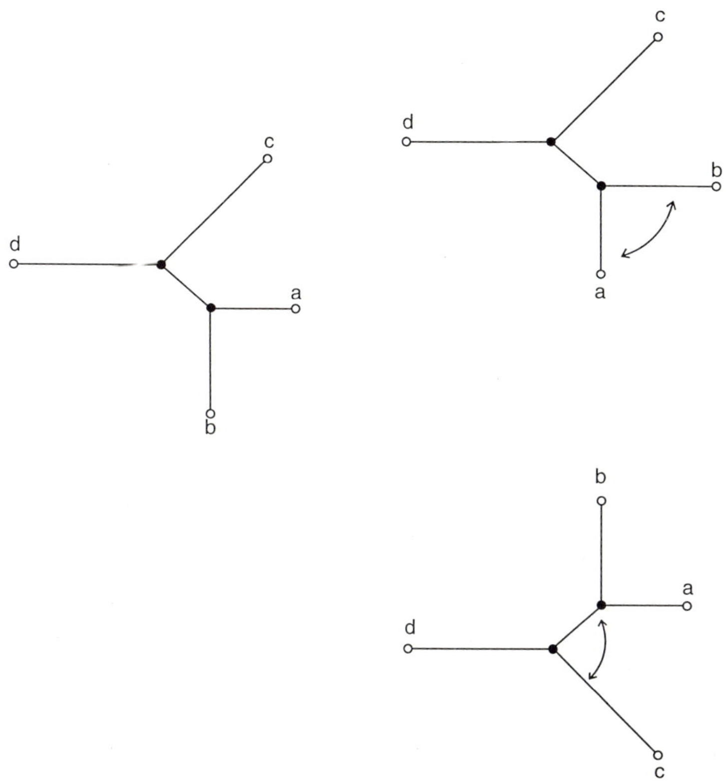

Figure 11. Topologically equivalent trees (shown right) that can be derived from an initial tree configuration (shown left). The number of equivalent trees that can be constructed with n leaves is 2^{n-1}.

sponds maximally with the arrangement of the 26 points on the map of Fig. 6, as viewed from the centroid of the pattern. Overall, the agreement between the tree and the map is fair. Only the branch leading to D4 cannot be moved to a position in the tree which agrees with its placement on the map without crossing other branches of the tree. The canonical form of the radial tree is the one that corresponds maximally with the corresponding factor map. It is a unique representation of the radial tree (apart from reflections about the horizontal and vertical axes).

3.3.5 Discussion

We have analyzed the observed dissimilarities between amino acid sequences of G-protein coupled receptors from two points of view. The first approach by Principal Coordinates Analysis produced a static picture in the form of a map which shows the dissimilarities between receptors by means of their distances in low-dimensional factor space. This map represents the situation as it is today. The second approach by phylogenetic clustering yields a dynamic picture in the form of a tree which shows the evolution from the more primitive ancestors to the present-day manifestations of the receptors. In philosophical terms, one may refer to the factor mapping as an idealistic Platonic representation, and to the phylogenetic tree as an evolutionary Aristotelian view. It is not uncommon in the history of science that emphasis shifts from one point of view to another, and back again [21]. Here, we adopt a synthetic position by combining the geometric properties of the mapping with the topological structure of the tree. We have shown that both views are complementary, in the sense that the branches of the radial tree can be arranged in the order as they appear on the map. Conversely, it is possible to superimpose the topological structure of the tree on the map, except for the D4 receptor which does not seem to fit well into a two-dimensional map [20]. The fact that such a high degree of complementarity exists between the two-dimensional projection and the clustering tree of an apparently high-dimensional pattern is remarkable. It seems as if nature constrains viable and functional random mutations within a low-dimensional subspace of a vastly high-dimensional mutation space [22]. It thus appears as if G-protein coupled receptors have been bound over several hundred millions of years by a low-dimensional attractor which may have forced them into a space of fractal dimensionality, possibly between 3 and 4.

In the mapping of Fig. 5 we highlighted the two discrepancies between the apparent clusters and the corresponding pharmacological classifications in the case of the Turkey B1 and Bovine A1A sequences. It is not clear yet whether these anomalies are due to experimental errors in the sequences, or to misclassification by the algorithm. A possible explanation may be found in a so-called horizontal gene translation by which a relatively large segment of the DNA sequence is moved from one place to another [23].

The canonical form of the tree and the factor map, on which it is based, allow a much easier comparison of phylogenetic results obtained within and between laboratories. The only geometric arbitrariness of the canonical form lies in a possible reflection of the horizontal and vertical axes. This results in three topologically equivalent mirror images of the tree and the map, which cannot be superimposed by means of a rotation in the plane of the diagram. A pictorial illustration of this concept is given in Fig. 12. In the maps of Figs. 6 and 10 we have placed, rather arbitrarily, the muscarinic cholinergic receptors at the left, and the D2, D3 subtypes of the dopaminergic receptor in the upper right corner. Other researchers may decide otherwise, but this geometric arbitrariness is substantially less significant than the topological one.

The radial tree of Fig. 10 reflects the presumed ancestral relationships between the 26 subtypes of G-protein coupled receptors, which can be summarized as follows: Earliest divergence is observed between the muscarinic acetylcholine receptors (M1 to M5). The histamine H1 receptor is also clearly different from all the others and

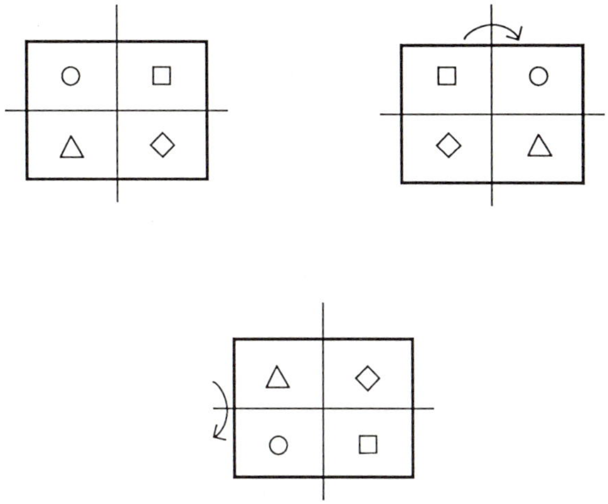

Figure 12. Symbolic illustration of the topologically equivalent maps that can be obtained by means of horizontal and vertical reflections (shown right) from an initial configuration (shown left).

especially from H2, as H1 branches off at a short distance from the computed root (indicated by an open circle). Likewise, there appears to be only a remote kinship between the D1, D2, D3, D4 subtypes of the dopaminergic receptor, on the one hand, and the D1, D5 subtypes, on the other hand. The α-adrenergic A2 receptors appear to be only remotely related to the A1 ones. Finally, the serotonergic receptors 5HT1C and 5HT2 seem to belong to a different lineage than the 5HT1A, -B, -D and -E receptors. This is in line with phylogenetic analyses of alpha-adrenergic receptors by Harrison, et al., [24] and of serotonin receptors by Goethert [25] and by Kim, et al., [26] which confirm the interpretation derived from the factor map in Fig. 6.

We also must address the problem of selecting the most appropriate metric for representing distances between receptor sequences. The present analyses are based upon dissimilarities, $\boldsymbol{D}$, as obtained from the pairwise comparisons of similarities, $\boldsymbol{L}$, of amino acid sequences by means of the VGAP computer program of Moereels, et al. [10], which accounts for variable gaps in the sequence alignments and for variable lengths of the sequences (Eq. (1)). Our Principal Coordinates Analysis has not shown any relevant violations of the assumptions for a Euclidean metric. Alternative metrics can be defined by means of the square root and the logarithmic [23] transformations:

$$\boldsymbol{D} = (100 - \boldsymbol{L})^{1/2} \tag{44}$$

$$\boldsymbol{D} = -\ln \boldsymbol{L} \tag{45}$$

We have observed no substantial difference between our linear metric (Eq. (1)) and the logarithmic one (Eq. (45)).

Branch lengths of the phylogenetic tree are expressed in percent changes in amino acid sequences. It is convenient and often justifiable to convert these branch lengths into evolutionary times [27]. Current estimates yield an upper bound of 10 million years for the production of a one percent change in amino acid sequences by random mutations [4]. Since the average distance of the leaves of the tree to the computed root is about 37 percent, this would suggest that first divergence took place some 370 million years ago. The latter calculation rests upon the assumption that the rate of mutation has been constant and homogeneous in all the subtypes of receptors [28]. The primordial ancestral receptor is thought to be much older, between 600 million and one billion years [4]. The discrepancy may be due to the incorrect assumption that there was a constant and homogeneous mutation rate over the past billion years. Moreover, some amino acid residues must have mutated several times, which leads to an underestimation of the number of mutations applied to any particular sequence. Many mutations may have led to non-functional receptors and, hence, have produced unviable or less competitive organisms. These are not represented in the phylogeny as they have become extinct. Finally, it must be understood that the phylogenetic tree of Fig. 10 is fundamentally unrooted. This means that there is no observational evidence for the existence of a primordial ancestor. As has been explained above, the root of the tree has been inferred from a statistical standpoint, such as to minimize the root mean square deviation of its distances from the tips (Eq. (34)).

A significant correlation is observed between the lengths of the amino acid sequences (number of AAs) and the evolutionary distances from the root (percent dissimilarity). These values are compiled in Table 1 and are displayed in Fig. 13. The product-moment (Pearson) coefficient of correlation is 0.61 ($P<.001$). At present, we cannot offer an explanation for this correlation.

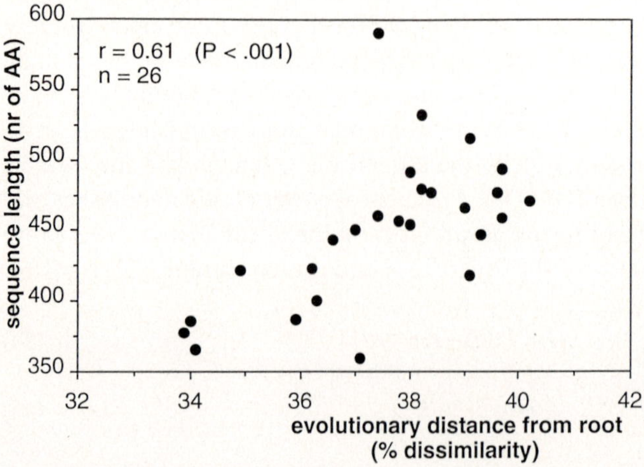

Figure 13. Relationship between sequence length (number of amino acid residues) and evolutionary distance from the computed root (percent dissimilarity).

We expect that our approach of combining factor mappings and phylogenetic trees may be helpful in bringing other tentative correlations to light. For example, one may look for patterns in the properties of the various G-protein coupled receptors, such as their type of signalling pathway which has been established to be either via activation of phospholipase C or of adenylate cyclase [10].

Acknowledgement

The authors thank B. Van den Poel, K. Van Reusel for assistance with computer programs, J. Van Hoof for programing SPECTRAMAP, D. Adriaensen for programing the Fitch-Margoliash algorithm, K. Sprangers and A. Biermans for typing the manuscript, T. Amery for proof-reading and J. Van Mierlo for the preparation of the figures.

References

[1] Janssen, P.A.J. and Moereels, H., *Three-Dimensional Structure of G-Protein Coupled Serotonin Receptors,* Inaugural lecture, Dr. Paul Janssen, Chair of Cell Biology, UIA, Antwerpen, Belgium, 20 November, 1992

[2] Starke, K., *Arzneim. Forsch./Drug. Res.* **42**, 182–183 (1992)

[3] Sternweis, P.C. and Smrcka, A.V., *TIBS (Trends in Biochemical Sciences)* **17**, 502–506 (1992)

[4] Yokoyama, S,. Isenberg, K.E., and Wright, A.F., *Mol. Biol. Evol.* **6**, 342–353 (1989)

[5] Goodson, H.V. and Spudich, J.A., *Proc. Natl. Acad. Sci.* **90**, 659–663 (1993)

[6] Janssen, P.A.J. and Moereels, H., *Serotonin Receptors: from Ligands to Sequence.* In: *From Cell Biology to Pharmacology and Therapeutics.* Paoletti, R., Vanhoutte, P.M. and Saxena, P.R., (eds.), Kluwer and Fondazione Giovanni Lorenzini, Houston, 1993

[7] Gower, J.C., *The Statistician* **17**, 13–28 (1967)

[8] Fitch, W.M. and Margoliash, E., *Science* **155**, 279–284 (1967)

[9] Dayhoff, M.O., Schwartz, R.M. and Orcutt, B.C., *Atlas of Protein Sequence and Structure,* Vol. V, Sup. 3, Dayhoff, M.O., ed., Nat. Biomed. Res. Found., Washington DC, 1978, pp. 352

[10] Moereels, H., De Bie, L. and Tollenaere, J., *Comp. Aided Mol. Des.* **4**, 131–145 (1990)

[11] Lewi, P.J. and Moereels, H., *TRAC (Trends in Analytical Chemistry)* **10**, 283–290 (1991)

[12] Tufte, E.R., *Envisioning Information.* Graphics Press, Cheshire, CONN, 1990

[13] Joliffe, J.T., *Principal Components Analysis,* Springer Verlag, New York, 1986

[14] Householder, A.S., *The Theory of Matrices in Numerical Analysis,* Blaisdell, New York, 1964

[15] Lewi, P.J., *SPECTRAMAP, Introduction to Multivariate Analysis of Rectangular Data Tables, with Special Emphasis on Biplots,* Janssen Pharmaceutical, Beerse, Belgium, 1993

[16] Yoshida, H., Kakuchi, J., Guo, D.-F., Furuto, H., Iwai, N., van der Meer-de Jong, R., Inagami, T. and Ichikawa, I., *Biochem. Biopyhs. Res. Comm.* **186**, 1042–1049 (1992)

[17] Alberts, B., Bray, D., Lewis, J., Raff, M., Roberts, K. and Watson, J.D., *Molecular Biology of the Cell,* Garland, New York, 1983, pp. 215

[18] Sneath, P.H. and Sokal, R.R., *Numerical Taxonomy. The Principles and Practice of Numerical Classification,* Freeman, San Francisco, 1973

[19] Felsenstein, J., *PHYLIP – Phylogeny Inference Package. Version 3.3,* Dept. of Genetics, University of Washington, Seattler, WA, 1991

[20] Lewi, P. J., Moereels, H. and Adriaensen, D., *Chemom. Intell. Lab. Syst.* **16**, 145–154 (1992)

[21] Oldroyd, D., *The Arch of Knowledge. An Introductory Study of the History of the Philosopyh and Methodology of Science,* Methuen, New York, 1986

[22] Eigen, M., *Sci. Amer.* **269**, 42–49 (1993)

[23] Smith, M. W., Feng, Da-Fei and Doolittle, R. F., *TIBS (Trends in Biochemical Sciences)* **17**, 489–493 (1992)

[24] Harrison, J. K., Pearson, W. R. and Lynch, R., *TIPS (Trends in Pharmacological Sciences)* **12**, 62–67 (1991)

[25] Goethert, M., *Arzneim. Forsch./Drug Res.* **42**, 238–246 (1992)

[26] Kim, J., Huang, K. N., Livelli, T. J., Axel, R. and Jesell, T. M., *Proc. Natl. Acad. Sci, USA* **87**, 928–932 (1990)

[27] Felsenstein, J., *Evolution* **38**, 16–24 (1984)

[28] Lewin, R., *Science* **239**, 561–563 (1988)

4 Advanced Statistical Techniques

4.1 Continuum Regression: A New Algorithm for the Prediction of Biological Activity

Jonathan A. Malpass, David W. Salt, Martyn G. Ford, E. Watcyn Wynn, and David J. Livingstone

Abbreviations

QSAR	Quantitative Structure-Activity Relationship
SAR	Structure-Activity Relationship
MLR	Multiple Linear Regression
PCR	Principal Components Regression
PLS	Partial Least Squares Regression
CR	Continuum Regression
CV	Cross-validation
PRESS	Predictive Error Sum of Squares
LOO	Leave-One-Out

Symbols

n	number of cases/observations
p	number of descriptor variables
$\dot{y}$	generic response variable vector
$\dot{X}$	generic descriptor variable matrix
$\dot{x}_i$	generic descriptor variable vector
y	mean-centered response variable vector
X	mean-centered descriptor variable matrix
x_i	mean-centered descriptor variable vector
$\hat{y}$	estimated value of response
r_{xy}	correlation coefficient of x_i and y

cov_{xy}	covariance of x_i and y
var_x	variance of x_i
T	Continuum Regression Generalized Criterion Function
α	Continuum Regression adjustable parameter
$\boldsymbol{c}$	latent variable/component coefficient vector
$\boldsymbol{c}'\boldsymbol{X}$	latent variable/component
$\boldsymbol{S} = \boldsymbol{X}'\boldsymbol{X}$	variance–covariance matrix
$\boldsymbol{s} = \boldsymbol{X}'\boldsymbol{y}$	
m	maximum number of non-zero eigenvalues of $\boldsymbol{S} = \boldsymbol{X}'\boldsymbol{X}$
v_i	eigenvectors of $\boldsymbol{S} = \boldsymbol{X}'\boldsymbol{X}$
e_i	eigenvalues of $\boldsymbol{S} = \boldsymbol{X}'\boldsymbol{X}$
z_i	set of coefficients used in defining $\boldsymbol{c}$
L	Lagrangian function
λ	Lagrangian multipliers
τ	algebraic term
ϱ	algebraic term
a_{ij}	algebraic term
D	algebraic term
A	algebraic term
d	algebraic term
σ	algebraic term
M	algebraic term
k	number of constraints in the Lagrangian function
R	dimension of Bordered Hessian
r	number of rows and columns that "border" the Hessian matrix
Δ	determinant of Bordered Hessian
M_i	class of models
D	complete data set
D^-	reduced data set
G	number of cross-validation groups
g_j	individual cross-validation group
E	Wold's cross-validatory statistic
F_{OST}	Osten's cross-validatory statistic
I	Stone and Brooks' cross-validatory statistic
q	number of non-zero partial regression coefficients
ω	number of components in model
t	regression *t*-statistic
R^2	multiple coefficient of determination
$\bar{R}^2$	adjusted R^2
I_{max}	maximum value of I
$-\log IC_{50}$	log of the inverse concentration giving median inhibition
I_{4OH}	indicator variable for the presence of a *para*-hydroxy substituent

π_{345}	sum of the lipophilicity at the *meta* and *para* positions in the phenyl ring
MR_{345}	sum of the molar refractivities at the *meta* and *para* positions in the phenyl ring
F_{345}	sum of the field and inductive effects at the *meta* and *para* positions in the phenyl ring
R_{345}	sum of the resonance effects at the *meta* and *para* positions in the phenyl ring

4.1.1 Introduction

Formulation of a structure-activity relationship (SAR) for a series of biologically active compounds is an important step in the molecular design process. Since Hansch's first use of multiple linear regression (MLR) [1], a number of multivariate statistical techniques have been employed in efforts to produce more accurate SARs. The use of Principal Components Regression (PCR) [2, 3] or Partial Least Squares regression (PLS) [4, 5], for example, can be successful. However, Quantitative Structure-Activity Relationship (QSAR) practitioners are often confused about the choice of method to employ. Another problem, which is inherent in all multivariate statistical techniques, is how to define the optimal model. Although the objective is to produce as parsimonious a model as possible while maintaining accuracy, it is not always clear how many components or original variables should be included in the final model specification. This problem is further confounded when data sets consisting of relatively few cases or objects, *n*, compared to the number of physico-chemical properties or descriptor variables, *p*, are to be analyzed. The plethora of multivariate statistical techniques has presented a further dilemma to the model specification stage and the choice of criteria to be considered, when constructing the components or latent variables, often appears arbitrary. However, the underlying structure of the data should always be considered when choosing the analytical procedure, since an inappropriate choice may violate the assumptions of that method and lead to severe limitations in the predictive power of the resulting model. This chapter addresses the complex problem of identifying the most accurate and most predictive QSAR model using relatively small samples of compounds to represent the population of properties.

Continuum Regression (CR), which was introduced by Stone and Brooks [6] to address several of these problems, allows the component construction stage of utilize fully the information in both the response variable, $\dot{y}$, and the descriptor variable set, $\dot{X} = \dot{x}_1, \dot{x}_2, \ldots, \dot{x}_p$. Furthermore, a criterion function, *T*, and the number of components to be included in the model are optimized using a cross-validation criterion for model specification. A new formulation of CR has now been developed [7] which offers further advantages over Stone and Brooks' original formulation. These include an analytical solution to identify an optimum, *T*, for each component, avoidance of infinite T, computational efficiency, and a reliable model specification procedure which does not require cross-validation. This newly developed CR is therefore, not subject to assumptions which have no theoretical basis and which are difficult to justify.

4.1.2 Equivalence of Continuum Regression with MLR, PLS, and PCR

Continuum Regression is a general regression technique which embraces the three popular procedures of Multiple Linear Regression (MLR), Partial Least Squares (PLS), and Principal Components Regression (PCR). The concept of relating MLR, PLS, and PCR may appear paradoxical at first. However, a closer inspection identifies the criterion function maximized in MLR as the correlation between the response and the descriptor variable set. In PLS the covariance of the response and the descriptor set is maximized, and in PCR it is the variance of the descriptor variables.

In vector notation, the correlation coefficient of the mean-centered variables x and y can be defined as:

$$r_{xy} = \frac{\boldsymbol{x}'\boldsymbol{y}}{\sqrt{||\boldsymbol{y}||^2 \boldsymbol{x}'\boldsymbol{x}}} \tag{1}$$

and, hence, the square of the correlation coefficient is:

$$r_{xy}^2 = \frac{(\boldsymbol{x}'\boldsymbol{y})^2}{||\boldsymbol{y}||^2 \boldsymbol{x}'\boldsymbol{x}} \; . \tag{2}$$

The covariance of the response and descriptor variable is:

$$\mathrm{cov}_{xy} = (\boldsymbol{x}'\boldsymbol{y})^2 \tag{3}$$

and the variance of the descriptor variable is:

$$\mathrm{var}_x = \boldsymbol{x}'\boldsymbol{x} \tag{4}$$

It is the similarity between these criterion functions that leads to the formulation of a generalized criterion function, T. If $\boldsymbol{c}$ is the vector of coefficients such that $\boldsymbol{c}'\boldsymbol{X}$ forms one component, then Stone and Brook's generalized criterion function is:

$$T = (\boldsymbol{c}'\boldsymbol{X}'\boldsymbol{y})^2 (\boldsymbol{c}'\boldsymbol{X}'\boldsymbol{X}\boldsymbol{c})^{(\alpha/1-\alpha)-1} \; . \tag{5}$$

The new formulation has the generalized criterion function:

$$T = (c'X'y)^{(2+2\alpha-4\alpha^2)}(c'X'Xc)^{(-1+2\alpha)} \tag{6}$$

In both functions the parameter, α, has the range $(0,1)$ and varying α adjusts the balance between the covariance of the response and the descriptor variables and the variance of the descriptor variable set.

Continuum Regression is so named because α is allowed to take any value in the continuum $(0,1)$. Substituting $\alpha = 0$ into Eqs. 5 and 6 yields the following:

$$T = \frac{(c'X'y)^2}{(c'X'Xc)} \tag{7}$$

which is the correlation between the response and the constructed component. By substituting $\alpha = 0.5$ into T, the function becomes:

$$T = (c'X'y)^2 \tag{8}$$

i.e., the covariance of the response and the new component. When $\alpha = 1$, T becomes

$$T = (c'X'Xc) \tag{9}$$

or the variance of the new component. By comparing Eqs. (2), (3) and (4) with Eq. (7), (8) and (9), it is apparent that the methods of MLR, PLS, and PCR are achieved by Continuum Regression when α is set to 0, 0.5, and 1, respectively. Each of the methods can be achieved to yield orthonormal components assuming that the set of vectors $c_1, c_2, \ldots, c_m$, where $m = \min[n-1, p]$, are constructed such that they are maxima, of unit length and are orthogonal to each other. The algorithm for constructing such components is outlined in Sec. 4.1.3.

4.1.3 Construction Algorithm

Research undertaken at the University of Portsmouth has led to the development of a robust and reliable CR algorithm. The details of this procedure, including its mathematical features, are presented below and identify the strategy adopted to overcome deficiencies in Stone and Brooks' CR procedure highlighted in Sec. 4.1.1.

4.1.3.1 A New Formulation of Continuum Regression

Let $v_1, v_2, \ldots, v_m$ be the orthonormalized eigenvectors of $S = X'X$ that correspond to the eigenvalues $e_1, e_2, \ldots, e_m$ such that $0 < e_1 < e_2 < \ldots < e_m$. The vectors $c_j, j = 1, 2, \ldots, m$ then take the form:

$$c_j = z_1 v_1 + z_2 v_2 + \ldots + z_m v_m \tag{10}$$

subject to the following constraints:

$$c_j' S c_k = 0 \qquad j \neq k \tag{11}$$

and

$$||c_j|| = 1 \qquad i = 1, 2, \ldots, m \tag{12}$$

and such that c_j maximizes T. If T is now expressed as a logarithmic function, ln T at $c = c_j$ can be written as:

$$\ln T = (2 + 2\alpha - 4\alpha^2) \ln (\tau) + (-1 + 2\alpha) \ln (\varrho) \tag{13}$$

where

$$\tau = d_1 z_1 + d_2 z_2 + \ldots + d_m z_m \tag{14}$$

$$\varrho = e_1 z_1^2 + e_2 z_2^2 + \ldots e_m z_m^2 \tag{15}$$

$$d_i = s' v_i \tag{16}$$

and $s = X'y$. The orthogonality and unit length constraints (Eqs. (11) and (12)) can now be written as:

$$z_1^2 + z_2^2 + \ldots + z_m^2 = 1 \tag{17}$$

$$a_{1j} z_1 + a_{2j} z_2 + \ldots + a_{mj} z_m = 0 \tag{18}$$

where $a_{ij} = e_i c_j' v_i$. The Lagrangian equation, L, which maximizes ln T is, in general,

$$L = (2 + 2\alpha - 4\alpha^2) \ln (\tau) + (-1 + 2\alpha) \ln (\varrho) - \lambda_0 (z_1^2 + z_2^2 + \ldots + z_m^2 - 1)$$
$$- \lambda_1 (a_{11} z_1 + a_{21} z_2 + \ldots + a_{m1} z_m) - \ldots - \lambda_k (a_{1k} z_1 + a_{2k} z_2 + \ldots + a_{mk} z_m) \tag{19}$$

where λ_i, $i = 0, 1, 2, \ldots, k$ are the Lagrangian multipliers.

Differentiating with respect to z_i yields

$$\frac{\delta L}{\delta z_i} = d_i(1+\alpha-2\alpha^2)\tau + e_i z_i(-1+2\alpha)\varrho - \lambda_0 z_i - \lambda_1 a_{i1} - \ldots - \lambda_k a_{ik} \tag{20}$$

To obtain a set of z_i that maximize $\boldsymbol{c}_j$ which in turn maximize ln T, set Eq. (20) to zero. This yields $\lambda_0 = 3\alpha - 2\alpha^2$, and the maximizing of z_i for $\boldsymbol{c}_1$ are found by solving

$$z_i = \frac{(1+\alpha-2\alpha^2)d_i}{\alpha\varrho\tau(3-2\alpha) - e_i\tau(-1+2\alpha)} \tag{21}$$

The remaining vectors, $\boldsymbol{c}_2, \ldots, \boldsymbol{c}_m$ are found by solving the matrix equation for $\boldsymbol{z}$, where

$$\begin{bmatrix} \boldsymbol{D} & A \\ A' & 0 \end{bmatrix} \begin{bmatrix} \boldsymbol{z} \\ \sigma \end{bmatrix} = \frac{\varrho(1+\alpha-2\alpha)}{\tau} \begin{bmatrix} \boldsymbol{d} \\ 0 \end{bmatrix} \tag{22}$$

$$\boldsymbol{D} = diag\,[e_1(-1+2\alpha) + \alpha\varrho(3-2\alpha), \ldots, e_m(-1+2\alpha) + \alpha\varrho(3-2\alpha)] \tag{23}$$

$$\boldsymbol{A} = (a_{ij}) \tag{24}$$

$$\sigma_i = \varrho\lambda_i \tag{25}$$

This can be shown to be equivalent to the iterative solution of the following:

$$\boldsymbol{z} = \frac{\boldsymbol{Md}}{||\boldsymbol{Md}||} \tag{26}$$

where

$$\boldsymbol{M} = \boldsymbol{D}^{-1} - \boldsymbol{D}^{-1}\boldsymbol{A}'(\boldsymbol{A}'\boldsymbol{D}^{-1}\boldsymbol{A})^{-1}\boldsymbol{A}'\boldsymbol{D}^{-1} \tag{27}$$

As the determination of z_i coefficients is an iterative technique, a starting vector must be estimated. In effect choosing a random vector will yield maximal results providing the method incorporates a maximum solution detection technique, such as that of the Bordered Hessian, outlined in Sec. 4.1.3.2.

4.1.3.2 Maximizing T

The Bordered Hessian method [8] tests whether the solutions of the optimization process are maxima, minima or saddle points. The Hessian matrix consists of all partial second derivatives of L with respect to z_i, i.e., $\delta^2 L/\delta z_i z_j$, and is "bordered" by the partial second derivatives of L with respect to the conditions λ_j, i.e. $\delta^2 L/\delta\lambda_i\lambda_j$.

If p is the number of variables and k is the number of conditions, then the Bordered Hessian is of dimension $(p+k)\times(p+k)$. The Bordered Hessian can now be considered as a series of submatrices of dimension R, $R = 2k+1,...,k+n$, which consist of the $k\times k$ common matrix and are supplemented by $r = k+1,...,n$ rows and columns of partial derivatives of z_i. If the determinant of each of these submatrices is evaluated, the condition for z_i to yield a maximizing solution is:

$$(-1)^r \times \Delta_H > 0 \ , \quad r = k+1,...,n \tag{28}$$

where Δ_H is the determinant of the matrix evaluated.

If the solution fails to yield a maximum, the starting vector of z_i is re-estimated and a new solution obtained.

4.1.3.3 Optimizing *α*

The new formulation of Continuum Regression yields an analytical solution of α which maximizes the information in the data set by optimizing the balance between the covariance of the response and descriptor variables and the variance of the descriptor variable set.

If the Lagrangian function Eq. (19) is differentiated with respect to α, i.e.,

$$\frac{\delta L}{\delta \alpha} = (-2+4\alpha)\ln(\tau) + 2\ln(\varrho) \tag{29}$$

then α is found as

$$\alpha = \frac{\ln(\tau) + \ln(\varrho)}{4\ln(\tau)} \tag{30}$$

This estimate of α is optimal and, hence, should produce a regression equation that yields more reliable and accurate predictions.

4.1.4 Model Specification

Determination of the final model depends on two criteria; the model should be as parsimonious as possible, i.e. it should contain as few components/original variables as possible whilst maintaining accurate prediction. Because cross-validation (CV) is recommended for model specification when using PLS, PCR, and Stone and Brooks' CR [9–11], a brief summary of the cross-validation procedure is presented below. Later sections will describe an alternative approach using the Portsmouth formulation of CR and compare the results obtained by the different procedures.

4.1.4.1 The Cross-Validation Procedure

For a class of models, M_i, where i is the number of components in the model, the full data set, $D = (y, X)$, is divided into G groups, where $2 \leq G \leq n$. Each group g_j, $j = 1,\ldots,G$ is then deleted, in turn, from the full data set to obtain a reduced set, D^-. The model parameters are then estimated on the reduced set and estimate(s), $\hat{y}$, for the response(s) in the deleted group obtained using the corresponding descriptor variable values. By squaring and summing over the differences $(y-\hat{y})$, where y is the original response a "partial *PRESS*" value is calculated and the procedure is repeated for all g_j. The Predictive Error Sum of Squares (*PRESS*) is obtained by summing over the partial *PRESS* values.

This procedure can then be repeated for the next model, M_{i+1}, in the class.

If G is chosen to equal n then each of the groups contain one case. This is commonly referred to as a "leave-one-out" (LOO) method and has been widely adopted in QSAR studies. The alternative approach is to split the data set, $D = (y, X)$, into two equal groups, i.e. $G = 2$. These two approaches represent two extremes and in practice G can vary between the limits $(2, n)$.

4.1.4.2 Model Specification using Cross-Validation

Once the *PRESS* scores for the series of models have been obtained, various criteria can be employed to determine the optimal model. The first of these criteria is to choose the model which yields the lowest *PRESS*. Since *PRESS* is a function of residuals, the optimal model will be the model which minimizes predictive errors. A second choice is to select the model that yields a local minimum. The model is chosen to contain the fewest components/variables whilst minimizing *PRESS*. Fig. 1a and 1b illustrate these two cases. A third technique is to set some a priori threshold value of *PRESS*; the optimal model is then chosen to be the first model to yield a *PRESS* score below this threshold (Fig. 2). These criteria, particularly the threshold method, are somewhat subjective.

Wold [9], Osten [12] and Stone and Brooks [6] have all employed different numerical criteria to determine the optimum model.

Wold's E-test criterion

Wold's *E*-test employs the *PRESS* scores as follows:

$$E = \frac{PRESS_{i+1}}{PRESS_i} \qquad i = 0, 1, 2, \ldots \tag{31}$$

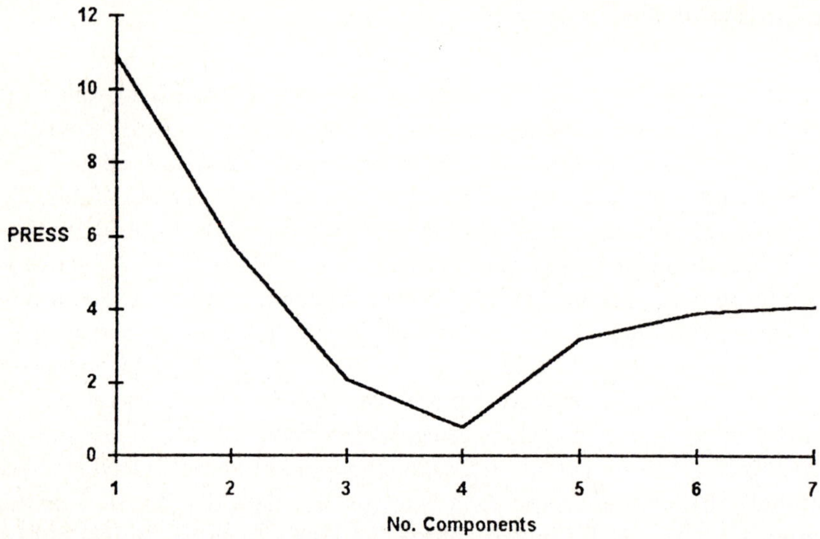

Figure 1a

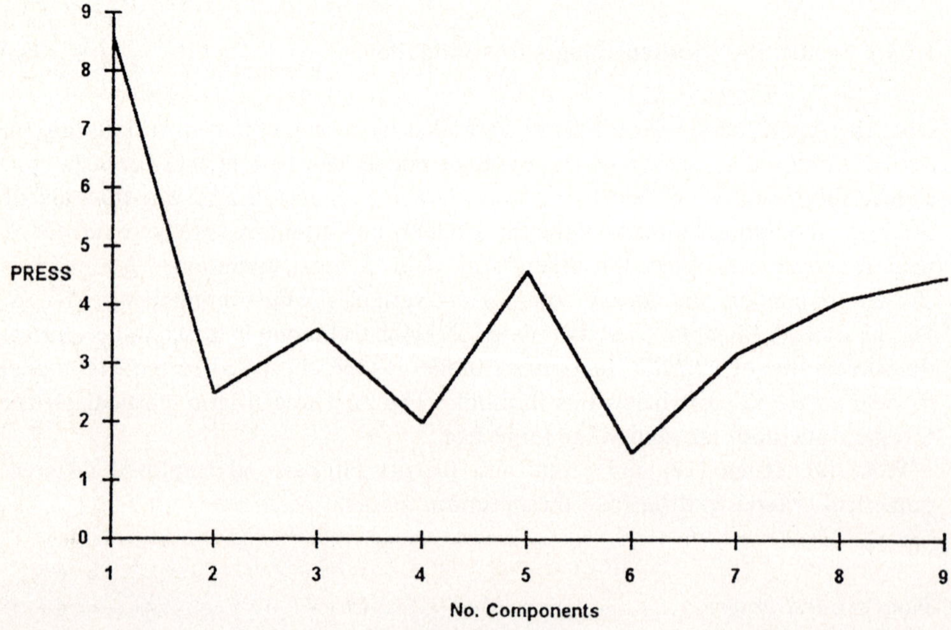

Figure 1b

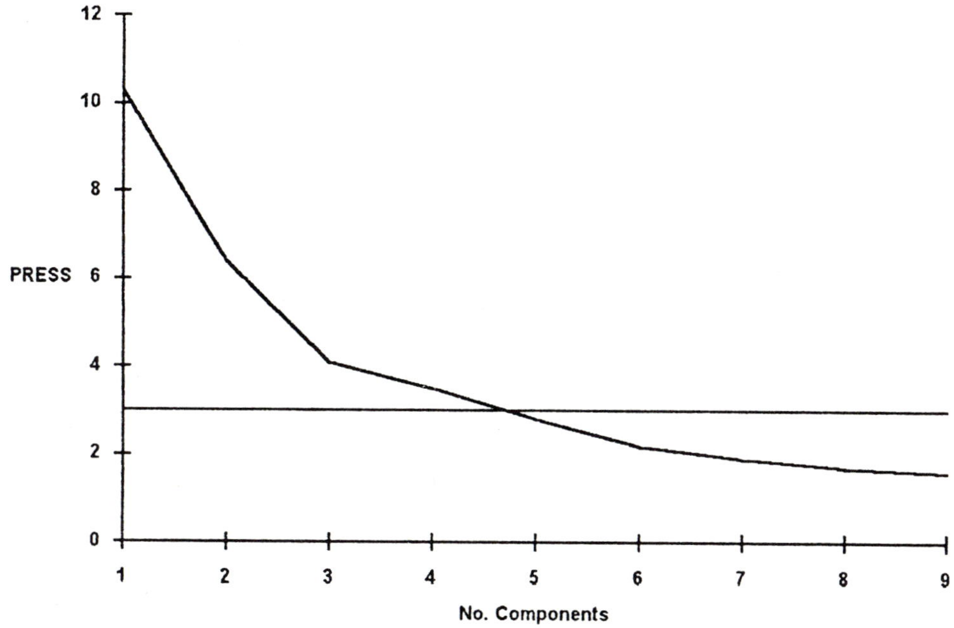

Figure 2

$PRESS_0$ is calculated from the model based on using the mean of the response, and $PRESS_i$, $i = 1, 2, \ldots$ is calculated from the models containing $1, 2, \ldots$ components. Wold originally suggested that components with $E < 1.0$ were significant and should be included in the prediction model, but he now recommends a more conservative value. If $E < 0.4$ then the model containing $i+1$ components is considered to be significant and Wold considers that the process should terminate when the E exceeds this value [13]. The choice of $E < 0.4$ appears to be somewhat arbitrary. However, in our experience 0.4 corresponds approximately to a 95% significance test.

A Cross-Validation Variance Statistic

Osten has produced an F-statistic similar to that used in regression for model comparison. Osten's F is defined as:

$$F_{OST} = \frac{(PRESS_i - PRESS_{i+1})/q}{PRESS_{i+1}/(np-(i+1)q)} \tag{32}$$

where q is the number of partial regression coefficients which do not equal zero. F_{OST} follows an F-distribution on q and $(np-(i+1)q)$ degrees of freedom. A component is deemed significant, at say the 95% significance level, if $F_{OST} > F_{TAB}$.

The I statistic

Stone and Brook's original formulation of Continuum Regression employs CV to determine the optimal values of α and ω, the number of components in the model. Their algorithm does not allow for an analytical solution of α, and so they employ the cross-validatory index, I, for this purpose where

$$I_{\alpha,\omega} = 1 - \frac{PRESS_i}{PRESS_0} \qquad i = 1, 2, \ldots \tag{33}$$

The values of α and ω which yield the highest value of I are deemed to produce the optimal model.

4.1.5 Model Specification without Cross-Validation

Wold [9], Osten [12], and Stone and Brooks [6] recommend that CV should be employed for the dual purpose of model specification and predictive assessment. However, there is no rigorous theoretical basis for this procedure, which in any case tends to yield over-optimistic results. Moreover, if CV is used to design a good predictive model, the same procedure cannot be used to provide an unbiased and independent assessment of predictive accuracy. For these reasons, a method of model specification which is not based on cross-validation, is desirable.

An appropriate procedure is outlined below.

1. Using the Portsmouth formulation of Continuum Regression, obtain the components and associated component scores based on the analysis at optimal α.
2. Regress the component scores on to the response variable, y, and obtain the regression t-statistics for each component.
3. Formulate the model by including components that have significant t-statistics.
4. Calculate the PRESS statistic for the specified model.
5. Calculate I, where

$$I = \left(1 - \frac{PRESS_{\text{mod}}}{PRESS_0}\right) \times 100\% \tag{34}$$

I is a cross-validatory R^2 statistic, i.e. it yields a value that cannot exceed 100%, and represents the percentage of predictive accuracy achieved by the model specified at Stage 3.

4.1.6 Properties and Performance of the Continuum Regression Algorithm

The problems encountered when attempting to specify a QSAR prediction model, based on relatively small sets of compounds, have been outlined in Sects 4.1.1 and 4.1.2. Such considerations raise a number of questions which do not appear to have been fully investigated. Should the choice of analytical procedure used to specify a model, for example, be based on the covariance structure of the data set, and how will an inappropriate choice affect the predictive power of a particular model? It would also of interest to establish whether a robust model can be specified without recourse to cross-validation. Finally, to what extent do the procedures introduced by the Portsmouth formulation of CR protec against spurious correlation, and how are standard procedures such as MLR, PLS, and PCR susceptible to this problem?

4.1.6.1 Does the Correlation Structure of a Data Set Affect the Choice of Analysis Method Used to Specify a Prediction Model?

To investigate whether the choice of method could affect the accuracy of the model, and whether the nature of the data set should govern the choice of technique, replicated data sets of known correlation structure were constructed prior to analysis by MLR, PLS, PCR and the new algorithm of CR described in Sec. 4.1.3. Because the size of the data set could also affect the choice of technique, replicate data sets of varying size were generated for each correlation structure to provide a set of random samples from a population of known characteristics.

The procedure employed to produce and analyze simulated data sets of known structure is presented below:

1. For a given correlation structure, sets of multivariate vectors were generated using a random number generator. NAG FORTRAN Library routines GO5CCF, GO5EAF and GO5EZF [14] were used to yield vectors of normally distributed random numbers using a non-repeatable seed. This system with a seed generated by using the time and date has the effect of yielding unique vectors of random numbers. Data sets were generated using population correlation structures (see Appendix 1) and are subject to the usual sample errors, i.e. the correlation structure of a data set will not be identical to the stated population correlation structure. The data sets varied in size from 4 to 6 descriptor variables and 10 to 50 cases (or objects).
2. Each data set was then analyzed using an in-house CR routine with α set to 0, 0.5 and 1, as well as allowing α to be optimized analytically. Preliminary studies using real data sets had shown that the CR routine, with $\alpha = 0$, 0.5, and 1, gave identical results to those yielded by MLR, PLS, and PCR routines.

3. For each data set R^2, $\bar{R}^2$ (adjusted R^2), and I_{max} were calculated and recorded. The first component α was also recorded for the CR run for which optimal α was estimated.
4. The procedure was repeated several times on replicate data sampled from each structured multivariate specification.

The experiments were performed on the University of Portsmouth Science Faculty VAX 6310 using codes written in FORTRAN 77. The population correlation structures of each multivariate population used in the study are given in Appendix 1.

A summary of the results identifying the population correlation structure, the size of the data set, the number of replicate analyses, and a series of summary statistics of the distribution of the parameter, α, is presented in Table 1. For data sets with a population structure where there is no association between any of the variables (population A1), the mean value of α tends towards the higher end of the continuum, i.e. values greater than 0.6. The minimum and maximum values indicate that the range of values tend to be [0.5, 1.0], with 50% of the values (between the lower quartile, Q1, and the upper quartile, Q3 lying approximately in the range [0.6, 0.7]. The second population, A2, is characterized by uncorrelated descriptor variables of which two exhibit a weak association with the response variable. The results show that α tends towards 0.5 (PLS) although the range of α values is large [0.01, 0.92]. The third population structure, A3, exhibits a stronger association between the descriptor variables and the response, although the descriptors remain uncorrelated. The tendency here is for α to be calculated as approximately 0.6, and the range of α is again large [−0.80, 1.53]. The characteristic of the fourth series of data sets, generated using population A4, is that whilst some of the descriptor variables are highly correlated with the response, there is some association between the descriptor variables. In this series of analyses, the first component α values exhibit much smaller ranges and tend towards 0.5, suggesting that for this data structure, PLS is an appropriate method of analysis.

A data set was generated with a structure in which two orthogonal descriptor variables were equally correlated ($r = 0.7$) to the response variable (population A5, Appendix 1). In this example, both descriptors were uniformly distributed about a zero mean. For this data, the optimum value of α (0.35) approached that of MLR ($\alpha = 0$), a satisfactory result since this data structure tends to satisfy some of the assumptions required for ordinary least squares regression. Thus, α can be used to adjust the criterion function, T, so that it matches the data structure and is a useful diagnostic.

As the ratio of the number of cases to the number of variables increases, α tends towards 0.5. Thus for population A2, the mean value of α for the three descriptor variable data sets decreases from 0.58 to 0.53 as the number of cases increases from 10 to 30.

There are a number of instances where the calculated value of α falls outside the continuum (0, 1), e.g. population A1, which comprises random variables (the three

Table 1. Results from simulated data sets of varying size and population correlation structure.

Population	No. Vars.	No. Cases	No. runs	1st component α					
				mean	st. dev.	min.	max.	Q1	Q3
A1	2	40	20	0.79	0.19	0.61	1.31		
	3	10	47	0.61	0.26	−0.93	0.91	0.56	0.71
	3	20	50	0.67	0.10	0.55	0.97	0.60	0.70
	3	30	45	0.67	0.09	0.52	0.97	0.62	0.72
	4	20	50	0.65	0.08	0.54	0.91	0.58	0.68
	4	30	67	0.66	0.09	0.55	0.97	0.61	0.69
	4	40	50	0.65	0.09	0.55	1.00	0.58	0.69
	4	50	20	0.68	0.09	0.56	0.93	0.61	0.74
	5	20	49	0.62	0.08	0.51	0.86	0.56	0.66
A2	3	10	49	0.58	0.12	0.01	0.92	0.53	0.63
	3	20	46	0.55	0.04	0.50	0.71	0.53	0.56
	3	30	50	0.53	0.02	0.50	0.59	0.52	0.54
	4	10	48	0.59	0.08	0.49	0.80	0.53	0.61
	4	20	48	0.56	0.05	0.50	0.69	0.53	0.58
	4	30	50	0.54	0.03	0.50	0.64	0.52	0.55
	4	40	40	0.53	0.02	0.49	0.57	0.52	0.54
	5	20	49	0.62	0.08	0.51	0.86	0.56	0.66
A3	2	10	46	0.60	0.18	−0.37	1.04	0.54	0.66
	2	20	49	0.60	0.09	0.51	0.95	0.55	0.61
	2	30	50	0.58	0.05	0.53	0.76	0.55	0.59
	3	10	46	0.67	0.15	0.54	1.18	0.57	0.69
	3	20	49	0.61	0.08	0.54	0.97	0.55	0.63
	3	30	51	0.59	0.06	0.54	0.99	0.56	0.60
	4	10	48	0.64	0.23	0.00	1.44	0.57	0.71
	4	20	50	0.62	0.25	−0.80	1.53	0.58	0.67
	4	30	50	0.60	0.03	0.54	0.72	0.57	0.61
A4	2	10	25	0.60	0.17	0.53	1.41	0.55	0.59
	2	30	25	0.54	0.01	0.53	0.57	0.54	0.55
	2	50	25	0.58	0.07	0.49	0.80	0.53	0.63
	4	10	25	0.53	0.01	0.51	0.55	0.52	0.55
	4	30	25	0.53	0.01	0.52	0.55	0.53	0.54
	6	30	25	0.55	0.02	0.50	0.62	0.55	0.56

variable, ten case data set with a value of −0.93, and the two variable, forty case data set with a value of 1.31). Although in Sec. 4.1.2 the range of the value of α was specified to be (0, 1) there is no actual theoretical bound to this continuum. Once again, the results suggest that α may be a useful diagnostic for this type of data structure. However, this property requires further investigation before it can be implemented.

4.1.6.2 Does the Choice of Method Affect the Predictive Capability?

The simulated data sets described in Sec. 4.1.6.1 were also used to investigate whether predictive capability is affected by the method of analysis employed to specify the model. By analyzing each data set with MLR, PLS, PCR, and the new formulation of CR and calculating the prediction criterion, I_{max}, the predictive capability of the models specified by each method of analysis can be compared. I_{max} represents the ability of each model to predict the response of known data. A selection of results describing the performance of the models specified by MLR, PLS, PCR, and the new formulation of CR is given in Table 2 which identifies the population structure and the maximum I-statistic for each of these methods; the reported α value is that calculated for the first component extracted by the CR analysis. The I_{max} statistics are those which represent the best model as specified by each of the methods, regardless of how many components constitute the model.

The results highlight a number of interesting features. First, whenever the calculated value of α is significantly different from 0, 0.5, or 1, CR clearly outperforms the other three methods, e.g. for population A2 where the three variable, ten

Table 2. A selection of summary results comparing the predictive performance of MLR, PLS, PCR and the new formulation of CR when analyzing simulated data sets.

Population	No. Vars.	No. Cases	I			CR	
			MLR	PLS	PCR	1st comp. α	I
A2	3	10	−26.15	41.46	30.94	0.70	53.62
	3	10	69.27	72.46	62.44	0.54	72.28
	4	10	−79.97	58.87	59.46	0.54	65.24
	4	10	5.88	54.54	34.21	0.58	53.95
	4	10	29.89	70.53	55.89	0.50	69.83
	4	20	16.34	21.96	21.17	0.55	37.26
	5	10	10.22	24.93	21.63	0.71	36.50
A3	2	10	38.85	51.23	38.85	0.54	50.66
	3	10	−23.52	39.57	28.14	0.59	40.74
	3	20	55.83	55.69	57.46	0.60	61.44
	4	10	−144.32	22.36	9.30	0.62	23.18
	4	10	54.70	77.00	67.23	0.56	76.20
	4	10	−29.58	35.79	20.95	0.59	35.25
	4	20	38.56	56.10	46.96	0.61	56.42
	4	20	17.14	39.88	34.51	0.56	40.18
A4	2	10	22.86	41.02	23.52	0.63	42.63
	4	10	12.09	65.78	60.58	0.58	66.12
	4	10	18.44	78.08	71.60	0.55	77.92

case example produces an α value of 0.70, and CR has an I-statistic far better than any of the other methods. Second, when α is approximately equal to 0.5, CR and PLS produce very similar results, e.g. population A3, four variables and twenty cases. This suggests that it is important to match the method of analysis to the data set since the predictive capability can be improved by adjusting α to give the most appropriate criterion function, T. Although for the majority of samples investigated, CR yields a larger prediction statistic, I_{max}, than the other methods, there are a few cases when MLR, PLS and PCR produce slightly better results. Such differences were relatively small, however, and may have arisen from chance effects. Furthermore, as the ratio case number/variable number increases, the results yielded by the four methods become very similar, e.g. for any of the four descriptor variable data sets (Table 2), the I-statistics yielded by MLR, PLS, PCR, and CR tended towards equality.

These observations confirm the importance of choosing the appropriate regression procedure, and, hence, the optimum criterion function, T, in order maximize the cross-validatory index, I, which reflects the robustness of the prediction model.

4.1.6.3 Can Robust Models be Specified Without Recourse to Cross-Validation?

To address the question of whether robust models can be specified without recourse to cross-validation, a number of data sets taken from the literature were analysed by CR, employing the model specification method of Stone and Brooks [6] described in Sec. 4.1.4.2 and that proposed by the Portsmouth group method (Sec. 4.1.5). Once again, the I statistic of Stone and Brooks' (Eq. (34)) was used as a basis for assessing predictive power. The values of I calculated for the models specified by each method were compared (Table 3) in order to indicate whether models can be specified without the use of cross-validation, i.e. using the procedure developed at the University of Portsmouth. In all but one, of the examples investigated, the cross-validatory statistic, I, obtained when the model is specified without recourse to cross-validation is as good as, or better, than that obtained when cross-validation is employed. This confirms that robust QSAR models can be constructed without reference to cross-validation. The one exception to this observation [21] still yields a satisfactory result ($I = 74.63\%$), although in this example, I falls substantially below the value calculated by the method of Stone and Brooks ($I = 86.21\%$). However, as mentioned earlier, estimates of I_{max} calculated using QSAR models specified using cross-validation as a construction criterion are likely to yield biased estimates which overstate the power of prediction (Sec. 4.1.5). The most notable improvement in I_{max}, obtained by switching from the procedure of Stone and Brooks to that of the Portsmouth group, being an increase from 0 to 30% [26]. This result demonstrates that a model specified when using cross-validation with no predictive value whatsoever, can be improved by basing component construction on the optimum value

Table 3. Summary of results from the experiment comparing model specification methods.

Ref. No.	No. Vars.	No. Cases	With Cross-Validation			Without Cross-Validation		
			I_{max}	No. comps.	$\bar{R}^2$	No. sig. comps.	I	$\bar{R}^2$
15	3	8	29.55	1	46.57	1	30.43	46.57
16	3	12	75.00	2	84.61	2	74.98	86.15
17	3	14	47.68	2	51.15	2	48.13	55.23
18	3	16	86.21	3	89.98	3	86.04	91.40
19	3	25	84.73	2	86.91	2	84.74	88.00
20	4	13	95.96	3	97.51	3	97.08	97.16
21	4	25	82.71	2	78.87	2	74.63	79.79
22	4	40	91.19	4	94.95	3	92.02	94.91
23	5	22	88.30	2	89.54	2	85.46	90.07
24	5	23	95.02	5	96.51	5	97.17	97.30
25	5	25	72.58	3	77.42	2	74.19	76.21
26[a]	6	11	−1.67	2	37.69	3[b]	31.81	72.48
27	6	21	92.75	4	94.86	3	93.96	95.58
27	6	28	90.52	6	98.08	6	96.79	98.38
28	7	15	81.55	6	90.96	5[c]	82.98	94.38

[a] reduced data set
[b] significant components were component numbers 1, 2, and 4
[c] significant components were component numbers 1, 3, 4, 5, and 6

of α. However, an I value of 31.81% indicates that the predictive capability of this model is limited.

It is interesting to note the very strong association between the I values obtained using LOO CV (Portsmouth algorithm) and the $\bar{R}^2$ values which express the degree to which the model fits the data used in its specification. A plot of I vs $\bar{R}^2$ (Fig. 3) illustrates this strong, linear relationship ($r = 0.91$).

4.1.6.4 Does Continuum Regression Protect Against Spurious Correlations?

An additional concern to QSAR practioners is the inclusion of non-significant variables, or components, into regression models. These Type 1 errors cause non-significant variables to be included in a QSAR model, when, in fact, they have little or no significant information to contribute. Spurious correlations have been investigated by Topliss and Costello [29], Topliss and Edwards [30] and, more recently, Wakeling and Morris [31]. These three studies address the problem using data sets constructed of random numbers that are uniformly distributed; the problem does not appear to have been investigated for normally distributed data.

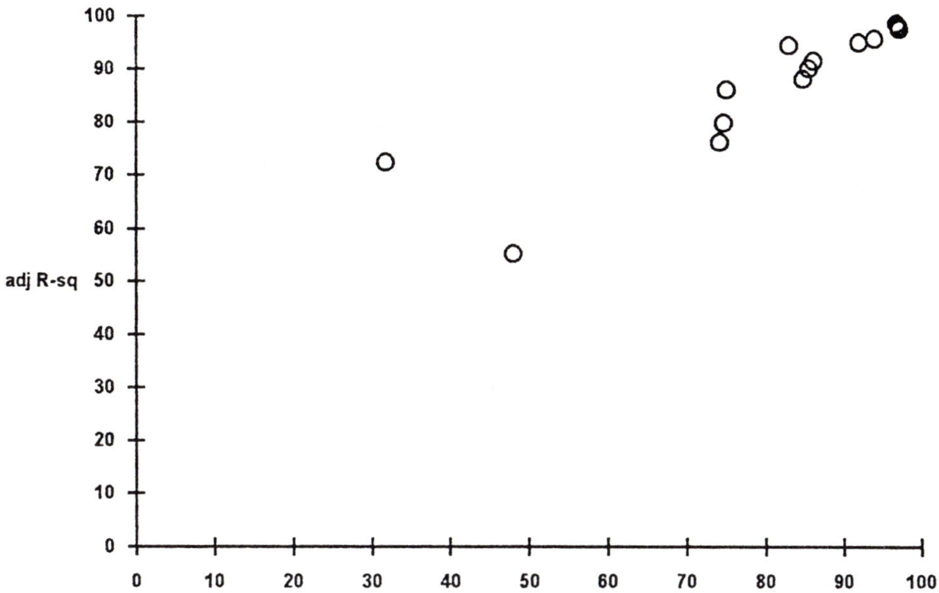

Figure 3

To investigate the extent to which models specified by the Portsmouth algorithm of CR are prone to chance effects, a limited study was undertaken. The study involved generating replicate data sets, of equal size (three descriptor variables and twenty cases), with three population correlation structures (Appendix A6). The population structures were chosen such that there would be ample opportunity for the analysis to give spurious results. The "uncorrelated" population (A6i) yielded random vectors that were independent of each another, and, hence, any significant result would have been a chance effect. The second population (A6ii) introduced a significant amount of correlation between two of the independent descriptors and the response and retained a third, random descriptor. The third population (A6iii) introduced multicollinearity into the data by constructing two of the descriptors to be dependent whilst still being highly correlated with the response. Once again, a third, random descriptor variable was included.

The appearance of chance effects in any of the analyses are easily detected. For population A6i, any component deemed significant by Wold's *E*-test (Eq. (31)) or the *t*-distribution (Sec. 4.1.5) is a chance effect; population A6ii and A6iii, where the populations comprise two significant descriptor variables, one-component models are more minimalistic and three-component models less minimalistic than those which would have been obtained by MLR. If the third component is deemed significant, it may indicate a spurious result arising from sampling errors. From the limited set of results presented in Table 4, there is some evidence to suggest that the Ports-

Table 4. The number of significant components included in prediction models constructed using simulated samples from populations of known condition structure.

Population		A6i			A6ii			A6iii		
No. components	Sig. test	1	2	3	1	2	3	1	2	3
Model specification by cross-validation	$E<1.0$	3	0	0	0	6	4	1	3	6
	$E<0.4$	0	0	0	3	7	0	8	0	0
Portsmouth Algorithm	$p<0.05$	1	0	0	1	6	3	4	4	2

mouth algorithm offers some protection against chance effects. Thus, only one in ten of the samples derived from population A6i gave rise to a significant CR model; in contrast, all twenty samples derived from the populations with correlation between y and $x1$, and y and $x2$ (A6ii, A6iii) gave a significant model. Of the samples from population A6ii, seven out of ten gave models with the expected number of components or less; from population A6iii, four out of ten gave one-component models and four gave two-component models. These results have been based on tail probabilities of the t-distribution derived using the Portsmouth model specification procedure (Sec. 4.1.5). Specifying the CR model using cross-validation and Wold's E-statistic leads to more conservative models when $E<0.4$, but less conservative models when $E<1.0$.

4.1.6.5 How CR Predictions Compare with those of other Regression Procedures

A comparison of the performances of MLR, PLS, PCR, and CR was undertaken using the data of Kruse et al. [25] for multisubstrate inhibitors of dopamine β-hydroxylase (DBH) acting at the phenethylamine binding site. The potency of the twenty-five 3,4,5-substituted DBH inhibitors tested and five chemical parameters used to quantify their chemical features are presented in Table 5, and the associated correlation matrix in Table 6. Five significant interparameter correlations ($r>0.38$, $p<0.05$) were observed. Furthermore, the data exhibited multicollinearity ($\lambda_{max}/\lambda_{min} = 451.6$). Thus, this data violates the assumption of that independent descriptor variables are required in order to perform ordinary least squares and multiple linear regression. It, therefore, provides a useful example to highlight the advantages, if any, of PLS, PCR, and CR.

The "best" MLR equation reported by Kruse [25] and his colleagues contained four variables,

$$-\log IC_{50} = 1.28(\pm 0.22)I - 0.14(\pm 0.02)MR + 0.65(\pm 0.16)\pi + 1.42(\pm 0.33)F - 1.26 \quad (35)$$

Table 5. Chemical parameters of 3,4,5-substituted DBH inhibitors. *I* is indicator variable for the presence of a 4-OH, π is the sum of the lipophilicity at the 3-, 4-, and 5-positions, *MR* is the molar refractivity at the 3-, 4-, and 5-positions, *F* is the sum of the inductive effects at the 3-, 4-, and 5-positions and *R* is a measure of the resonance effects.

Substituent	$-\log IC_{50}$	I_{4OH}	π_{345}	MR_{345}	F_{345}	R_{345}
3-NO_2, 4-OMe	−2.55	0	−0.30	16.23	0.93	−0.35
4-OMe	−2.30	0	−0.02	9.87	0.26	−0.51
3-OMe	−2.19	0	−0.02	9.87	0.26	−0.51
3-OH	−2.17	0	−0.67	4.85	0.29	−0.64
4-Cl	−1.98	0	−0.71	8.03	0.41	−0.15
4-NO_2	−1.72	0	−0.28	9.36	0.67	0.16
3-Me, 4-OH	−1.69	1	−0.11	9.50	0.25	−0.77
4-F	−1.67	0	0.14	2.92	0.43	−0.34
3,5-Cl_2, 4-OMe	−1.67	0	1.40	19.93	1.08	−0.81
3,5-F_2, 4-OMe	−1.55	0	0.26	9.71	1.12	−1.19
H	−1.50	0	0.00	3.00	0.00	0.00
3-NO_2, 4-OH	−1.49	1	−0.95	11.21	0.96	−0.48
3,4-Cl_2	−1.44	0	1.42	13.06	0.82	−0.30
3-Br, 4-OH	−1.07	1	0.19	12.73	0.73	−0.81
3-Cl	−1.07	0	0.71	8.03	0.41	−0.15
3-F	−0.74	0	0.14	2.92	0.43	−0.34
4-OH	−0.41	1	−0.67	4.85	0.29	−0.64
3,5-Cl_2	−0.38	0	1.42	13.06	0.82	−0.30
3,4-$(OH)_2$	−0.34	1	−1.34	6.70	0.58	−1.28
3-Cl, 4-OH	−0.30	1	0.04	9.88	0.70	−0.79
3-F, 4-OH	−0.17	1	−0.53	4.77	0.72	−0.98
3,5-F_2	−0.07	0	0.28	2.84	0.86	−0.68
3,5-Cl_2, 4-OH	−0.16	1	0.75	14.91	1.11	−0.94
3,5-F_2, 4-OH	1.13	1	−0.39	4.69	1.15	−1.32
3,4-$(OMe)_2$	−2.75	0	−0.04	16.74	0.52	−1.02

and had an $R^2 = 0.83$, which indicated a good fit to the data. The data was analyzed using the CR algorithm reported in Sec. 4.1.3., with α fixed at 0, 0.5, and 1, or optimized analytically. This resulted in four models corresponding to MLR, PLS, PCR, and CR, respectively. Because cross-validation is recommended for use during model construction, the models specified for PLS and PCR have been based on this procedure. However, the CR model was constructed according to the procedure outlined in Sec. 4.1.5. The MLR procedure ($\alpha = 0$) gave β coefficients within the tolerances of their standard errors, which were identical to those reported by Kruse et al. [25] thus, validating the CR algorithm.

Table 7 gives comparisons of the four methods. The equation giving the best fit to the data (i.e. the largest R^2) was obtained using MLR; R^2 values of 0.77 were obtained for PLS, PCR and CR. The goodness of fit obtained using MLR is probably

Table 6. Correlation matrix of chemical parameters of 3,4,5-substituted DBH inhibitors. Significant pairwise correlation are in bold.

	$-\log IC_{50}$	I_{4OH}	π_{345}	MR_{345}	F_{345}	R_{345}
$-\log IC_{50}$	1.00					
I_{4OH}	0.58	1.00				
π_{345}	−0.03	**−0.45**	1.00			
MR_{345}	−0.35	−0.06	**0.43**	1.00		
F_{345}	0.40	0.21	0.23	**0.45**	1.00	
R_{345}	−0.40	**−0.55**	0.28	−0.10	**−0.43**	1.00

Table 7. Summary of statistics comparing the fit and prediction of CR, MLR, PLS and PCR models for 3,4,5-substituted DBH inhibitors.

Regression Method	1st Component α	No. of Sig. Comps.	I	R^2
CR	0.67	2	74.19	76.77
MLR	0.0	1	71.65	83.37
PLS	0.5	2	73.13	77.31
PCR	1.0	3	68.39	77.48

Table 8. Comparative partial regression coefficients (β) of the original variables used to construct QSAR models of 3,4,5-substituted DBH inhibitors.

Regression Method	I_{4OH}	π_{345}	MR_{345}	F_{345}	R_{345}
CR ($\alpha = 0.67$)	1.4077	0.5076	−0.1274	0.6852	−0.6617
MLR	1.1699	0.6398	−0.1497	1.2990	−0.0730
PLS	1.0509	0.5298	−0.1292	0.6972	−0.6561
PCR	0.9617	0.6804	−0.1387	0.6275	−0.7594

due to overfitting, since it is not reflected in the predictive capability of the MLR equation. The I values obtained for the different equations is in fact ranked as the following: CR > PLS > MLR < PCR. Table 8 suggests that the electronic effects of substituents at positions 3, 4, and 5 of the phenyl ring account for these discrepancies; the parameters F_{345} and R_{345} which are significantly correlated (Table 6) are given with contrasting influences on $-\log IC_{50}$ depending on which method of regression was used for model specification. The best prediction was obtained using the model constructed by the CR algorithm developed at the University of Ports-

Table 9. Loadings of the components deemed significant by CR, MLR, PLS, and PCR analyzes. Significant loadings ($p<0.05$) are in bold.

Regression Method	Comp. Index	I_{4OH}	π_{345}	MR_{345}	F_{345}	R_{345}
CR	1	0.1099	−0.0148	**−0.9915**	0.0380	−0.0566
	2	**0.6945**	0.3399	−0.0346	**0.4560**	**−0.4393**
MLR	1	**0.6259**	0.3423	−0.0801	**0.6950**	−0.0391
PLS	1	0.1556	0.0093	**−0.9817**	0.0689	−0.0854
	2	**0.6906**	0.3518	−0.0348	**0.4600**	**−0.4318**
PCR	1	0.0067	−0.0677	**−0.9972**	−0.0309	0.0082
	2	**−0.5384**	**−0.5639**	0.0504	−0.3752	**0.4989**
	3	**0.6102**	**−0.7495**	0.0471	0.2523	−0.0011

mouth, where equal weighting was given to both electronic terms. The low I value using a three-component PCR model probably reflects the lack of supervision during the construction of the components. The equations obtained using PLS ($\alpha = 0.5$) and CR ($\alpha = 0.67$ for this example) were constructed criteria functions based on different degrees of supervision (i.e. dependence on $-\log IC_{50}$).

The loadings of the original variables onto the significant components (Table 9) suggest that molar refractivity (MR_{345}) is described by a single component, whereas the remaining variables describing the partition and electronic properties of the *meta* and *para* positions of the phenyl ring and the presence or absence, of a *para*-hydroxy substituent combined to produce a molecular feature which is useful for predicting novel compounds of high biological activity. A tentative interpretation of the structure-activity relationship as given by the various models is that bulk and electronic character were both important in determining the potency of DBH inhibitors. The electronic character for the set of substituents used in this study is supplemented by the partition properties and the presence of the hydroxy group to provide a predictor variable orthogonal to that describing the bulk of the substituents. Inhibitor activity seems to be associated with small substituents which tend to withdraw electrons by induction and donate electrons to the π-bond system delocalized over the phenyl ring. The bulk and inductive effects have already been identified by Kruse et al. [25] who suggest that the latter may decrease the p*Ka* of any hydroxyl group present in the ring. However, the resonance term included in the most successful prediction models was not recognized by these workers, possibly because it is a source of multicollinearity (Table 6). Resonance will be reinforced by the presence of dissociated hydroxyls, i.e. low p*Ka* values, since the anion has strong electron donating properties. Inductive withdrawal and mesomeric donation of electrons would also be possible with the 3,5-dihalogen substitution pattern associated with potent inhibitors [25]. The independence of bulk and electronic effects is not easily recognizable in the model obtained using MLR for which the loading pattern (onto a single component) is less clear.

4.1.7 Concluding Remarks

The Continuum Regression algorithm was developed to overcome the problems associated with the construction of parsimonious components and model specification. Refinements to the original method of Stone and Brooks [6] which allow the data structure to determine the nature of the general criterion function, *T*, have led to the development of an algorithm which is proving to be a useful research tool. This algorithm is available commercially as part of the University of Portsmouth Enterprise Limited's Drug Design Software, PARAGON^{TM}. Further work is required to establish whether Continuum Regression can improve the prediction of biological activity based on sets of molecular properties which are characterized by multicollinerarity.

Appendix

The population correlation structures used to generate the simulated data sets are given below.

A1. Data sets of all sizes were generated with the correlation structure shown below. The four descriptor variable set is used as an example.

	y	$x1$	$x2$	$x3$	$x4$
y	1.0				
$x1$	0.0	1.0			
$x2$	0.0	0.0	1.0		
$x3$	0.0	0.0	0.0	1.0	
$x4$	0.0	0.0	0.0	0.0	1.0

A2. The data sets were generated so that variables $x1$ and $x2$ were correlated with y for all set sizes, e.g. four descriptor variables.

	y	$x1$	$x2$	$x3$	$x4$
y	1.0				
$x1$	0.25	1.0			
$x2$	0.25	0.0	1.0		
$x3$	0.0	0.0	0.0	1.0	
$x4$	0.0	0.0	0.0	0.0	1.0

A3. Data sets of all sizes were generated so that three descriptor variables, $x1$, $x2$, and $x3$ were highly correlated with the response variable, e.g. four descriptor variables.

	y	$x1$	$x2$	$x3$	$x4$
y	1.0				
$x1$	0.5	1.0			
$x2$	0.5	0.0	1.0		
$x3$	0.5	0.0	0.0	1.0	
$x4$	0.0	0.0	0.0	0.0	1.0

A4. Some correlation amongst the descriptor variables was introduced, e.g. four descriptor variables.

	y	$x1$	$x2$	$x3$	$x4$
y	1.0				
$x1$	0.5	1.0			
$x2$	0.5	0.3	1.0		
$x3$	0.5	0.4	0.6	1.0	
$x4$	0.5	0.2	0.3	0.2	1.0

A5. This data set was generated so that two orthogonal descriptor variables were equally correlated with the response variable. The sample correlation structure is:

	y	$x1$	$x2$
y	1.0		
$x1$	0.705	1.0	
$x2$	0.700	0.0	1.0

A6. The population correlation structures used in the investigation into spurious correlations were defined as follows:

(i) Uncorrelated:

	y	*x*1	*x*2	*x*3
y	1.0			
*x*1	0.0	1.0		
*x*2	0.0	0.0	1.0	
*x*3	0.0	0.0	0.0	1.0

(ii) Some Correlation:

	y	*x*1	*x*2	*x*3
y	1.0			
*x*1	0.7	1.0		
*x*2	0.7	0.0	1.0	
*x*3	0.0	0.0	0.0	1.0

(iii) Multicollinearity:

	y	*x*1	*x*2	*x*3
y	1.0			
*x*1	0.8	1.0		
*x*2	0.8	0.4	1.0	
*x*3	0.0	0.0	0.0	1.0

References

[1] Hansch, C., Maloney, P.P., Fujita, T. and Muir R.M., *Nature* **194**, 178–180 (1962)
[2] Krzanowski, W.J., *Biometrics* **43**, 575–584 (1987)
[3] Mager, P.P., *Med. Res. Rev.* **2**, 93–121 (1982)
[4] Hellberg, S., Wold, S., Dunn III, W.J., Gasteiger, J. and Hutchings, M.G., *Quant. Struct.-Act. Relat.* **4**, 1–11 (1985)
[5] Nilsson, L.M., Carter, R.E., Sterner, O. and Liljefors, T., *Quant. Struct.-Act. Relat.* **7**, 84–91 (1988)
[6] Stone, M. and Brooks, R.J., *J.R. Statist. Soc. B* **52**, 237–269 (1990)

[7] Malpass, J.A., Salt, D.W., Ford, M.G., Wynn, E.W. and Livingstone, D.J., *Prediction of Biological Activity Using Continuum Regression.* In: *Trends in QSAR and Molecular Modelling 92*, Wermuth, C.G., ed., ESCOM, Leiden, 314–316, (1993)
[8] Koutsoyiannis, A., *Theory of Economics*, Macmillan, London, 1973
[9] Wold, S., *Technometrics* **20**, 397–405 (1978)
[10] Cramer III, R.D., Bunce, J.D., Patterson, D.E. and Frank, I.E., *Quant. Struct.-Act. Relat.* **7**, 18–25 (1988)
[11] Norden, B., Edlund, U., Johnels, D. and Wold, S., *Quant. Struct.-Act. Relat.* **2**, 73–76 (1983)
[12] Osten, D.W., *J. Chemometrics* **2**, 39–48 (1988)
[13] Wold, S., *Quant. Struct.-Act. Relat.* **10**, 191–193 (1991)
[14] *NAG Fortran Library Manual, Mark 15,* The Numerical Algorithms Group Limited, 1991
[15] Young, R.C., Durant, G.J., Emmett, J.C., Ganellin, C.R., Graham, M.J., Mitchell, R.C., Prain, H.D. and Raontree, M.L., *J. Med. Chem.* **29**, 44–49 (1986)
[16] Diana, G.D., Oglesby, R.C., Akullian, V., Carabateas, P.M., Cutcliffe, D., Mallamo, J.P., Otto, M.J., McKinlay, M.A., Maliski, E.G. and Michalec, S.J., *J. Med. Chem.* **30**, 383–388 (1987)
[17] Hasegawa, K., Miyashita, Y., Sasaki, S.-I., Sonoki, H. and Shigyou, H., *Chem. Intell. Lab. Sys.* **16**, 69–75 (1992)
[18] Snee, R.D., *Technometrics* **4**, 415–428 (1977)
[19] Clark, M.T., Coburn, R.A., Evans, R.T. and Genco, R.J., *J. Med. Chem.* **29**, 25–29 (1986)
[20] Draper, N.R. and Smith, H., *Applied Regression Analysis,* Wiley, New York, 1981
[21] Cantelli-Forti, G., Guerra, M.C., Barbaro, A.M., Hrelia, P., Biagi, G.L. and Borea, P.A., *J. Med. Chem.* **29**, 555–561 (1986)
[22] Li, R., Hansch, C., Matthews, D., Blaney, J.M., Langridge, R., Delcamp, T.J., Susten, S.S. and Freisheim, J.F., *Quant. Struct.-Act. Relat.* **1**, 1–7 (1982)
[23] Hopfinger, A.J., *J. Med. Chem.* **26**, 990–996 (1983)
[24] Leo, A., Hansch, C. and Church, C., *J. Med. Chem.* **12**, 766–771 (1969)
[25] Kruse, L.I., Kaiser, C., DeWolf, Jr., W.E., Frazee, J.S., Ross, S.T., Wawro, J., Wise, M., Flaim, K.E., Sawyer, J.L. Erickson, R.W., Ezekiel, M., Ohlstein E.H. and Berkowitz, B.A., *J. Med. Chem.* **30**, 486–494 (1987)
[26] Ijzerman, A.P., Bultsma, T. and Timmerman, H., *J. Med. Chem.* **29**, 549–554 (1986)
[27] Wilson, L.Y. and Famini, G.R., *J. Med. Chem.* **34**, 1668–1674 (1991)
[28] Wold, S., Dunn III, W.J. and Hellberg, S., *Pattern Recognition as a Tool for Drug Design.* In: *Drug Design: Fact or Fantasy?* Jolles, G. and Wooldridge, K.R.H., eds., Academic Press, London, 95–117, (1984)
[29] Topliss, J. and Costello, R.J., *J. Med. Chem.* **15**, 1031–1033 (1972)
[30] Topliss, J. and Edwards, R.P., *J. Med. Chem.* **22**, 1238–1244 (1979)
[31] Wakeling, I.N. and Morris, J.J., *J. Chemometrics* **7**, 291–304 (1993)

4.2 Molecular Taxonomy by Correspondence Factorial Analysis (CFA)

Jean-Christophe Doré and Tiiu Ojasoo

Abbreviations and Symbols

CFA	Correspondence Factorial Analysis
MCA	Multiple Correspondence Analysis
PCA	Principal Component Analysis
AC	Absolute Contribution
RC	Relative Contribution ($= \cos^2 \theta$)
RBA	Relative Binding Affinity
ER	Estrogen Receptor
AR	Androgen Receptor
PR	Progesterone Receptor
GR	Glucocorticoid Receptor
MR	Mineralocorticoid Receptor

(i)	Molecules ($\Sigma i = n$)
(j)	Tests ($\Sigma j = p$)
R^i	Multidimensional space into which the (i) molecules are projected as regards to the (j) tests
R^j	Multidimensional space into which the (j) tests are projected for the (i) molecules
k_{ij}	Basic relationship between molecule (i) and test (j)
f_{ij}	Relationship between molecule (i) and test (j) in terms of probability
$f_{i.}$	Marginal relative frequency of molecule (i) for the (j) tests
$f_{.j}$	Marginal relative frequency of test (j) for the (i) molecules
$\{S\}'_{jj}$	Semi-matrix of the distances between the (j) tests
φ_α	Factorial axis α
$\varphi_{\alpha i}$	Coordinate of variable (i) along α-axis
$\varphi_{\alpha j}$	Coordinate of variable (j) along α-axis
V	Eigenvector
λ	Eigenvalue
τ	Percent variance accounted for by the axis

4.2.1 Introduction

One of the tools required by the medicinal chemist is a simple, objective, pictorial method of representing the biochemical and biological data obtained from families of chemical compounds. The pictorial representation can form the common ground for a cogent interpretation of the results with the biologist and for the design of future tailor-made molecules for use as markers or drugs. The need for such a descriptive multivariate tool is essential at the interface between Chemistry and Biology because of the infinite permutations of molecular descriptors in chemical synthesis and the complexity of living organisms. The multivariate approach should serve several purposes, such as sifting through information without imposing any a priori determinate classification, and enable the development of a model for the prediction of activity profiles of as yet unsynthesized molecules. However, under no circumstances, should the tool be an end in itself, but should open up new areas for further experiments and analyses by other complementary multivariate methods. In the present chapter, we propose that these requirements are aptly met by Correspondence Factorial Analysis (CFA).

4.2.1.1 The Need for an Interface Between Chemistry and Biology

There exists the one extreme, where large numbers of molecules are blindly screened, that too often is wasteful of resources, intellectually unsatisfying and of uncertain outcome. At the other extreme, is the design of a handful of highly specific and, hopefully non-toxic ligands which is based on the structural information obtained from the active sites of proteins by sophisticated "know-how" and analytical tools. Between these two extremes, there exists a middle pathway that makes the most of all available information and neither rejects screening nor depends entirely on state-of-the-art knowledge. The number of new molecules that need be synthesized is minimized by objectively analyzing existing data obtained from the interactions between molecules and effectors in the relevant biological and pharmacological tests and from toxicological studies. This is a desirable alternative to the design of compounds by the overexploitation of methods of synthesis already mastered, or by searching for loopholes in patents.

4.2.1.2 Concept of a Multivariate System

Medicinal chemistry has to deal with the interface between two systems (organized self-consistent wholes), one composed of molecules and the other of biological parameters. A molecule, whether a small chemical entity or a macromolecule, should not be considered as an isolated entity but as an integral part of a greater whole (system). The uniqueness of each molecular structure is gauged with respect to its peers. A given biological response occurs in the context of an environment of multiple complex interactions and interferences. When considering the interfacing of the two systems, a multivariate method that encompasses both and attaches equal importance to each is required.

The data take the form of matrices (table $i \times j$ of the responses of (i) molecules on (j) tests). Although such tables can be analyzed by a univariate approach, this is extremely laborious and also biased because of a natural tendency to search for maxima and minima and to rank rather than to structure. Nor are an iterative bivariate approach (regression), which considers the variables in pairs, or multiple regression analysis, which preselects a dependent variable, suitable for a preliminary analysis. These methods are also laborious and their use only postpones the finding of an all-embracing solution. On a less serious note, one could describe uni- and bivariate approaches as "*molecular psychology*" and multivariate approaches as "*molecular sociology*" which seeks the whole gamut of correlations between a population of molecules and a population of receptors.

4.2.1.3 The Choice of Correspondence Factorial Analysis (CFA)

CFA surpasses the notion of rank by adopting the concept of structure or organization and, in so doing, can be considered as an abstract form of Pattern Recognition. It is governed by the laws of probability and, thus, yields an overview that is an approximation. However, it dispenses with artificially determined mathematical probability levels (p values indicating the significance of correlations), that are less crucial in multiparametric relationships than bivariate relationships and which may be of little relevance in the biological context. Instead CFA highlights legitimate and meaningful correlation levels by eliminating redundant information, filtering out noise and minimizing artefacts. In this respect, it belongs to the realm of fuzzy logic (expert systems, neural networks, etc.).

The likelihood of establishing imaginary unfounded relationships between totally independent variables by CFA is minimal since it seeks convergence of indicators and yields a theoretical index (λ) of the quality of the factorial representation. Moreover, a CFA requires expert interpretation. In the results of a CFA, both chemists and biologists will not only find affirmation of the tried and tested (a sign of the strength and validity of the approach) as well as of some of their untested hypotheses, but also a number of similarities and contrasts that can lead to new ideas and that may cast doubt upon certain preconceptions. CFA is no substitute for human decision-making powers, but is an appropriate and effective tool for interpretation and decision-making. The factorial plots of CFA, read in conjunction with tables giving the contributions of the variables to the factorial axes, reveal the most probable correlations among the extraordinarily vast number of possible combinations in an ordered fashion. To do so, like other multivariate methods, it relies upon a technique of data reduction, more specifically, the computing of vectors from a diagonal matrix by an eigenfactor–eigenvector routine.

4.2.1.4 Multivariate Data Reduction by χ^2-Metrics in CFA

In factorial analysis, one of the oldest methods of data reduction is covariance as applied, for instance, in Principal Component Analysis (PCA). CFA, on the other hand, is based on the use of χ^2-metrics on the assumption that the data table $(i \times j)$ is a frequency table, i.e., there is a probability of a statistical link (kinetic interaction) between molecule (i) and test (j) under defined conditions (as in the statistics of thermodynamics). In a way, CFA can be regarded as a two-fold unified PCA performed on the molecules, on the one hand, and on the tests, on the other. This allows the direct representation of both molecules and tests simultaneously on single factorial axes or plots and the disclosure of relationships among molecules, tests, and between molecules and tests within each successive factorial plot required, by decreasing order of variance, to describe the total variance (inertia) of the experimental system in a stepwise fashion. As in PCA, the first factorial plot ($\varphi_1\,\varphi_2$) reveals the strongest correlations among the variables, the $\varphi_3\,\varphi_4$ plot weaker correlations, and so on, thus highlighting the principal organization but also substructures of the system. However, unlike in the case of PCA, the lower order correlations are as relevant as the higher order correlations. In CFA, there is not theoretical limit to the number of variables (molecules or tests) that can be analyzed. This could be as low as 2×2 (if this were meaningful), but is more usually as high as 100×200. In practice, the upper limit is governed by the computation time, the degree of graphic resolution (approx. 200 items per page), and ease of interpretation.

Owing to the use of χ^2-metrics, CFA obeys the principle of distributional equivalence which offers several advantages: (a) clustered items (e.g, molecules) can be grouped together into barycenters in the factorial plots to improve clarity. This means that, for the analysis of very large data matrices, a stepwise approach can be adopted, whereby items are preassembled into groups (by selection or by a partition technique) for group analyses. These groups can be subsequently split into their individual components. (b) The normalized profiles of supplementary variables, either molecules or tests, can be routinely introduced into a CFA that is used as a mathematical model. The process can be stretched to test hypotheses on other types of data that are considered pertinent, such as molecular descriptors (e.g., physicochemical properties, spectroscopic data) etc. Relative contributions establish whether the new data are truly relevant to the existing analysis.

A CFA can be validated by the use of other cluster techniques applied to the χ^2-distance square matrix such as independent classification of the study fields (molecules and tests) by nearest-neighbor analysis, partitioning around mobile target items, minimum spanning trees, or ascending hierarchical clustering in order to obtain an overview of the projections of each field over several or all of the factorial axes describing the experimental system. A CFA can also be a basis for prospective modeling by applying a qualitative method of category assignment, such as discriminant analysis, or a quantitative method of evaluating a dependent variable, such as stochastic regression, to selected factorial axes. This variable, usually a biological test, is considered to be dependent upon the structural descriptors of the molecules. In this case, the use of regression analysis is totally justified since correlations are not based on the initial variables, but on superlative variables constituted by the factorial axes which are orthogonal and, therefore independent. In multivariate statistics, no single method is allowed to dominate other methods and the results of an analysis will be all the more reliable, if several methods point to the same conclusion.

4.2.2 Applications and Methodology of CFA

CFA was first developed by Benzécri and coworkers [1–4] and is reviewed in [5–8] with up-to-date discussion by van der Heijden et al. [9] and Goodman [10]. This method has long been a specialist area of French scientists (viz. the journal "*Les Cahiers de l'Analyse des Données* which was founded in 1976 and is entirely devoted to CFA) and has recently gained a much wider audience in an ever increasing number of disciplines. CFA is now common in studies relating to geology and ecology [11–15], evolution [16], analytical chemistry (e.g., chromatography [17, 18], electrophoresis [19], the analysis of electron micrographs of macromolecules [20]), sensor chemistry [21] and increasingly in medicine [22–29]. (The aforementioned references are just but a few selected from a long list for illustrative purposes). To our knowledge, the application of CFA to metallurgy [30, 31], olfactory substances [32], and, most importantly, to structure–activity relationships of chemical substances has yet to be addressed by others. The list of the different types of chemical compound, either natural or synthetic, whose structures and properties have been analyzed by CFA, is fairly extensive and relate to both agronomy (pesticides [33], insecticides [34], pheromones [35, 36]) and medicine (antibiotics [37], antiparasitic agents [38], flavonoids [39], steroids [40–42], triphenylethylene analogs of estrogens [43–48] and others [49, 50]).

A brief description of the CFA method will follow to aid in the understanding of the application to steroid receptor binding data which is presented in the second half of this chapter.

4.2.2.1 The Data Matrix

Direct application of CFA to a set of data is possible on four conditions: the data do not comprize negative values, they are homogeneous (i.e, the measurements correspond to intensive variables), exhaustive (no variables have been knowingly omitted), and relevant (not purely independent variables). The data can be made homogeneous by the appropriate scaling (logarithmic transformation, centering, normalizing, centering+normalizing etc.). A limited number of missing values for any given variable can be extrapolated. On the whole, these are relatively drastic conditions since, outside industrial screening programs, it is rare to encounter data that are not the result of a linear step-by-step reasoning.

The crude data for analysis by CFA may take several forms: contingency and frequency tables, rankings, "yes" or "no" responses (0/1), experimental values with or without calculated antivalues in order to account not only for the specificity, but also the amplitude of response of the test molecules, tables of disjunctive variables used as such or to construct frequency tables (e.g., the Burt matrix [51]), etc.

4.2.2.2 Statistical Procedure

Let us consider the case of steroids interacting with different hormone receptors. A data matrix of the (i) molecules ($\Sigma i = n$) (rows) by the (j) tests ($\Sigma j = p$) (columns) is constituted. The data table is subjected to an appropriate transformation procedure to allow representation of both molecules and tests on single-display factorial maps. The j tests are, thus, projected into the multidimensional space made up of the i molecules R^{i} and vice versa the i molecules are projected into the j-dimensional space of the tests R^j. The position of each point representing a single test result for a given molecule in the R^i space is given by the probability that test (j) has an amplitude of f_{ij} for molecule (i) and is defined by the ratio

$$f_{ij}/f_{.j} \tag{1}$$

where

$$f_{ij} = k_{ij}/\sum_{ij} k_{ij} \tag{2}$$

and where

$$f_{.j} = \sum_{j} f_{ij} \tag{3}$$

A symmetrical calculation defines the position of each molecule for each test ($f_{ij}/f_{i.}$) in the R^j space. This dual procedure yields comparable normalized profiles (histograms) for the rows and columns, thus, enabling their comparison by a technique that can be considered as a form of pattern recognition.

To represent these two sets of points, principal projection axes are established as in PCA by determining eigenvalues (λ) and eigenvectors (V_{x}). A symmetrical matrix is constituted of the distances $S_{jj'}$ between pairs of molecules (χ^2-distance) as follows:

$$R = \{S_{jj'}\} = \left[\sum_{i=1}^{n} \frac{1}{f_{i.}} \frac{f_{ij}f_{ij'}}{\sqrt{f_{.j}f_{.j'}}} \right] \tag{4}$$

The calculation is performed by solving equations of the type $[\boldsymbol{R}] - \lambda\,[x] = 0$ and $[\boldsymbol{R}][V_x] = \lambda_x [V_x]$ (diagonalization of the symmetric matrix). This procedure is simpler in CFA than in PCA because one of the sets of points is given by the matrix $[\boldsymbol{R}] = [\boldsymbol{M}] \cdot [\boldsymbol{M}']$. The permutation of the indices is, thus, equivalent to transposing the matrix onto the other set of points $[\boldsymbol{M}'] \cdot [\boldsymbol{M}]$ with the same eigenvalues as those of R.

The coordinates φ_j of the tests for factorial axis α are calculated by using the formula:

$$\varphi_{\alpha \mathrm{j}} = \lambda_{\alpha}^{1/2} V_{\alpha j} / f_{.j}^{1/2} \tag{5}$$

where $\lambda_{\alpha}^{1/2}$ is the square root of the non-trivial eigenvalue λ_{α}, $V_{\alpha j}$ the corresponding eigenvector, and $f_{.j}^{1/2}$ the square root of the marginal relative frequency of test (j) for the (i) molecules. The correspondence between the molecules and tests is given by the transition formulae:

$$\varphi_{\alpha i} = (1/\lambda_{\alpha}^{1/2}) \sum_{j=1}^{p} (f_{ij}/f_{i.}) \varphi_{\alpha j} \qquad \text{for the molecules} \; - \tag{6}$$

$$\varphi_{\alpha j} = (1/\lambda_{\alpha}^{1/2}) \sum_{i=1}^{n} (f_{ij}/f_{.j}) \varphi_{\alpha i} \qquad \text{for the tests} \; - \tag{7}$$

The factorial axes φ_{α} are ranked by their order of importance in accounting for the total variance of the system (φ_1, φ_2, $\varphi_3 \ldots \varphi_{n-1}$). Factorial maps are then drawn by plotting any two of these orthogonal axes and displaying the projection of points.

If a large part of the total variance is not accounted for by the two principal factorial axes φ_1 and φ_2, i.e, the true points are not close to their projections onto the $\varphi_1 \varphi_2$ map, it is necessary to refer to the absolute contribution (AC) and relative contribution (RC) of each variable to all factorial axes, in order to assess how well a particular axis represents the variance of the system (ACs of the variables) and how a variable is dispersed across all the axes (RCs of the variables). For test j,

$$\mathrm{AC}\alpha\,(\mathrm{j}) = f. \varphi_{\alpha i}^{2} / \lambda_{\alpha}\; 100 \quad (\Sigma\mathrm{ACs} = 100\% \text{ for any axis } \alpha) \tag{8}$$

and

$$\mathrm{RC}\alpha\,(\mathrm{i}) = \varphi_{\alpha i}^{2} / d_{p}^{2}(\mathrm{i}, G) \quad (\Sigma\mathrm{RCs} \text{ of each variable to all axes} = 1) \tag{9}$$

where G is the distance from the center of gravity of the points. RC is in fact the square of the cosine of the test j for axis α.

4.2.2.3 CFA Program Availibility

Computations for the writing of CFA programs are presented in several textbooks [8, 52, 53]. Standard CFA programs are commercially available from French sources including ADDAD (Association pour le Développement et la Diffusion des Données, Laboratoire de Statistique, Tour 45 – 55, 4 Place Jussieu, 75005 Paris), ITCF (Institut Technique des Céreales et des Fourrages, 8 ave du Président Wilson, 75116 Paris), and DP Tool Club (Program ADSO [54], BP 745 59657 Villeneuve d'Ascq, France), but also in other countries, Kovach Computing Service (Wales, U.K.), SimCA Version 2 (Prof. M. J. Greenacre, P.O. Box 567, Irene, 1675 South Africa), BMDP Statistical Software Inc., BMDP PC-90 user's Guide, 1990 (Los Angeles), SPSS Categories Reference Guide 1990, SPSS Inc. (Chicago), ANOVA-FREQ Version 6, SAS/STAT User's Guide Vol. 1, 1990, SAS Institute Inc. (Cary, North Carolina). These standard programs can be run on mainframe and personal computers (PC/AT, Macintosh [55]) as well as on UNIX work-stations. Computing time for matrices smaller than 200×50 is at present reasonable, especially if the computer is equipped with an arithmetic processor.

In the example given below, calculations were performed on a microcomputer (16 – 32 bits of 655K of central memory, Hewlett-Packard 9836) with an in-house program written especially for the analysis of structure-activity relationships. The program has many subroutines and, unlike the above commercial programs, is flexible and not restrictive, enabling many variations in input and output (calculations as well as graphs). This flexibility is essential if CFA is to be considered as an important route to other techniques. The program was adapted for BASIC (Microsoft Language) from FORTRAN ANACOR software.

4.2.3 Application of CFA to the Analysis of Steroid-Receptor Relationships

One chapter cannot illustrate all the facets and advantages of CFA in the field of QSAR, and the reader may need to refer to published examples for further details. The present illustration is an unpublished analysis of the binding of 187 steroids to 5 steroid hormone receptors (see Appendices): estrogen receptor (ER), progesterone receptor (PR), androgen receptor (AR), mineralocorticoid receptor (MR) and glucocorticoid receptor (GR) present in "cytosol", a "high speed" supernatant, obtained from different target tissues and species (187×5 matrix). The crude data was taken from several publications [56 – 60]. The steroid population was primarily, but not solely, characterized by differences in ring saturation and by the presence or absence of alkyl and hydroxyl groups and of a C-17-α ethynyl substituent (see appendix). There were no bulky substituents or lengthy side chains in strategic positions that to our knowledge might preferentially interact with amino-acids outside the binding site of the natural hormone. For each receptor, the tests were performed under identical experimental conditions. The results are expressed as relative binding affinities (RBAs) with respect to the competition of the natural hormone (estradiol, progesterone, testosterone, aldosterone) or of a biologically active synthetic analog (dexamethasone) with an RBA taken to be equal to 100. The range of RBAs for each receptor was small enough not to warrant logarithmic transformation. Although the vast majority of the results consists of the means of at least three replicate experiments, some are

Table 1. Coded RBAs of steroids **20** to **50** for five receptors. An excerpt from the converted data table describing the relative binding affinities (RBAs) of 187 steroids (see Appendix) for five steroid receptors (ER = estrogen receptor, PR = progesterone receptor, AR = androgen receptor, MR = mineralocorticoid receptor, GR = glucocorticoid receptor). The RBAs for each receptor (R), measured as previously described [56], were coded as follows: R0 = RBA < 1, R1 = 1 – 3, R2 = 3 – 10, R3 = 10 – 15, R4 = 15 – 25, R5 = 25 – 50, R6 = 50 – 75, R7 = 75 – 100, R8 = 100 – 125, R9 = 125 – 150, R10 = 150 – 250, R11 = 250 – 600, thus, yielding a total of 54 categories.

No.	ER	PR	AR	MR	GR
	R 0 1 2 3 4 5 6 7 8	0 1 2 3 4 5 6 7 8 9 10 11	0 1 2 3 4 5 6 7 8 9 10	0 1 2 3 4 5 6 7 8 9 10	0 1 2 3 4 5 6 7 8 9 10
20	------◆--	--◆---------	-◆---------	◆----------	-◆---------
21	--------◆	----◆-------	-◆---------	◆----------	-◆---------
22	---◆-----	◆-----------	◆----------	◆----------	--◆--------
23	---◆-----	◆-----------	◆----------	◆----------	◆----------
24	--◆------	◆-----------	◆----------	◆----------	◆----------
25	------◆--	--◆---------	◆----------	◆----------	--◆--------
26	--------◆	--◆---------	◆----------	◆----------	◆----------
27	◆--------	--------◆---	--◆--------	--◆--------	◆----------
28	◆--------	------◆-----	-◆---------	--◆--------	◆----------
29	◆--------	---------◆--	----◆------	--◆--------	◆----------
30	◆--------	---------◆--	--◆--------	---◆-------	--◆--------
31	◆--------	---------◆--	----◆------	--◆--------	-◆---------
32	◆--------	---------◆--	--◆--------	----◆------	◆----------
33	◆--------	------◆-----	-◆---------	◆----------	◆----------
34	◆--------	--------◆---	------◆----	◆----------	◆----------
35	◆--------	---------◆--	----◆------	◆----------	--◆--------
36	◆--------	---------◆--	----◆------	◆----------	--◆--------
37	◆--------	----------◆-	--◆--------	-◆---------	◆----------
38	◆--------	---------◆--	◆----------	-◆---------	--◆--------
39	◆--------	---------◆--	--◆--------	◆----------	-◆---------
40	◆--------	--------◆---	◆----------	◆----------	-◆---------
41	◆--------	-----------◆	--◆--------	◆----------	--◆--------
42	◆--------	---------◆--	-◆---------	◆----------	---◆-------
43	◆--------	-----------◆	--◆--------	--◆--------	---◆-------
44	◆--------	---◆--------	◆----------	◆----------	--◆--------
45	◆--------	-----------◆	----◆------	--◆--------	-◆---------
46	◆--------	---------◆--	-----◆-----	----◆------	◆----------
47	◆--------	---------◆--	-----◆-----	--◆--------	----◆------
48	◆--------	-----------◆	--------◆--	-------◆---	------◆----
49	◆--------	-----------◆	-◆---------	--◆--------	◆----------
50	◆--------	--------◆---	◆----------	◆----------	◆----------

single determinations with a much higher margin of error. For this reason, we decided not to analyze the true means or experimental values but to distribute the RBAs into 9 to 12 categories according to receptor class as indicated in Table 1 for steroids **20** to **50**. This resulted in a 187×54 matrix.

A crucial feature of this table that justifies analysis by CFA is the lack of specificity of the majority of compounds including natural hormones such as progesterone (**27**). One of the objectives of the pharmaceutical industry has been, and still remains, the design of progestational drugs with minimal androgenic side effects (AR binding). However, the most specific progesterone analog, although often a very useful tool for studying mechanisms of action, is not necessarily the most suitable drug. The necessity for an anti-mineralocorticoid activity component (MR binding), for instance, may be expressed, and the object then is to identify the compounds with the appropriate mix of activities.

In this study, we have limited the biological variables to just the 5 classes of hormone receptor but, in previous studies on diverse populations of steroids or steroid analogs, we have taken into account other or additional columns such as RBA determinations under different incubation conditions [41] or on cytosols of different origins [41, 42, 45], inhibition of kinase activation [46–48], growth responses and even cytotoxicity [45, 47, 48]. Additional columns such as descriptors of chemical structure, the cost of synthesis, and so on, could also be included.

We shall now proceed to illustrate several correspondence factorial analyses of the data set represented by the excerpt in Table 1.

4.2.3.1 Multiple Correspondence Analysis (MCA)

Each cell K_{ij} of the 187×54 matrix as already defined presents the probability (0/1) of association of steroid (i) with response (jx). This is an example of multiple correspondence analysis (MCA), since each test is divided into subcolumns [33, 34, 43, 44]. A major advantage of MCA compared to binary CFA is that MCA does not assume a linear relationship between the two types of variable, in this case molecules and tests. Table 1 can be analyzed directly by MCA (see below) or, more simply, after conversion into a Burt matrix [51, 61].

Analysis of Relationships Among Receptors via a Burt matrix

Prior to factorial analysis, the data set, of which Table 1 is an excerpt, was converted into a Burt matrix describing the frequency of co-occurrence of the various subclasses of receptor interaction (symmetrical 54×54 square matrix). This matrix is illustrated by a chequer board representation in Fig. 1. Each cell $K_{jj'}$ of the matrix – or minisquare of the board – represents the frequency of association of the receptor

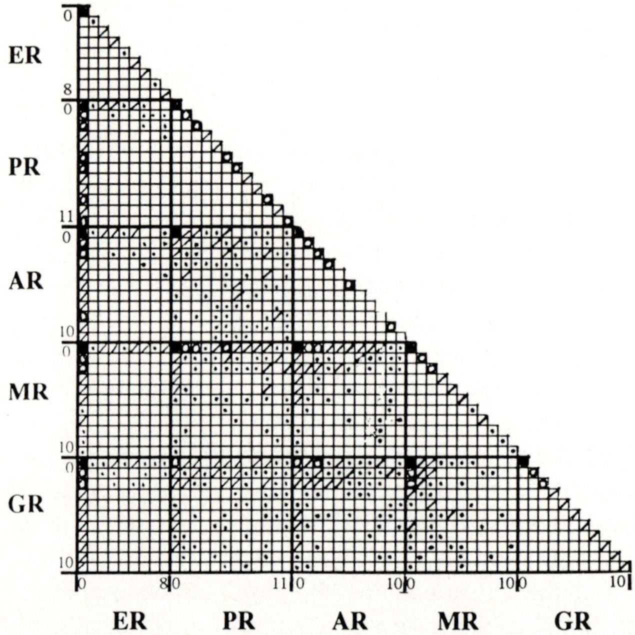

Figure 1. Chequer representation of a 54×54 Burt semi-matrix describing the frequency of occurrence of each combination of binding categories. (Frequency: □ = 0, ⊡ = 1–2, ▨ = 3–9, ◘ 10–49, ■ ≥50).

binding levels jx_n and $j'x_{n'}$ for a molecule. The sum total of the frequencies per large "receptor square" corresponds to the population of test molecules, namely, 187. The frequency of occurrence of each binding category, i.e., the diagonal of this matrix, is shown by the histograms in Fig. 2. The sum of the frequencies along the diagonal per large square is, of course, 187. This type of approach is convenient for a factorial analysis of highly heterogeneous data but is not recommended here since, apart from necessitating lengthy calculations, it only gives a rough picture of the experimental system partly because of high background noise as a result of having to subdivide the test of columns. Indeed, as indicated in the factorial plot of Fig. 3 a, the principal projection axes describe less than 18% of the total variance (10.1% for φ_1 and 6.8% for φ_2).

In Fig. 3 a, we have linked the highest binding levels for each receptor (5 to 11) into polygons and highlighted the zero binding levels by encircled crosses. AR_0 and PR_0 are close, as are MR_0 and GR_0. All four are located within a restricted area distant from ER_0 (near the origin) but close to high ER binding levels. It is, thus, possible to conclude that the binding capability of the steroids is correlated negatively as follows $ER/(PR+AR+GR+MR)$, i.e., that steroids that bind to ER do not bind well to the other receptors and vice versa, or that ER is the most atypical of the five receptors. The high-binding-level polygons establish the isolation of ER (top right-hand quadrant) and reveal some degree of overlap between all the other receptors with the

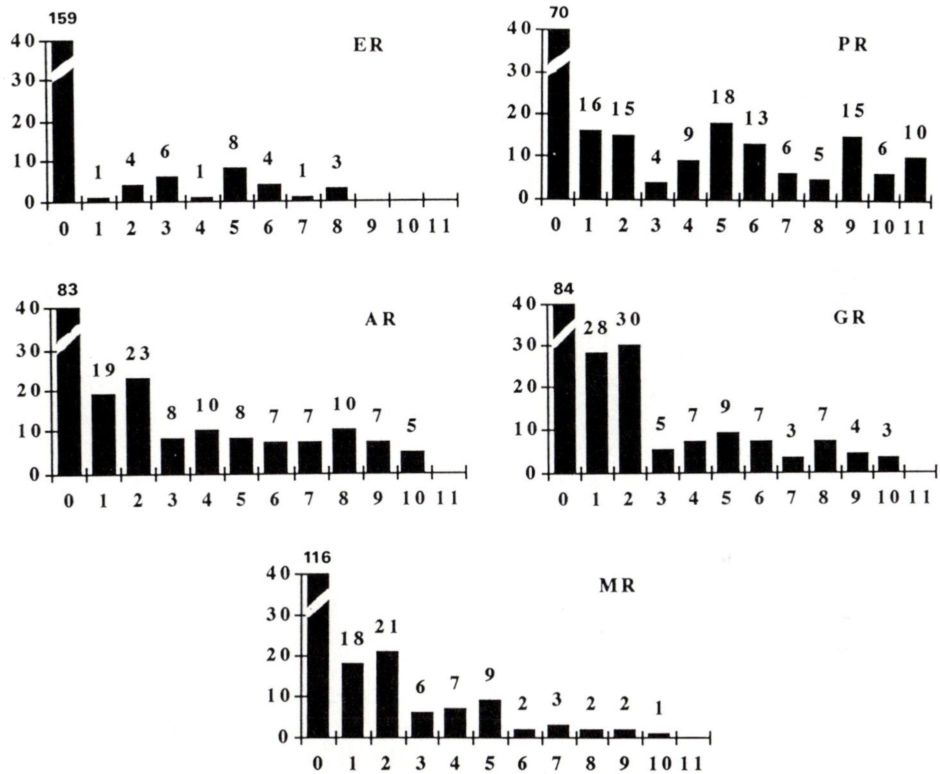

Figure 2. Frequency of occurrence of each binding category for each receptor as given by the diagonal in Fig. 1.

most overlap occurring between AR and PR (bottom quadrants) and between GR and MR (left-hand quadrants). However, some overlap is also evident between GR and PR, even AR, and also between AR and MR.

Analysis of Relationships Among Steroids and Receptors by MCA

MCA of the data set represented by Table 1 (187×54 matrix) yields a $\varphi_1 \varphi_2$ factorial plot (Fig. 3b) that is directly superimposable upon Fig. 3a. For the sake of clarity, only the highest binding level for each class of hormone receptor is shown, whereas all steroids are represented. The distribution of the steroids in relation to the receptor categories is not uniform, but an examination of the individual position of each steroid within this plot would be like trying to find a needle in a haystack. For this reason, and because the plot accounts for only a small proportion of the total

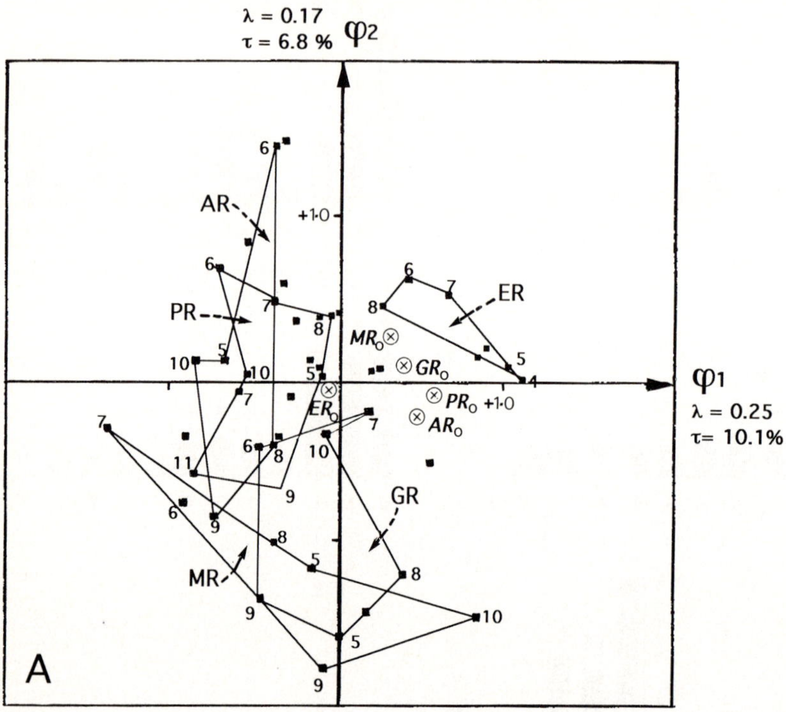

Figure 3a. Correspondence factorial plot depicting the two main axes ($\varphi_1 \varphi_2$ (16.9% of the total variance) obtained by analysis of the Burt matrix illustrated in Fig. 1. (λ = eigenvalue of the factorial axis, τ = percent variance accounted for by the axis). For each receptor, the binding categories ≥ R5 have been joined to form a polygon. The location of the zero binding levels (ER_0, MR_0, GR_0, PR_0 and AR_0) are indicated by ⊗.

variance of the system, we decided to discard this approach for one that would be more fruitful.

4.2.3.2 CFA of Binding Profiles (Probability Scales) to Determine Specificities

The K_{ij} results in Table 1 can be considered as probabilities of each steroid molecule (i) binding to each receptor molecule (j) which are expressed according to a probability scale based on the binding affinity (interaction kinetics) of the natural hormone (= 100). Establishing a profile (pattern) with respect to a norm, that temporarily sets aside the absolute value of the response (amplitude), is a common and practical way of setting out data in a CFA. In this instance, in the absence of the true experimental values in Table 1, the mean RBA of each binding category was attributed to each steroid, as appropriate (187×5 matrix of Table 2). A CFA of the data in Table 2 supplies the following information.

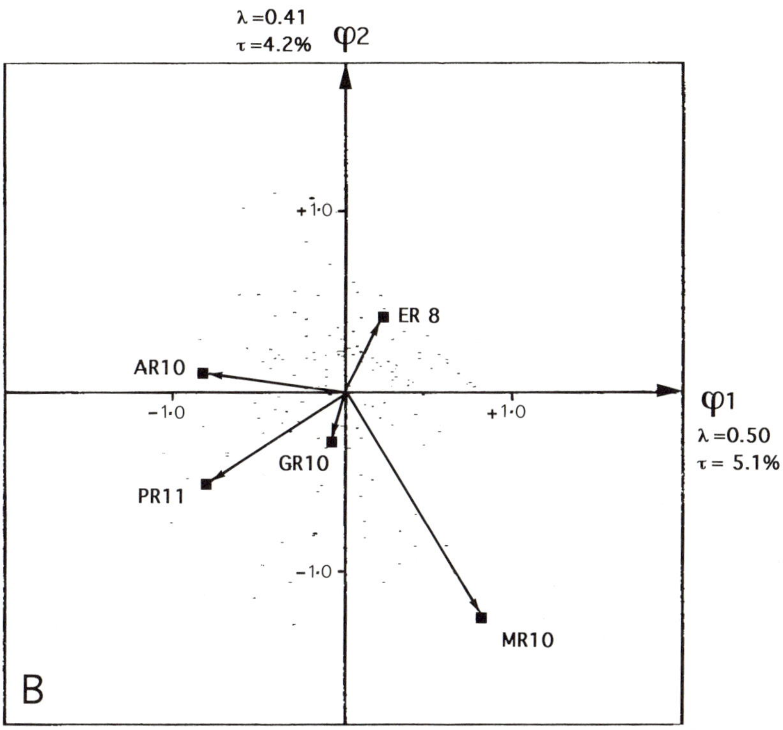

Figure 3b. $\varphi_1\varphi_2$ Factorial plot (9.3% of total variance) obtained by multiple correspondence analysis of the data set represented by the excerpt from Table 1. For each receptor, only the location of the highest binding category is shown (8 for ER, 10 for AR, GR, MR, 11 for PR) (■ = receptor, – = steroid). The plots in (a) and (b) superimposable.

Marginal Weights

The mean marginal weight of each receptor, which denotes the overall level of binding to the different receptors for all steroids, is as follows: ER (5%), PR (47%), AR (23%), GR (16%), MR (9%). This bias toward PR might be explained by the ease of synthesis of certain analogs, or by an instinctive selection of molecules with a progestin component for biochemical study. The bias might reflect the aims of a pharmaceutical research program and could be minimized by increasing the steroid population and/or by the use of techniques of random sampling.

Distances from the Center of Gravity

The distances of the receptors from the center of gravity (centroid) of the multivariate cloud (ER (16.4), MR (3.56), GR (2.56), AR (1.38), PR (0.48)) clearly indicate that ER is the most atypical receptor in its behavior toward this population of molecules and is very different from the other receptors that fall into an irregular

Table 2. Mean RBAs of steroids **20** to **50** according to binding category.

No.	ER	PR	AR	MR	GR
20	62.5	6.5	1.5	0	1.5
21	112.5	20	1.5	0	1.5
22	12.5	0	0	0	6.5
23	112.5	0	0	0	0
24	6.5	0	0	0	0
25	62.5	6.5	0	0	6.5
26	112.5	6.5	0	0	0
27	0	112.5	6.5	6.5	0
28	0	62.5	1.5	6.5	0
29	0	137.5	20	6.5	0
30	0	137.5	6.5	12.5	6.5
31	0	137.5	20	6.5	0
32	0	137.5	6.5	20	0
33	0	62.5	1.5	0	0
34	0	112.5	62.5	0	0
35	0	137.5	20	0	6.5
36	0	137.5	20	0	6.5
37	0	200	6.5	1.5	0
38	0	137.5	0	1.5	6.5
39	0	137.5	6.5	0	1.5
40	0	112.5	0	0	1.5
41	0	400	6.5	0	6.5
42	0	137.5	1.5	0	12.5
43	0	400	6.5	6.5	12.5
44	0	12.5	0	0	6.5
45	0	400	20	6.5	1.5
46	0	137.5	37.5	20	0
47	0	137.5	37.5	6.5	20
48	0	400	112.5	87.5	62.5
49	0	400	1.5	6.5	0
50	0	112.5	0	0	0

Each steroid in Table 1 was allocated a mean RBA (R0 = 0, R1 = 1.5, R2 = 6.5, R3 = 12.5, R4 = 20, R5 = 37.5, R6 = 62.5, R7 = 87.5, R8 = 112.5, R9 = 137.5, R10 = 200, R11 = 400).

step-wise progression (MR → GR → AR → PR) away from the center. PR is situated nearest to the center of the multivariate cloud.

Receptor and Steroid Profiles

As mentioned above, CFA operates a multiple comparison of patterns. Typical receptor binding patterns for two steroids **129** and **132** are illustrated in Fig. 4. The sum

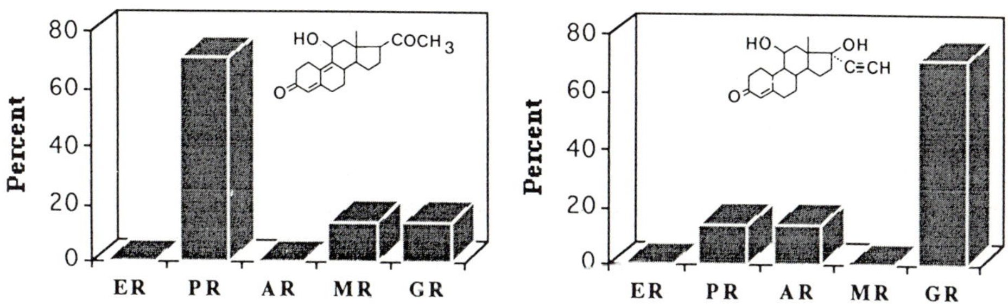

Figure 4. Comparative binding patterns of steroids **129** (shown left) and **132** (shown right).

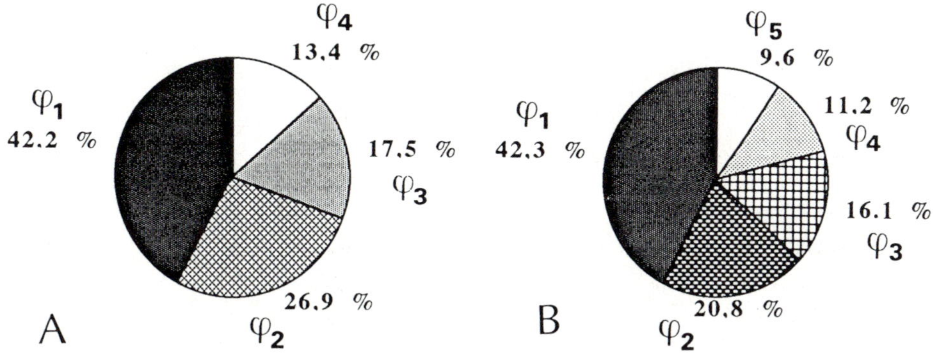

Figure 5. (a) Distribution of the total variance over the four factorial axes required to describe the 187×5 matrix of Table 2 (Sec. 4.3.3.2: Analysis of binding specificity). (b) Distribution of the total variance over the five factorial axes (φ_1 to φ_5) required to describe the 187×10 matrix obtained after splitting the data columns of Table 2 into mean values and anti-values (Sect. 3.3: Analysis of binding specificity and amplitude). Axes $\varphi_6 - \varphi_9$: $<0.001\%$ of the total variance.

of each profile is 100%. Corresponding patterns for the binding of all steroids to each receptor are not shown.

Distribution of Variance over the Factorial Axes

There are 5 biological tests in this study and, therefore, 4 factorial axes since, unlike in the case of PCA, the first latent root is trivial (= 1). The total variance (inertia) of the system is distributed over these axes as shown in Fig. 5a.

Absolute and Relative Contributions

The absolute contribution (AC) is the extent to which a factorial axis is representative of the variance of the system ($\Sigma AC = 100\%$). The relative contribution (RC)

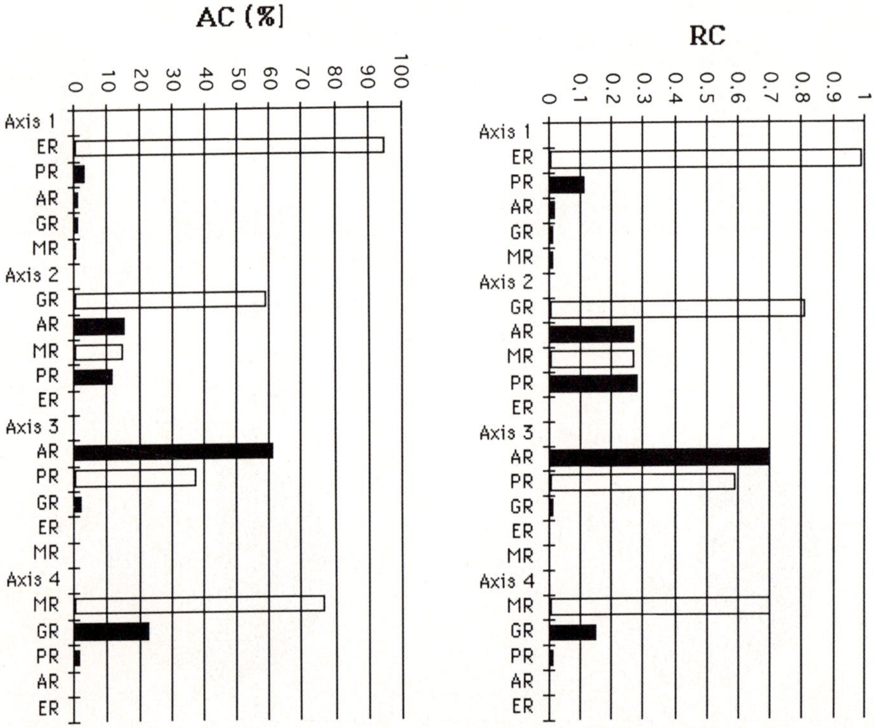

Figure 6. Absolute (AC) and relative contributions (RC) of the receptors to the four factorial axes required to describe the 187×5 matrix of Table 2. Solid histogram bars indicate that the coordinate of the receptor is positive with respect to the axis, hollow bars reflect negative coordinates. (ΣACs = 100% for all variables per factorial axis, ΣRCs = 1 per variable over all factorial axes).

($\cos^2 \theta$) is the dispersion of a variable over all the factorial axes (ΣRCs of each variable to all axes = 1). The ACs and RCs of the receptors to all 4 factorial axes are given in Fig. 6 and reveal the meaning that can be attached to the factorial axes. The φ_1 axis describes the distinctive specificity of ER which is the most striking feature of the steroid population. Once this aspect has been dealt with, φ_2 subsequently indicates that GR and MR can be considered as a group that contrasts to the AR-PR group with the exception of a few molecules that can further distinguish between AR and PR (φ_3 axis) and between MR and GR (φ_4 axis). The molecules at the root of these relationships can be identified from the corresponding AC and RC tables (not shown).

Correspondence Factorial Plots

Figure 7a is the plot of the principal factorial axes ($\varphi_1 \varphi_2$) obtained by CFA of the data set represented by Table 2 and describes a substantial proportion (68.6%) of the

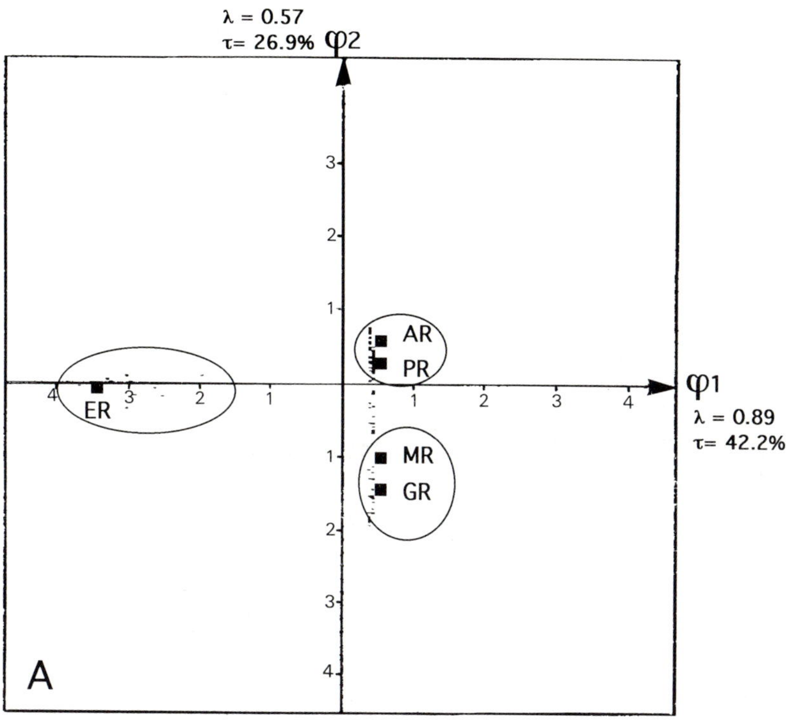

Figure 7 a. $\varphi_1 \varphi_2$ correspondence factorial plot of the analysis of the mean RBAs of the 187×5 matrix of Table 2. (■ = receptors, – = steroids).

information content of the experimental system (42.2% for φ_1; 26.4% for φ_2). Steroids near to the center of a factorial plot are unrelated to the others (e.g. inactive steroids, steroids with a mean profile, or steroids that are described by lower factorial axes). Clustering of variables within the plot signifies correlation, whereas diametrically opposed positions reflect an anti-correlation. Figure 7a clearly illustrates the fundamental features of this system already discussed above, i.e., first, an opposition between ER and the other steroid receptors (negative vs positive coordinates along the φ_1 axis) and, second, close analogies between AR and PR and between GR and MR which are, however, in opposition (above and below the φ_1 axis respectively). The compounds that account for this receptor distribution map are either dispersed around the ER pole in the left-hand quadrants (steroids with a phenolic A-ring) or along a vertical axis in the opposite right-hand quadrants (derivatives of 3-keto-4-enes), thus, reflecting the well-known specificity of these receptors. There is no cross-specificity between these two types of receptor for this population of molecules.

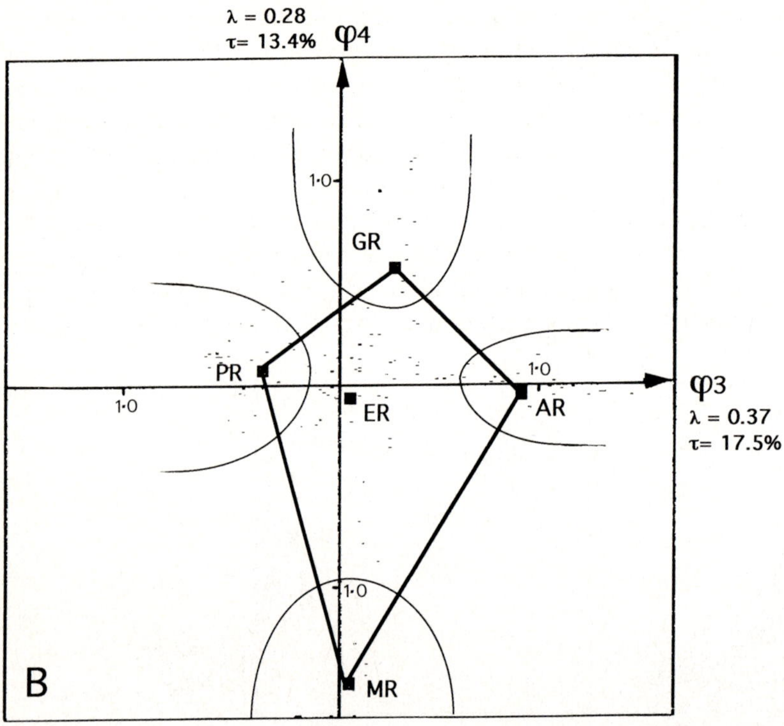

Figure 7 b. $\varphi_3\varphi_4$ correspondence factorial plot of the analysis of the mean RBAs of the 187×5 matrix of Table 2. (■ = receptors, – – = steroids).

The remaining information content (31%) is plotted in the $\varphi_3\varphi_4$ map (Fig. 7 b) which occults ER's participation in the total variance as indicated by its central position near the origin. This plot illustrates the finer distinctive features among the receptors. Whereas AR and PR are broadly similar in their antithesis to ER (Fig. 7 a), there are features that distinguish between them (Fig. 7 b). GR and MR differ to an even greater extent. The steroids at the extremes of the cloud in Fig. 7 b account for most of the observed differences in behavior. Those located between the PR and GR poles infer that interaction with these receptors is achieved by common structural features and that in other words, PR and GR have close similarities as regards to their hormone binding sites. A substantial number of steroids at the center of the cloud are attracted equally by all four receptors, a known property of most 3-keto-4,9,11-trienes [59].

Barycenters of Archetypal Steroids

Figures 7 a and 7 b account for the total variance of the system and are much easier to understand than Figs. 3 a and 3 b that only describe 10–20% of the total variance.

As already mentioned in Sec. 4.3.1.4, matters can be simplified even further by identifying either clusters of steroids with common descriptors in Fig. 7 or subfamilies of compounds within the original data bank. In each case, the profiles of the identified groups are introduced as a single supplementary variable into the analysis (see Sec. 4.3.5.1 below) [49, 62]. To illustrate this principle, we considered the following families of steroids (barycenters): steroids with a phenolic A-ring and 3-keto-4-enes with one of three functional groups at C-17: $-COCH_3$, $-\beta$-OH, or $-COCH_2OH$ (Fig. 8). Each plot in Fig. 8 represents the calculated profile of the family (designated total) and the profiles of selected subfamilies with distinctive substituents or combinations of substituents. These barycenters were introduced as supplementary vari-

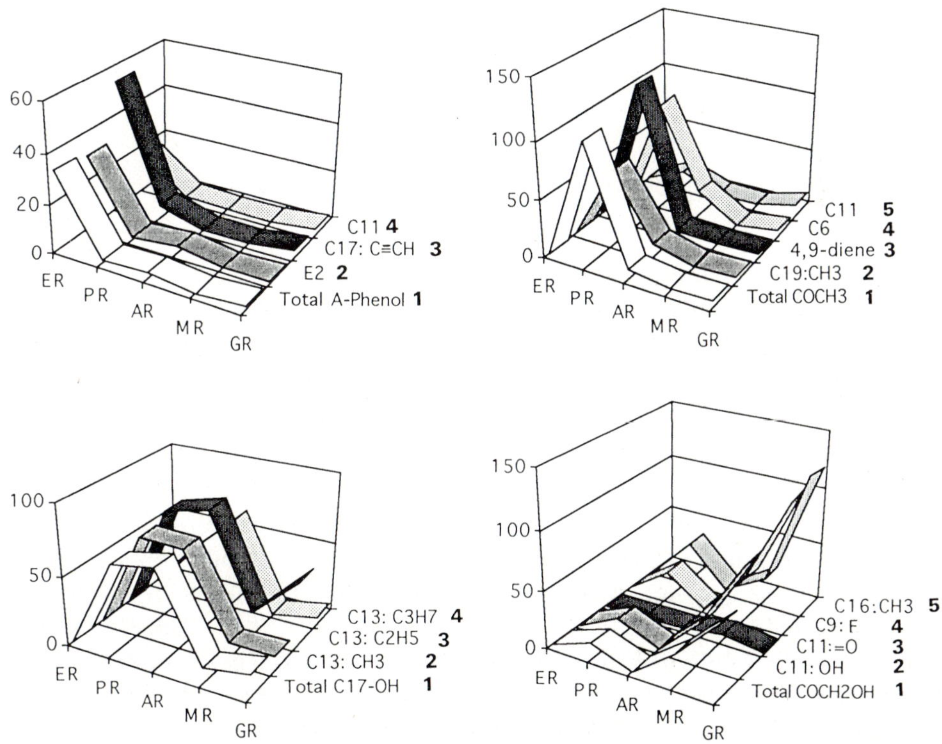

Figure 8. Calculated binding profiles of steroid families and subfamilies with distinctive structural features. Top left: *A-ring phenolic steroids* (**1**), with a C-13-methyl and C-17-β-hydroxy substituent as in estradiol (E_2) (**2**), with a C-17-α-ethynyl substituent (**3**), with a C-11-substituent (**4**). Top right: *Steroids with a C-17-$COCH_3$* substituent (**1**), with a C-19-methyl substituent as in progesterone (**2**), with double bonds at C-4 and C-9 (**3**), with a substituent or double bond at C-6 (**4**), with a substituent at C-11 (**5**). Bottom left: *Steroids with a C-17-β-hydroxy* substituent (**1**) with diverse alkyl groups at C-13 (**2, 3, 4**). Bottom right: *Steroids with a C-17-$COCH_2OH$* substituent (**1**), and a C-11–OH substituent (**2**), or carbonyl at C-11 (**3**), or fluorine at C-9 (**4**), or with a C-16-α-methyl substituent (**5**).

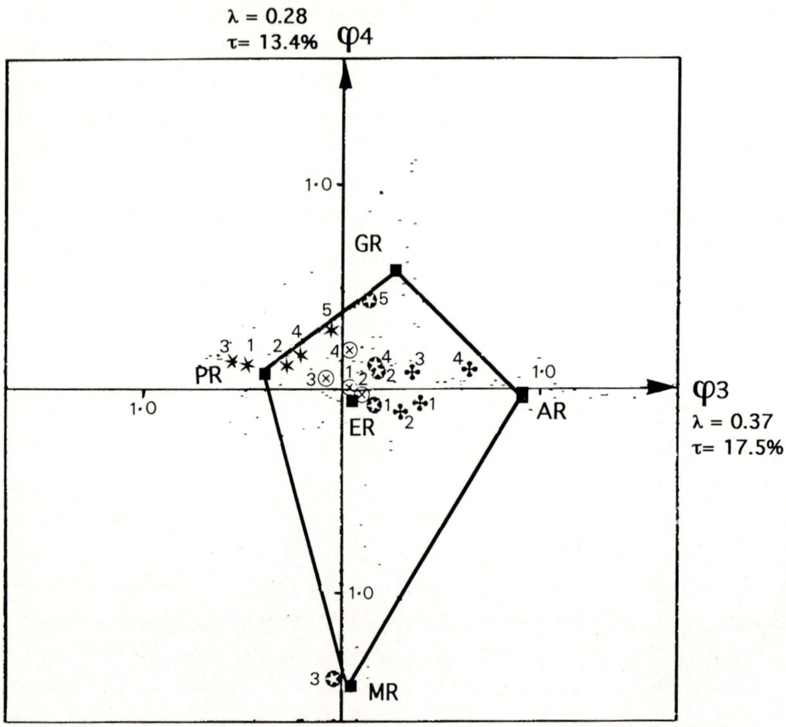

Figure 9. Introduction of the barycenters defined in Fig. 8 into the factorial plot of Fig. 7b. A-ring phenolic steroids (⊗), steroids with a C-17 – $COCH_3$ (★), a C-17 – $COCH_2OH$ (☆), a C-17-*β*-hydroxy (+) substituent. The numbering corresponds to the subfamilies defined in the caption to Fig. 8.

ables into the factorial plots of Figs. 7a and 7b which describe specificity of binding only and disregard amplitude of response (see Sec. 4.3.3.3) (Fig. 9).

The top left-hand plot in Fig. 8 shows that the A-ring phenolic molecules within the population with a C-13 – methyl and C-17-*β* – OH as in estradiol do not have a specificity profile and amplitude of response that are very different from those of the total population of ring A phenols. It also shows that the introduction of an ethynyl group at C-17-*α* enhances ER binding but also introduces a slight PR binding component. On the other hand, a C-11-substituent decreases ER binding under these experimental conditions and introduces a marginal GR binding component. These barycenters are obviously situated in the left-hand quadrants of Fig. 7a in the immediate vicinity of the ER pole (not shown). Their position in relation to Fig. 7b is given in Fig. 9 and reflects the corresponding induction of PR and GR binding by the ethynyl group at C-17 and by the substituent at C-11. Both barycenters move away from the locus of the total A-ring phenolic population toward these poles.

The introduction of a C-11 substituent into the 3-keto-4-ene population with a C-17 – $COCH_3$ diminishes binding to PR whilst introducing a noticeable GR binding component (top right-hand plot of Fig. 8 and quartile of Fig. 9). A C-6 substituent or double bond initiates a trend toward AR. On the other hand, an additional double bond at C-9 increases PR binding and decreases AR binding, leading to the barycenter that is most inherently PR-like. In the C-17 – $COCH_2OH$ series (bottom right-hand plot of Fig. 8 and quartile of Fig. 9), the presence of either a C-11 – OH or of a C-9 – F is favorable toward both MR and GR binding, with a slight preference for GR. Both these barycenters are situated between the GR and MR poles. The introduction of a C-16-α – CH_3 increases the GR binding specificity as shown by the position of this barycenter next to the GR pole. On the other hand, a C-11 ketone is manifestly highly unfavorable to GR binding as illustrated by the barycenter position in a zone which is diametrically opposed to GR. Only a minor MR component characterizes these C-11 keto compounds. In the C-17-β – OH series, we chose to investigate the effects of the alkyl at C-13 (bottom left-hand plot of Fig. 8 and quartile of Fig. 9). A C-13 ethyl group influenced the amplitude of GR binding and a C-13 propyl had an influence on AR binding [40].

It is not the purpose of the present chapter to detail all the structural information that can be gleaned from the creation of such barycenters. The above examples just illustrate how CFA brings to the fore, in pictorial terms, what the biochemists know by experience, intuition, and subjective analysis of data tables [56, 58] but also what they might still ignore. Differences that appear evident at high response values are often missed at lower response levels. Because this type of analysis is far more objective and, at all times, founded on a much wider information base than any analysis undertaken manually by the expert who can only master a limited amount of information at a time, its potential usefulness for studying the nature of hormone binding sites on receptor proteins is obvious [63, 64].

4.2.3.3 Dual CFA (Specificity and Amplitude of Binding)

The CFA described in Sec. 4.3.3.2 is based on a comparison of specificity profiles independently of the binding levels. To account for the amplitude of the binding interaction, each RBA column in Table 2 has to be split into two subcolumns corresponding, on the one hand, to the mean RBA values (analyzed above) and, on the other, to "anti-values" (not shown) obtained by subtracting each mean RBA from the maximum mean RBA value recorded in that column. This procedure gives a 187×10 matrix that can be analyzed as above (for distribution of variance, see Fig. 5) to give factorial plots where each receptor is no longer represented by a point but by a vector (Fig. 10). Yet again, the atypical behavior of ER is apparent as are the close similarities between AR and PR and the slightly less marked kinship between GR and MR. The position of the molecules in this map is, however, not only a function of their specificity but also of the affinity of binding. Other examples of dual CFAs are illustrated elsewhere [33, 36].

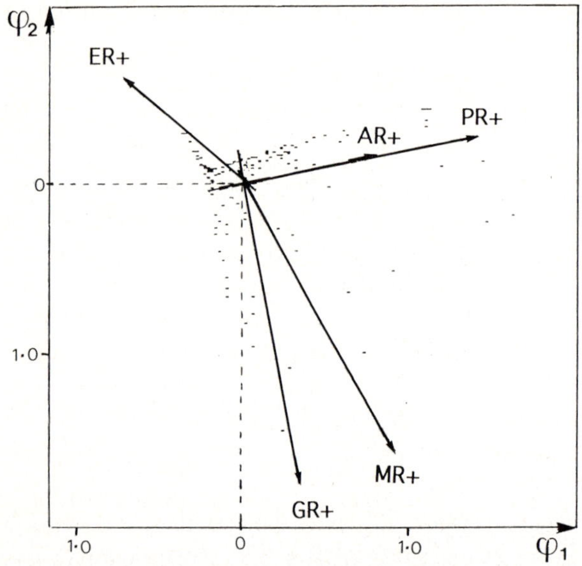

Figure 10. $\varphi_1\varphi_2$ Correspondence factorial plot obtained by analysis of the 187×10 matrix obtained after splitting the columns of Table 2 into mean RBA values and corresponding calculated anti-values. (tip of vector (+) = high RBA, origin of vector = low RBA). (φ_1: $\lambda = 0.20$, $\tau = 42.3\%$; φ_2: $\lambda = 0.10$, $\tau = 20.8\%$).

4.2.4 Post-CFA Analyses: Minimum Spanning Trees and Hierarchical Classifications

As already mentioned in the introduction, CFA is not an end in itself but can be complemented by the use of other algorithms applied to the χ^2-distance square matrix such as, for instance, algorithms for minimum spanning trees, that link variables into a shortest-distance network, or for hierarchical classifications, in which correlated variables (either receptors or steroids) are clustered beneath interconnected nodes of different heights [65–70]. These methods can be applied to describing the total variance of the system [47, 71] or of the variance of selected factorial axes so that, for example, correlation levels derived from the hierarchical classifications [48] or tree-like networks [39, 72] can be transposed onto the factorial plots.

4.2.4.1 Minimum Spanning Trees

A minimum spanning tree is the shortest route within the multidimensional space that links the steroids into a network on the basis of their responses toward the receptors (or the receptors into another network on the basis of their reactivity toward the steroids). The responses can be considered either in terms of specificity alone or of

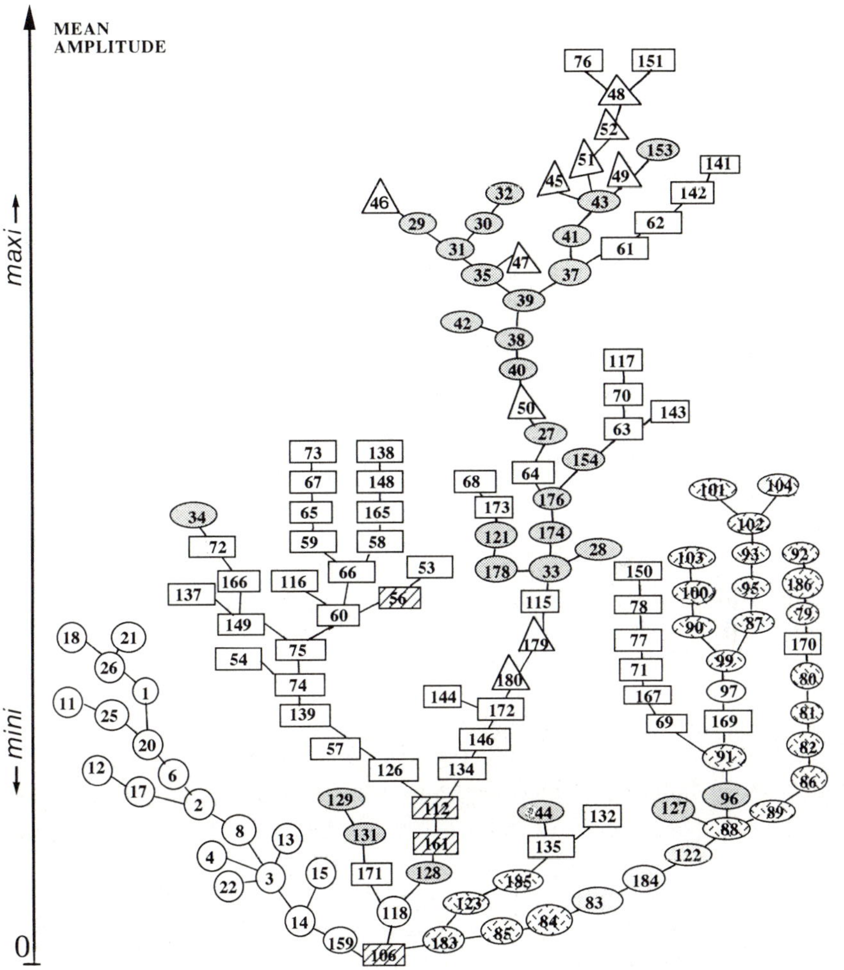

Figure 11. Minimum spanning tree of the 187 steroids drawn on the basis of their binding specificity and mean amplitude as deduced from the mean RBAs of Table 2 and corresponding calculated anti-values (187 × 10 matrix). Only 124 out of 187 steroids are shown since the remaining 63 coincide with illustrated steroids. Due to lack of space, the distances between steroids are not represented to scale. As an indication, the greatest interval is 0.87 (between **37** and **41**), the shortest 6.5×10^{-3} (between **3** and **4**). The majority of the steroids are separated by an interval of approx 5×10^{-2}. The symbols correspond to different chemical affiliations: ○ = A-ring phenolic steroids, □ = 3-keto-17-β-hydroxy steroids, ▩ = steroids with C-17-$COCH_3$ substituent, ⊠ = steroids with C-17-$COCH_2OH$, △ = lactones and oxathiolanes, ▨ = miscellaneous).

specificity and amplitude as shown in Fig. 11 which is an analysis of our 187 × 10 matrix using the algorithm of Prim [73]. (When steroids fall into identical positions, only one steroid is indicated). The distance along a path between two steroids reflects the degree of similarity in their behavior towards the battery of receptors. No loops or backtracking is permitted. The restriction that the route linking the steroids be as short as possible is important when one considers that there are nearly 500 million ways in which just 12 items can be arranged linearly. The minimum spanning tree is, thus, not only a logical way of arranging steroids but a true description of the inherent structure of the experimental system under study. Like factorial plots and hierarchical trees (see below), minimum spanning trees give a clear pictorial representation of the data. The position of any one steroid (e.g., part of a cluster, a milestone or at a branch end) describes its affiliation to the other steroids. The branch pathways describe how the cross-section of receptor binding properties evolves. Additional molecules can be introduced into the experimental system. The new molecule may be located at a branch extremity (enhanced selectivity), between two adjacent nodes (an intermediary), or between two non-adjacent nodes (creation of a new pathway).

In the minimum spanning tree of Fig. 11, the ground level, which is represented by steroid **106** (a progesterone derivative with a 5β configuration), corresponds to absence of affinity or very low affinity, whereas the branch ends correspond to a high mean binding affinity level. There is no common trunk. Instead, the system spreads out immediately into three main branches. The left-hand branch is composed solely of A-ring phenolic steroids that bind specifically to ER. The right-hand branch is mostly made up of molecules with a C-17 $-$ $COCH_2OH$, i.e., corticoids, with however, a few exceptions. Compounds **44, 96** and **127** all belong to the C-17 $-$ $COCH_3$ family but all have either a hydroxy or a methoxy substituent at C-11. Several molecules, which are nearly all an offshoot from the corticoid **91**, are derivatives of 3-keto-17-hydroxy compounds. Of these, **132, 135, 167, 169** are similarly substituted at C-11, **71, 77,** and **78** have a C-13-ethyl group, **150** and **170** have modified A-rings (A-*nor* and 2-*oxo*, respectively) and **167** has an electronically interesting substituent ($CH_2CH = CH_2$) at C-17-α.

The middle branch has a bifurcation at molecule **112**. The left-hand bifurcation is entirely composed of 3-keto-17-hydroxy derivatives except for molecule **34**, a progesterone derivative characterized by a C-6-α $-$methyl. The right-band bifurcation is more varied comprising the hard core of the C-17 $-$ $COCH_3$ derivatives, but also several 3-keto-17-hydroxy compounds, in particular those with a C-17 α-ethynyl (e.g. **61–64, 68, 70, 115, 117, 143, 144, 173**) or a C-7 α-methyl (**76, 141, 142**). The latter have a particularly high mean binding affinity as do the majority of the lactones and oxathiolanes (**45–52**) within the steroid bank.

This is only a broad description of the tree which contains many interesting features that cannot be detailed here, but which can form the basis for fruitful discussion between the medicinal chemist and the biologist.

4.2.4.2 Hierarchical Clustering

An ascending hierarchical tree describing the relationships among receptors (Fig. 12) was constructed by applying an algorithm for agglomerative hierarchical clustering that uses the aggregation criterion of Lance and Williams [74] and standard coefficients of $\alpha = 0.625$ and $\beta = -0.25$ to the χ^2-distance table. The matrix of mean RBA values (187×5 – Table 2) was converted into a matrix of the distances that separate the receptors (j), taken as pairs, when these are projected into the multidimensional space defined by the (i) steroids. These distances were arrayed into a symmetrical $j \times j$ semi-matrix in which the two closest receptors were united into a single group and the dissimilarity of the newly-formed group with each of the other receptors was calculated. The two closest receptors or groups of receptors were again united and the process was iterated $j-1$ times. An equivalent hierarchical steroid tree was also built (not shown).

This method applied to the total variance of the experimental system without filtering out lower axes highlights the characteristic binding specificity of ER. Yet again, MR and GR are seen to be affiliated as are PR and AR, but, as indicated by the lower node height, the AR/PR relationship is closer than the GR/MR relationship for this population of molecules. However, since the stem above the node that groups MR and GR is very short, there is no significant gap between the two pairs of receptors. This classification is yet another illustration of how different classes of steroid hormone receptor are apprehended by the interacting steroids and is an indirect comparison of the stereo-chemistry of ligand binding sites [42]. It has been compared to phylogenetic trees of receptors obtained by analysis of selected amino-acid

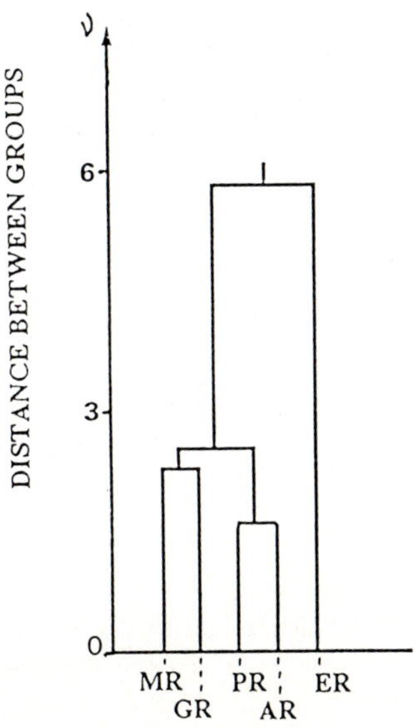

Figure 12. Hierarchical classification of the receptors on the basis of the mean RBA values in Table 2 (187×5 matrix).

sequences presumed to be involved in ligand-dependent regulation of transactivation and in receptor dimerization.

4.2.5 Simulation and Prediction Studies

4.2.5.1 Introduction of Additional Steroids and Tests into a CFA

The factorial plots in Figs. 7 and 10 can be considered as mathematical models. If the results for all receptors are available for a further compound (steroid or non-steroid) or for all steroids in a further test (receptor binding or other), it is possible to evaluate the position of the compound or test in the model by using the transition formula for the calculation of coordinates given in Sec. 4.3.2.2 [75]. As an illustration we have introduced the binding profiles of several C-11 substituted phenolic A-ring steroids (Table 3) [76, 77] into the $\varphi_1\ \varphi_2$ factorial maps of Fig. 7. Missing MR binding values were extrapolated on the basis of GR values in view of the affiliation be-

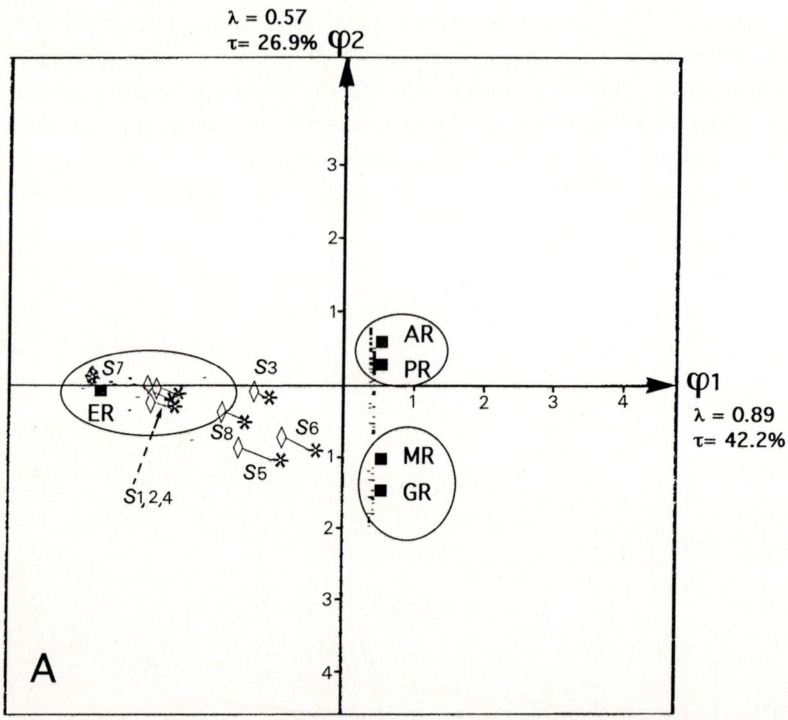

Figure 13a. Use of the $\varphi_1\varphi_2$ CFA plot in Fig. 7a as a mathematical model for introducing the binding profiles of eight steroids substituted at C-11 (Table 3). Missing MR values were extrapolated on the basis of GR (*) or AR (◇) values.

tween GR and MR in Fig. 7a and on the basis of AR values in view of the similar quadrant location of AR and MR in Fig. 7b. These compounds were omitted from the initial analysis, not only because of missing values, but because there is good evidence to believe that the introduction of bulky or long-chain substituents at C-11 of the steroid skeleton leads to interactions with amino-acid residues outside of the binding site of the endogenous hormone. These compounds would, therefore, not be directly relevant to a study of the stereochemistry of the binding site.

The introduction of certain bulky substituents at C-11 can not only dramatically affect the specificity of 3-keto-4-ene steroids [78], but also the specificity of phenolic A-ring steroids [76, 77] (Figs. 13a and b). Irrespective whether missing MR values were extrapolated on the basis of GR or AR, it is clear that for certain substituents (vinyl, thienyl, *m*- or *p*-methoxyphenyl), there is a distinct drift away from the ER pole towards the other receptors (Fig. 13a). In the plot of the lower factorial axes (Fig. 13b), steroid location naturally depends upon whether GR or AR values were used for the extrapolation of MR (above and below the φ_3 axis, respectively).

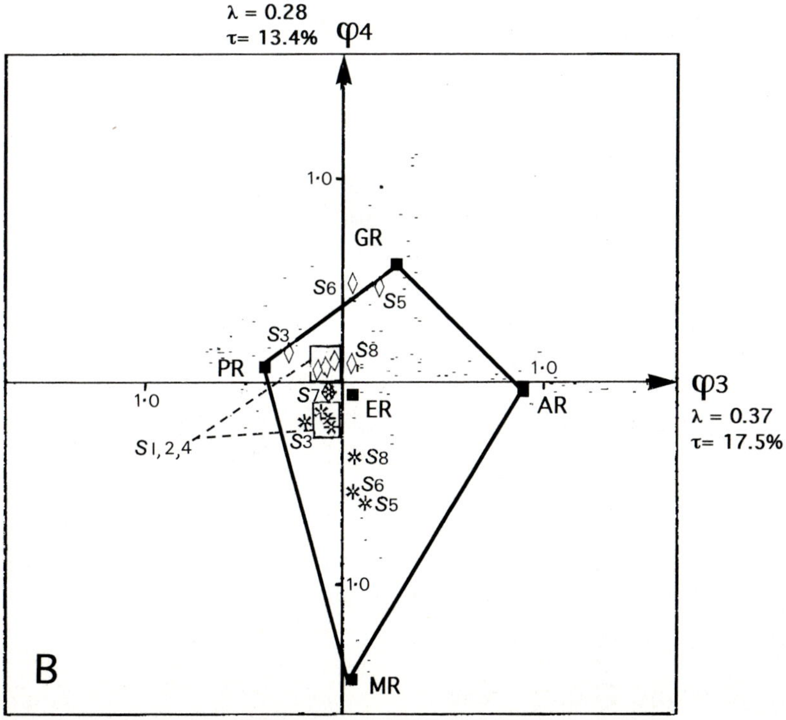

Figure 13b. Use of the $\varphi_3\varphi_4$ CFA plot in Fig. 7b as a mathematical model for introducing the binding profiles of eight steroids substituted at C-11 (Table 3). Missing MR values were extrapolated on the basis of GR (*) or AR (◇) values.

Table 3. Mean RBAs of supplementary steroids (C-11 substituted derivatives of ethynyl estradiol) according to binding category

	C-11	ER	PR	AR	MR	GR
S1	propyl	62.5	6.5	0	n.d.	6.5
S2	*n*-butyl	87.5	6.5	0	n.d.	12.5
S3	vinyl	62.5	62.5	1.5	n.d.	20
S4	allyl	87.5	20	0	n.d.	6.5
S5	*m*-methoxy-phenyl	20	0	0	n.d.	20
S6	*p*-methoxy-phenyl	62.5	37.5	0	n.d.	87.5
S7	benzyl	12.5	0	0	n.d.	0
S8	thienyl	87.5	20	7.5	n.d.	37.5

n.d.: not determined

4.2.5.2 Analyzing the Construction of a System

CFA describes the experimental system in terms of a series of factorial axes of decreasing variance. Projections onto the φ_1 axis, a sort of trunk road, gives the relative positions of the most important variables or by analogy, cities. Stepwise inclusion of additional factorial axes leads to an increasingly accurate, detailed and complex road map. Thus, a minimum spanning tree based on the $\varphi_2\varphi_3$ axes will still yield a rather large scale map, whereas further axes will give rise to an increasing number of ramifications. This progressive complexity induced by stepwise addition of factors is a useful simulation of the construction of an intricate experimental system. Ideally, at this stage, the CFA progam should be completely interactive.

4.2.5.3 Predicted Profiles of Hypothetical Steroids

If the chemical structures of the test compounds are coded according to a chosen system and possible permutations of these codes are on computer lists [79], mathematical modeling and multivariate analysis of the resultant computer-conceived molecules can help predict their relationships with the existing test compounds and thus their probable receptor binding profiles. In this way, the actual synthesis of a major proportion of computer-conceived molecules becomes unnecessary.

4.2.6 Conclusions and Future Trends

In conclusion, correspondence factorial analysis (CFA) of molecular screening data is an invaluable preliminary, but strategic tool that establishes how the screening data are organized and, thereby assists expert opinion on the design of new molecules. Without multivariate analysis, the correlations and governing trends within the initial data matrix may be difficult to grasp because of the sheer volume of the data, and because of the presence of much redundant information.

CFA is essentially an interface between areas of knowledge because it gives condensed pictorial views of large sets of data that can form the common ground for a much-needed dialogue between chemist and biologist, whilst preserving the essence of each discipline. The biologist can classify test-compounds without the aid of a theoretical chemist, design appropriate studies on the mechanism of action to explain unexpected analogies, and introduce further variables such as time [29] into the CFAs used as mathematical models in order to obtain a dynamic view of the experimental system. The chemist can include further variables relating to chemical structure. These may concern 3D-structures (e.g. crystalline coordinates or van der Waals spheres translated into contours, vectors, pixel densities, skeletons...), chemical formulae (compared by fitting or breaking up into fragments), spectral signatures (mass and NMR spectra or transformed UV and IR spectra), quantum variables (isopotential map, charge density, free valency index, π electron density...) etc. It is also possible to include information from other disciplines, e.g., economic factors, because CFA does not correlate items and properties but sets of properties [50].

CFA is also an interface between statistical methods because data reduction by the use of χ^2-metrics enables the subsequent application of other statistical methods. The choice of new molecules to be synthesized can, thus, be further optimized by methods that, unlike CFA, are no longer purely descriptive. The correspondence factorial axes are in fact a new set of variables of particular interest since they are independent. This warrants the legitimate use of methods such as stochastic regression [80] and discriminant analysis applied to one, several, or all of the factorial axes.

Finally, the use of an interactive CFA program is an elegant and relatively simple way of simulating complex situations since its rationale is based upon organizing a system in stepwise fashion.

References

[1] Benzécri, J.P. et al., *L'Analyse des Données. Tome I: La Taxinomie. Tome II: L'Analyse des Correspondances.* (1st edn) Bordas/Dunod, Paris, 1973

[2] Benzécri, J.P., *Histoire et Préhistoire de l'Analyse des Données,* Bordas/Dunod, Paris, 1982

[3] Benzécri, J.P. and Benzécri, F., *Pratique de l'Analyse des Données. I. L'Analyse des Correspondances, Exposé Elémentaire.* Bordas/Dunod, Paris, 1980

[4] Bastin, C., Benzécri, J.P., Bougarit, C., and Cazes, P., *Pratique de l'Analyse des Données. II. Abrégé Théorique – Etudes de Cas Modèle.* Bordas/Dunod, Paris, 1980

[5] Lebart, L., Morineau, A., and Warwick, K.M., *Multivariate Descriptive Statistical Analysis.* Wiley, Chichester, UK, 1984

[6] Jambu, M., *Exploratory and Multivariate Data Analysis,* Academic Press, New York, 1991

[7] Greenacre, M. J., *Theory and Application of Correspondence Analysis,* Academic Press, New York, 1983
[8] Greenacre, M. J., *Correspondence Analysis in Practice*, Academic Press, New York, 1993
[9] van der Heijden, P.G.M., de Falguerolles, A., and de Leeuw, *Appl. Statist.* **38**, 249–292 (1989)
[10] Goodman, L., *Am. Stat. Assn.* **86**, 1085–1138 (1991)
[11] Guillaume, A., *Introduction à la Géologie Quantitative,* Masson, Paris, 1977
[12] Mellinger, M., *J. Geochem. Explor.* **21**, 455–469 (1984)
[13] Avila, F. and Myers, D.E., *Chemom. Intell. Lab. Syst.* **11**, 229–249 (1991)
[14] Devillers, J. and Karcher, W., *Correspondence Factor Analysis as a Tool in Environmental SAR and QSAR Studies.* In: *Practical Applications of Quantitative Structure-Activity Relationships (QSAR) in Environmental Chemistry and Toxicology,* Vol. **I**. (Euro-courses: Chemical and Environmental Science Series), Karcher, W. and Devillers, J., eds., Kluwer Academic Publishers, Dordrecht, 1990, p. 181–195
[15] Miquel, J.F., Planchon, C., Barthélémy, M., Labia, R. and Doré, J.C., *C.R. Acad. Sc. Paris* **296** (Série II), 953–956 (1983)
[16] Manella, C.A., Frank, J., and Delihas, N., *J. Mol. Evol.* **24**, 228–235 (1987)
[17] Doré, J.C. and Jaubert, J.N., *Parfums, Cosmétiques, Arômes.* **61**, 79–85 (1985)
[18] Chrétien, J.R., Righezza, M., Hassani, A., and Meklati, B.Y., *J. Chromatogr* **609**, 261–267 (1992)
[19] Pun, T., Hochstrasser, D.F., Appel, R.D., Funk, M., Villars-Augsburger, V. and Pellegrini, C., *Appl. Theor. Electrophor.* **1**, 3–9 (1988)
[20] Bretaudière, J.P. and Frank, J., *J. Microsc.* **144**, 1–14 (1986)
[21] Avila, F., Myers, D.E., and Palmer, C., *J. Chemom.* **5**, 455–465 (1991)
[22] Sandor, G., Lenoir, P., and Kerbaol, M., *C.R. Acad. Sc. Paris* **272**, 331–334 (1971)
[23] O'Connor, K.P., Hallam, R., Beyts, J., and Hinchlife, R., *J. Psychosom. Res.* **32**, 291–302 (1988)
[24] Crichton, N.J. and Hinde, J.P., *Stat. Med.* **8**, 1351–1352 (1989)
[25] Ciampi, A., Schiffrin, A., Thiffault, J., Quintal, H., Weitzner, G., Poussier, P., and Lalla, D., *J. Clin. Epidemiol.* **43**, 701–715, 1990
[26] Doré, J.C., Lacroix, J., Lacroix, R., and Etienne, T., *J. Pharm. Clin.* **10**, 257–263 (1991)
[27] Greenacre, M., *Stat. Methods Med. Res.* **1**, 97–117 (1992)
[28] Ojasoo, T., Fiet, J., Raynaud, J.P., and Doré, J.C., *J. Steroid Biochem. Mol. Biol.* **46**, 183–193 (1993)
[29] Fiet, J., Doré, J.C., Le Gô, A., Ojasoo, T., and Raynaud, J.P., *Prostate* **23**, 291–313, 1993
[30] Faty, S., da Cunha Belo, M., and Doré, J.C., *C.R. Acad. Sc. Paris* **296** (Série II), 1055–1060 (1983)
[31] Kraiem, J., Doré, J.C., and Vu Quang, K., *C.R. Acad. Sc. Paris* **316** (Série II) 587–594 (1993)
[32] Doré, J.C., Gòrdon, G., and Jaubert, J.N., *C.R. Acad. Sc. Paris* **299** (Série II) 315–320 (1984)
[33] Doré, J.C., Marçot, B., Pillon, D., and Viel, C., *C.R. Acad. Agri. France* **70**, 649–657 (1984)
[34] Doré, J.C., Miquel, J.F., Mrlina, G., and Calmon, J.P., *C.R. Acad. Sc. Paris* **297** (Série II) 125–130 (1983)
[35] Doré, J.C. and Renou, M., *Acta Oecol/Oecol. Applic.* **6**, 269–284 (1985)
[36] Renou, M., Lalanne-Cassou, B., Michelot, D., Gordon, G., and Doré, J.C., *J. Chem. Ecol.* **14**, 1187–1215 (1988)
[37] Labia, R., Morand, A., Verchere-Beaur, C., and Doré, J.C., *J. Antimicrob. Chemother. (Suppl. A)* **11**, 147–152 (1983)
[38] Doré, J.C., Lacroix, J., Lacroix, R., and Viel, C., *Eur. J. Med. Chem.* **22**, 109–117 (1987)
[39] Limasset, B., Le Doucen, C., Doré, J.C., Ojasoo, T., Damon, M., and Crastes de Paulet, A., *Biochem. Pharmacol.* **46**, 1257–1271 (1993)
[40] Doré, J.C., Gilbert, J., Ojasoo, T., and Raynaud, J.P., *J. Med. Chem.* **29**, 54–60 (1986)
[41] Ojasoo, T., Doré, J.C., Gilbert, J., and Raynaud, J.P., *J. Med. Chem.* **31**, 1160–1169 (1988)
[42] Ojasoo, T., Raynaud, J.P., and Doré, J.C., *J. Steroid Biochem. Mol. Biol.* **48**, 31–46 (1994)
[43] Doré, J.C. and Miquel, J.F., *C.R. Acad. Sc. Paris.* **293** (Série II), 1061–1064 (1981)

[44] Doré, J.C., Gilbert, J., Crastes de Paulet, A., Michel, F., and Miquel, F., *C. R. Acad. Sc. Paris* **294** (Série III), 730–734 (1982)
[45] Bignon, E., Pons, M., Crastes de Paulet, A., Doré, J.C., Gilbert, J., Abecassis, J., Miquel, J.F., Ojasoo, T., and Raynaud, J.P., *J. Med. Chem.* **32**, 2092–2103 (1989)
[46] Bignon, E., Pons, M., Doré, J.C., Gilbert, J., Ojasoo, T., Miquel, J.F., Raynaud, J.P., and Crastes de Paulet, A., *Biochem. Pharmacol.* **42**, 1373–1383 (1991)
[47] Doré, J.C., Gilbert, J., Bignon, E., Crastes de Paulet, A., Ojasoo, T., Pons, M., Raynaud, J.P., and Miquel, J.F., *J. Med. Chem.* **35**, 573–583 (1992)
[48] Ojasoo, T., Bignon, E., Crastes de Paulet, A., Doré, J.C., Gilbert, J., Miquel, J.F., Pons, M., and Raynaud, J.P., *J. Steroid Biochem. Mol. Biol.* **44**, 239–250 (1993)
[49] Valla, A., Giraud, M., and Doré, J.C., *Pharmazie* **48**, 295–301 (1993)
[50] Doré, J.C., Viel, C., Lacroix, R., and Lacroix, J., *J. Pharm. Belg.* **45**, 101–110 (1990)
[51] Burt, C., *Br. J. Stat. Psychol.* **3**, 166–185 (1950)
[52] Lebart, L., Morineau, A., and Fénelon, J.P., *Traitement des Données Statistiques. Méthodes et Programmes*, Dunod, Paris, 1979
[53] Foucard, T., *Analyse Factorielle. Programmation sur Micro-ordinateurs.* Masson, Paris, 1982
[54] Dubus, A., *Méthodes et Pratique du Traitement Statistique en Sciences Humaines. Logiciel ADSO,* Atelier des Trois Monts, Lille, 1992
[55] Thioulouse, J., *Comput. Appl. Biosci.* **5**, 287–292 (1989)
[56] Ojasoo, T. and Raynaud, J.P., *Cancer. Res.* **38**, 4186–4198 (1978)
[57] Delettré, J., *Interaction "Hormone-Stéroïde-Protéine Cytoplasmique" au Travers d'une Approche Structurale des Stéroïdes Estrogènes. Progestogènes, Androgènes. Le Stéroïde à "l'Etat Isolé", Corrélations Paramètres Structuraux-Réponse Biologique,* Ph. D. Thesis, University of Paris VI, 1978
[58] Raynaud, J.P., Ojasoo, T., Bouton, M.M., and Philibert, D., *Receptor Binding as a Tool in the Development of New Bioactive Steroids.* In: *Drug Design Vol.* **VIII** Ariëns, E.J., ed., Academic Press, New York, 1979, p. 169–214
[59] Delettré, J., Mornon, J.P., Lepicard, G., Ojasoo, T., and Raynaud, J.P., *J. Steroid Biochem.* **13**, 45–59 (1980)
[60] Ojasoo, T., Delettré, J., Mornon, J.P., Turpin-Vandycke, C., and Raynaud, J.P., *J. Steroid Biochem.* **27**, 255–269 (1987)
[61] Devillers, J., Steiman, R., Seigle-Murandi, F., Sage, L., Benoit-Guyot, J.L., and Doré, J.C., *System Appl. Microbiol.* **14**, 196–204 (1991)
[62] Ojasoo, T., Raynaud, J.P., and Doré, J.C., *Steroids* (submitted for publication)
[63] Ojasoo, T., Mornon, J.P., and Raynaud, J.P., Steroid Hormone Receptors. In *Comprehensive Medicinal Chemistry* Vol. **III**, Emmett, J.C., ed., Pergamon Press, Oxford, 1990, p. 1175–1226
[64] Ojasoo, T., Doré, J.C., Mornon, J.P., and Raynaud, J.P., *Two Approaches to Structure-Activity Relationships in the Field of Sex-Steroids and their Analogs.* In: *Molecular Structure and Biological Activity of Steroids,* Bohl M., and Duax, W.L., eds., CRC Press, Boca Raton, 1992, p. 157–207
[65] Sokal, R.R. and Sneath, P.H.A., *Principles of Numerical Taxonomy,* Freeman, San Francisco, 1963
[66] Gower, J.C. and Ross, G.J.S., *Appl. Statist.* **18**, 54–64 (1969)
[67] Jambu, M., *Classification Automatique pour l'Analyse des Données: I. Méthodes et Algorithmes,* Dunod, Paris, 1978
[68] Roux, M., *Algorithmes de Classification,* Masson, Paris, 1985
[69] Barthélémy, J.P. and Guénoche, A., *Les Arbres et les Représentations des Proximités,* Masson, Paris, 1988
[70] Chatfield, C. and Collins, A.J., *Multidimensional Scaling and Cluster Analysis.* In: *Introduction to Multivariate Analysis*, Chapman and Hall, London, 1989
[71] Devillers, J. and Doré, J.C., *Ecotox. & Environ. Safety* **17**, 227–235 (1989)

[72] Gilbert, J., Doré, J.C., Pons, M., and Ojasoo, T., *QSAR* **13** (1994) in press
[73] Prim, R.C., *Bell Syst. Techn. J.* **36**, 1389–1401 (1957)
[74] Lance, G.N. and Williams, W.T., *Computer J.* **9**, 60–64 (1966)
[75] Pack, P. and Jolliffe, I.T., *Appl. Statist. J. Roy. Statist. Soc. (Series C)* **41**, 365–380 (1992)
[76] Raynaud, J.P., Ojasoo, T., Bouton, M.M., Bignon, E., Pons, M., and Crastes de Paulet, A., Structure-activity Relationships of Steroid Estrogens. In: *Estrogens in the Environment 2,* McLachlan J.A., ed., Elsevier, New York, 1985, p. 24–42
[77] Raynaud, J.P. and Ojasoo, T., *J. Steroid Biochem.* **25**, 811–833 (1986)
[78] Teutsch, G., Ojasoo, T., and Raynaud, J.P., *J. Steroid Biochem.* **31**, 549–565 (1988)
[79] Doré, J.C., Lacroix, J., Lacroix, R., and Viel, C., *J. Pharm. Belg.* **45**, 375–384 (1990)
[80] Devillers, J., Zakarya, D., Chastrette, M., and Doré, J.C., *Biomed. Envir. Sci.* **2**, 385–393 (1989)

Appendix

A1. Estradiol and Related Derivatives

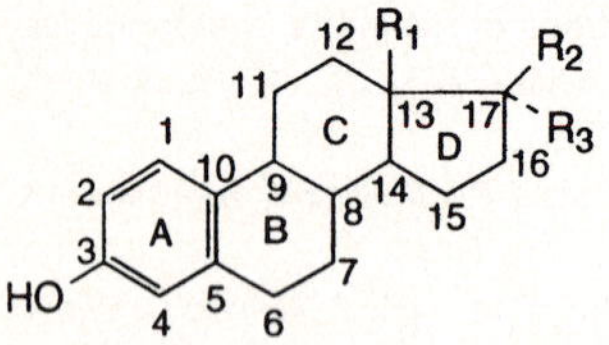

No.	R_1	R_2	R_3	Other
1	CH_3	OH		
2	CH_3	=O		
3	CH_3		OH	
4		OH(17a)		D-homo
5		OH		D-homo
6	CH_3	CH_2OH		
7	CH_3		CH_2OH	
8	CH_3	OH		C-16: β-OH
9	CH_3	OH		C-16: α-OH
10	CH_3		OH	C-16: α-OH
11	C_2H_5	OH		
12	C_3H_7	OH		
13	C_4H_9	OH		
14	CH_3	OH		C-11: β-OH
15	CH_3	OH		C-11: β-OCH_3

No.	R_1	R_2	R_3	Other
16	CH_3	OH		C-11 : β-OC_2H_5
17	CH_3	OH		C-2 : CH_3
18	CH_3	OH		C-7 : α-CH_3
19	CH_3	OH		C-9 : α-CH_3
20	CH_3	OH	CH_3	
21	CH_3	OH	C≡CH	
22	CH_3	OH	C≡CH	C-11 : β-OCH_3
23	CH_3	OH	C≡CH	C-11 : α-OCH_3
24	CH_3	OH	C≡CH	C-11 : β-OC_2H_5
25	CH_3	OH	C≡CH	C-11 : β-C_3H_7
26	CH_3	OH	C≡CH	C-12 : β-CH_3
118	CH_3			C-16 : β-C≡CH, α-OH
136	CH_3	OH		C-11 : β-OCH_3, C16 : α-OH
155	CH_3	OH		C-3 : methoxylated
156	CH_3	OCH_3		
157	CH_3	OCH_3		C-3 : methoxylated
158	CH_3	OH		C-9 : β(iso)
159	CH_3			

A2. Progesterone and Related Derivatives

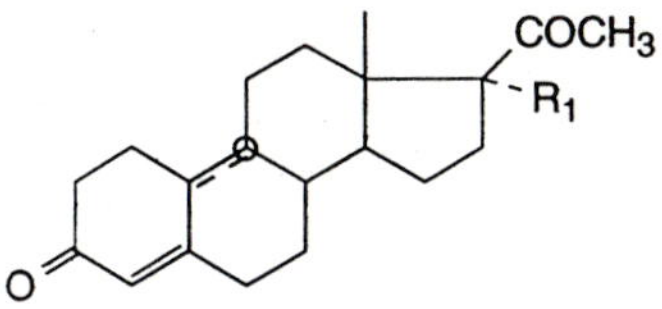

No.	Δ	C-19	R_1	Other
27		CH_3		
28	9 (11)	CH_3		
29	11	CH_3		
30	9			
31	9, 11			
32				
33		CH_3		C-16 : α-CH_3
34		CH_3	$OCOCH_3$	C-6 : α-CH_3
35	6	CH_3	$OCOCH_3$	C-6 : α-CH_3
36	6	CH_3	$OCOCH_3$	C-6 : Cl
37			CH_3	
38	9		CH_3	

No.	Δ	C-19	R_1	Other
39	9, 11		CH_3	
40	9		C_2H_5	
41	9		nC_3H_7	
42	9		CH_3	C-21 : CH_3
43	9		CH_3	C-13 : C_2H_5
44	9			C-11 : β-OCH_3
96	1	CH_3	OH	C-9 : F, C-11 : OH, C-16 : α-CH_3
119	9			C-2 : *gem*-$(CH_3)_2$
120		CH_3		C-16 : *gem*-$(CH_3)_2$
121		CH_3		C-6, 16 : α,α-$(CH_3)_2$
127		CH_3		C-11 : β-OH
128		CH_3		C-11 : α-OH
129	9			C-11 : β-OH
130/117		CH_3	OH	
131	9		OH	
153				C-16 : α-C_2H_5, C-21 : OH
154	6	CH_3	OH	C-2 : *gem*-$(CH_3)_2$, C-6 : Cl
174		CH_3	α-C-17 : acetate	
175		CH_3		C-8 : β-CH_3
176				β-C-13 : propyl
178	6	9β, 10α		

A 3. Lactones/oxathiolanes

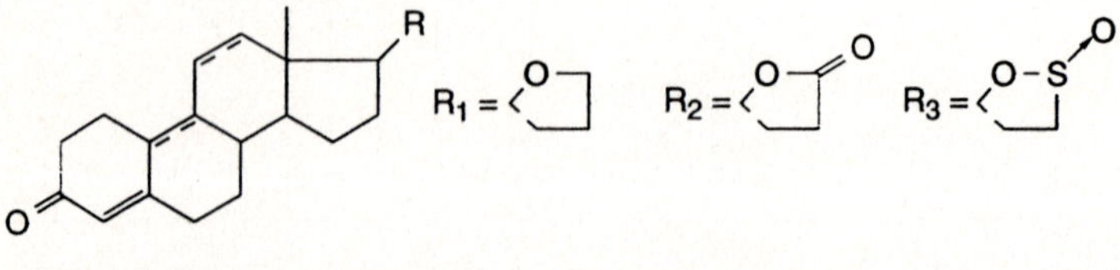

No.	Δ	C-19	R
45			R_1
46			R_2
47	9		R_2
48	9, 11		R_2
49			R_3
50	6		R_3
51	9		R_3
52	9, 11		R_3
179		CH_3	R_1
180		CH_3	R_2

A 4. Testosterone and Related Derivatives

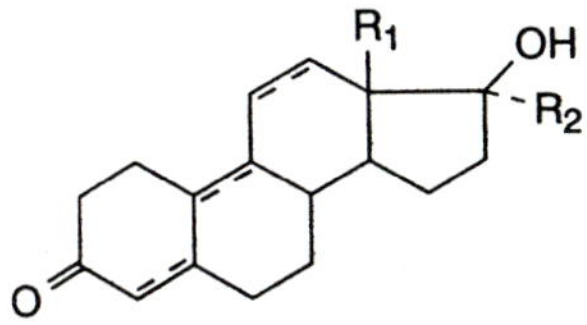

No.	Δ	R_1	R_2	Other
53	4			C-19 : CH_3
54	1, 4			C-19 : CH_3
57	5 (10)	CH_3		
58	4	CH_3		
59	4	C_2H_5		
60	4	C_3H_7		
61	4	CH_3	C≡CH	
62	4	C_2H_5	C≡CH	
63	4	C_3H_7	C≡CH	
64	4	CH_3	C≡CH	C-11 : β-OCH_3
65	4	CH_3	CH_3	
66	4, 9	CH_3		
67	4, 9, 11	CH_3		
68	4, 9	CH_3	C≡CH	
69	4, 9	CH_3	C≡CH	C-11 : β-C_6H_5
70	4, 9, 11	CH_3	C≡CH	
71	4, 9, 11	C_2H_5	C≡CH	
72/140	4, 9	CH_3	CH_3	
73	4, 9, 11	CH_3	CH_3	
74	4, 9, 11	CH_3	CH_3	C-2 : β-CH_3
75/149	4, 9, 11	CH_3	CH_3	C-4 : CH_3
76	4, 9, 11	CH_3	CH_3	C-7 : α-CH_3
77	4, 9, 11	C_2H_5	CH_3	
78	4, 9, 11	C_2H_5	CH_3	C-7 : α-CH_3
115	4		C≡CH	C-19 : CH_3
116	4, 9, 11	C_3H_7		
117	4, 9, 11	C_3H_7	C≡CH	
124	4	CH_3		C-2 : α-CH_3
125	4, 9, 11	CH_3	CH_3	C-2 : β-CH_3
126	4, 9, 11	CH_3	CH_3	C-2 : *gem*-$(CH_3)_2$
132	4	CH_3	C≡CH	C-11 : β-OH
133	4, 9	CH_3	C≡CH	C-11 : β-OH
134	4	C_2H_5	C≡CH	C-11 : β-OH
135	4, 9	CH_3	C≡CH	C-11 : β-OCH_3
137	4, 9	C_2H_5		
138	4, 9, 11	C_2H_5		
139	4, 9	C_3H_7		
141	4	CH_3	CH_3	C-7 : α-CH_3
142	4, 9	CH_3	CH_3	C-7 : α-CH_3
143	4, 9	C_2H_5	C≡CH	
144	4, 9	C_3H_7	C≡CH	

No.	Δ	R_1	R_2	Other
145	4, 9, 11	CH_3	CH_3	C-3 : deoxo
146	4, 9, 11	CH_3		C-17 : no OH
148	4, 9, 11	CH_3	CH_3	C-6 : *gem*-$(CH_3)_2$
150	4, 9, 11	CH_3	CH_3	A-*nor*
151	4, 9, 11	CH_3	CH_3	2-*oxo*
152	5 (10)	CH_3	$C{\equiv}CH$	
164	4, 9	CH_3	$C{\equiv}CH$	C-11 : keto
165	4, 9, 11	CH_3	CH_3	C-4 : Cl
166	4, 9, 11	CH_3	$(CH_2)_2CH_3$	
167	4, 9, 11	CH_3	$CH_2CH{=}CH_2$	
168	4	CH_3	$C{\equiv}CH$	C-11 : β-OH; C-19 : CH_3
169	4, 9	C_2H_5	$C{\equiv}CH$	C-11 : β-OCH_3
170	4, 9, 11	CH_3	$C{\equiv}CH$	2-*oxo*
171	4	CH_3	CH_3	C-17 : acetate; C-19 : CH_3
172	4, 9, 11	CH_3		C-17 : acetate
173	4, 9, 11	C_2H_5	$C{\equiv}CH$	C-17 : acetate

A5. Corticoids

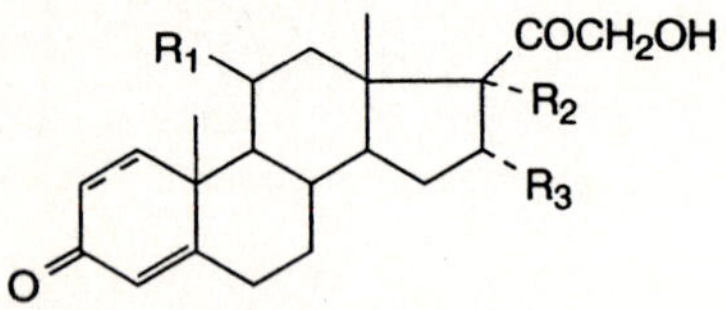

No.	Δ	C-9	R_1	R_2	R_3	Other
79			OH			C-18 : CHO
80						
81					C-21 : acetate	
82	6					
83						C-2 : CH_3
84				OH		
85			=O	OH		
86			OH			
87	1		OH			
88			OH	OH		
89	1		OH	OH		
90	1		OH		CH_3	
91	1		OH	OH	CH_3	
92		F	OH	OH		
93		F	OH	OH	CH_3	
94		F	OH	OH	CH_3	C-21 : acetate
95	1	F	OH	OH	CH_3	
97	1	F	OH	OH		
98	1	F	OH	OH	OH	

No.	Δ	C-9	R_1	R_2	R_3	Other
99		F	OH	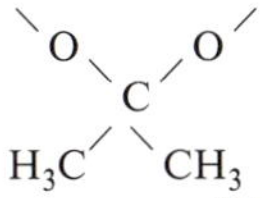		
100		F	OH	*	CH_3	
101	1	F	OH		CH_3	
102		F	OH		CH_3	
103		F	OH		CH_3	C-6 : F
104	1	Cl	Cl		CH_3	
122						C-2 : *β*-CH_3
123						C-2 : *gem*-$(CH_3)_2$
181	1		=O	OH		
182			=O			C-16 : *α*-CH_3
183	1		=O			C-16 : *α*-CH_3
184			OH	OH		C21 : acetate
185			OCH_3	OH		
186		F	OH	OH		C-21 : acetate
187		F	OH	*		

* Function at C-17: $-COCH_2O-CH-CH=CH_2$
O

A 6. Miscellaneous

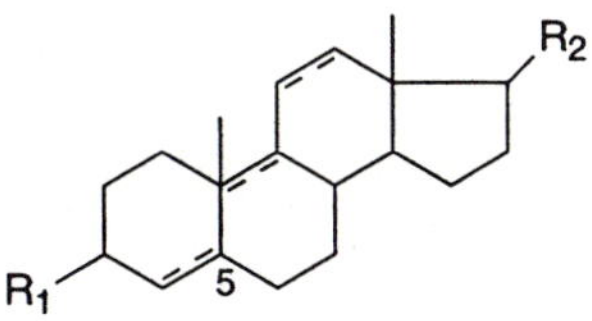

No.	R_1	Δ	C-5	R_2	Other
55	=O		*α*-5	*β*-OH	
56	=O		*α*-5	*β*-OH, *α*-CH_3	
105	=O		*α*-5	$COCH_3$	
106	=O		*β*-5	$COCH_3$	
107	*β*-OH		*α*-5	$COCH_3$	
108	*β*-OH		*β*-5	$COCH_3$	
109	*α*-OH		*α*-5	$COCH_3$	
110	*α*-OH		*β*-5	$COCH_3$	
111	=O		*β*-5	$COCH_2OH$	
112	=O		*β*-5	OH	
113	*β*-OH		*α*-5	OH	
114	*α*-OH		*α*-5	OH	
160	=O		*α*-5	=O	
161	=O	4		=O	
162/147	=O	4, 9, 11		=O	
163	=O	4, 9, 11		=O	C-13 = C_2H_5

4.3 Analysis Of Embedded Data: *k*-Nearest Neighbor and Single Class Discrimination

Valerie S. Rose, John Wood and Halliday J.H. MacFie

Abbreviations

*k*NN	*k*-Nearest Neighbor
SIMCA	Soft Independent Modeling of Class Analogy
CSA	Cluster Significance Analysis
PCA	Principal Component Analysis
CVA	Canonical Variate Analysis
SCD	Single Class Discrimination
GSCD	Generalized Single Class Discrimination
LLM	Linear Learning Machine

Symbols (equations)

$$d_{ij} = \sqrt{\sum_{h=1}^{m} (x_{ih} - x_{jh})^2} \tag{1}$$

$$d_{ij} = \sqrt{\sum_{h=1}^{m} f_{\mathrm{h}} (x_{ih} - x_{jh})^2} \tag{2}$$

$$w_i = 0.05 + 0.90 \left(\frac{c_i - min}{max - min} \right) \qquad \text{for } i = 1 \text{ to } n \tag{3}$$

$$v_i = 1 - w_i \qquad \text{for } i = 1 \text{ to } n \tag{4}$$

$$wm_{\mathrm{j}} = \frac{\sum_{\mathrm{i}=1}^{n} (x_{\mathrm{ij}}\, w_{\mathrm{i}})}{sumw} \qquad \text{for } j = 1 \text{ to } p \tag{5}$$

$$ws_j = \sqrt{\frac{\sum_{i=1}^{n} (x_{ij} - wm_j)^2\, w_{\mathrm{i}}}{sumw}} \qquad \text{for } j = 1 \text{ to } p \tag{6}$$

$$(XI^{\mathrm{T}}XI - \lambda_j^2 XA^{\mathrm{T}}XA)g_j = 0 \qquad \text{for } j = 1 \text{ to } p \tag{7}$$

$$(XV^{\mathrm{T}}XV - \lambda_j^2 XW^{\mathrm{T}}XW)g_j = 0 \qquad \text{for } j = 1 \text{ to } p \tag{8}$$

4.3.1 Embedded Data

Biological activity data is described as being "embedded" (or "asymmetric") if a set of active compounds occurs as a cluster in a diffuse cloud of inactive compounds, when the compounds are plotted in physico-chemical property space. Active compounds (actives) are, therefore, similar to each other with respect to certain properties, e.g. log *P*, molecular weight and p*K*a, while inactive compounds (inactives) lack this characteristic pattern of similarity for a variety of reasons. There is, thus, an optimum range of values for a subset of properties which promotes activity, and any deviation from this range results in inactivity. Identification of this informative subset can be problematic due to the presence of properties which are not relevant for determining activity and which, consequently, contribute noise to the system. Embedded data has been described by some authors, for example, Magee [1], McFarland and Gans [2, 3] and Dunn and Wold [4].

The main differences between "embedded" and "non-embedded" activity data are summarized in Fig. 1.

Fig. 1 clearly shows that the statistical methods developed for the analysis of linearly separable classes of activity, such as linear discriminant analysis, are generaly, inappropriate for embedded data as they depend on there being a difference in the mean values of the active and inactive class. In the case of embedded data, the

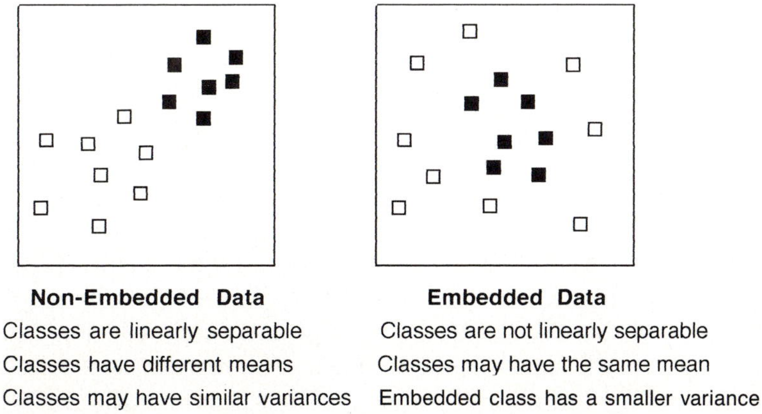

Figure 1. Diagram showing the main differences between non-embedded and embedded activity data. Filled squares denote active compounds and open squares denote inactive compounds.

active and inactive class may have the same mean value. In some situations, inclusion of transformations of the original property variables (e.g. squared variables) in the property matrix has enabled methods developed for linear discrimination of activity classes to be employed successfully, but at the expense of increasing the dimensionally of the descriptor array.

Multivariate embedded data have been successfully analyzed in QSAR using SIMCA (Soft Independent Modeling of Class Analogy) [5], Cluster Significance Analysis (CSA) [2, 3], k-Nearest Neighbor (kNN) [6] and Single Class Discrimination (SCD) [7, 8]. These methods are all suitable for the analysis of classified biological activity data; i.e. where the compounds are qualitatively labeled "active" or "inactive". SCD is also applicable to quantitatively defined activities. The methods of kNN and SCD are described below. SIMCA and CSA are discussed elsewhere in volume 2 of this series.

4.3.2 *k*-Nearest Neighbor Analysis

4.3.2.1 Methodology

k-Nearest Neighbor (*k*NN) [6] is a method for classifying unknown samples (test samples) based on their proximity to samples of known class (the training set) without actually fitting a model. This method is suitable for resolving classification problems associated with embedded and non-embedded activity data, and is quite simple when considered on a computational and conceptual basis. The distance of all samples in the training set from a test sample are determined in multivariate space. Generally, an Euclidean distance measure is used and the Euclidean distance, d, between two samples, i and j, is calculated as:

$$d_{ij} = \sqrt{\sum_{h=1}^{m} (x_{ih} - x_{jh})^2} \qquad (1)$$

where m is the number of properties and x_{ih} is the hth property value of sample, i. The class membership of k-nearest neighbors of the test sample is ascertained. The value of k is user-determined, e.g. such as 5. A simple "majority vote" on the classes of the k-nearest neighbors can then be used to predict the class of the test sample and this concept is illustrated in Fig. 2. The active compounds are embedded within the inactives and two unknowns are included (X and Y). It is apparent that X will always be predicted as inactive for values of k from 1 to 5, whereas the predicted class of Y depends on the value of k. Y is predicted as active for $k = 1$ or $k = 5$, but as inactive for $k = 3$.

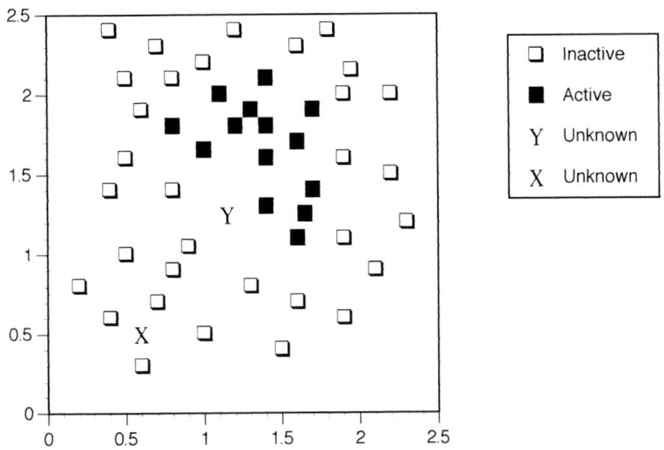

Figure 2. Plot showing the classification of 2 "unknown" samples by *k*NN in a two-dimensional property space.

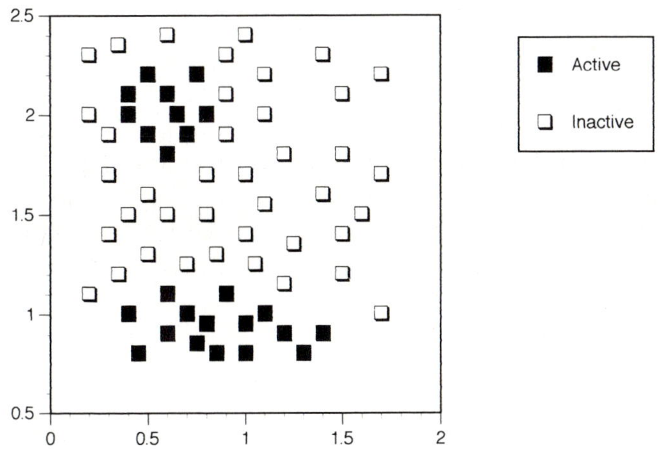

Figure 3. Plot showing 2 active cluster locations in a two-dimensional property space.

A particular advantage of *k*NN is that it is also appropriate for multiclass problems, or for classes with clusters situated at more than one location in property space as shown in Fig. 3. This situation may arise in QSAR when two different modes of action are occurring in the compound set, each requiring a different optimum property profile.

Further aspects of this method can be divided into two categories: the selection of *k*, and the scaling of the descriptors. These topics are discussed in the following sections.

4.3.2.2 Selection of *k*

A suitable value of *k* can be selected based on a "leave-one-out" principle. In order to perform this selection, a number of different *k* values are first selected, e.g. 1, 3, 5, 7, 9. Each member of the training set is used as a "test" compound and its *k*-nearest neighbors are used to predict its class for all values of *k*. The value of *k*, which best predicts the class of all members of the training set in this manner, is selected as the optimum value to be used for true unknowns.

*k*NN, in its simplest form may not perform well when a different number of samples exists in each class, particularly when one class occupies the property space with a higher density than the other class. The situation is further exacerbated with overlapping classes as depicted in Fig. 4. Only one of the actives would be correctly classified for $k = 1$ or 3, as the active class occupies the property space less densely than the inactive class.

Coomans and Massart [10] have proposed a modification to the "majority vote" classification procedure for such circumstances, termed the "alternative vote" method. This cpu-intensive approach requires the testing of many possibilities before the best one can be selected. They defined a decision rule which states: "For 2 classes, a test object is only classified as class 1 if, for its *k*-nearest neighbors, at least α of them are in class 1, otherwise the object belongs to class 2." The user must select and test several values of *k* and α. The optimum combination can be chosen using a leave-one-out approach, as described previously. This approach can be extended to choosing different (k, α) pairs for each class, and also allows prior probabilities of class membership to be incorporated.

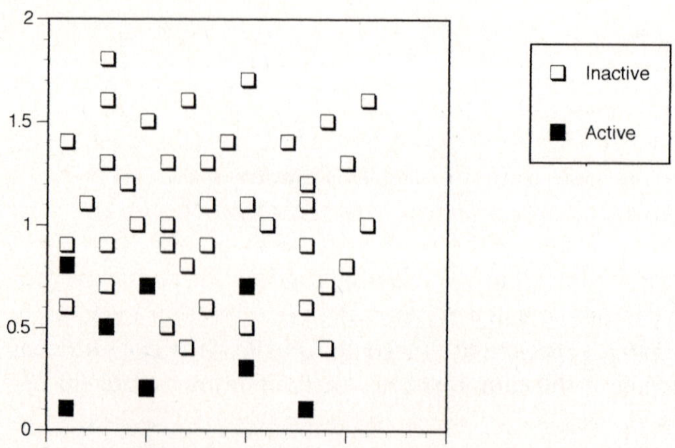

Figure 4. Plot showing the problem of classification for overlapping classes with different densities of points in a two-dimensional property space.

4.3.2.3 Scaling and Weighting

As kNN works on the premise that compounds which are similar in property space will possess similar activity, then its success is highly dependent on the set of properties chosen to portray similarity. Inclusion of properties which are not relevant to activity will introduce noise. Generally, important properties are not known prior to analysis, which can lead to property matrices containing considerable noise, resulting in a poor classification by kNN.

Once a particular set of properties has been chosen, the problem of whether or not to scale or weight the property data must be considered. Physico-chemical property data are usually on different scales, and, thus, a preliminary autoscaling of the properties is usually required.

Various ways of weighting the individual properties have been proposed with the aim of improving classification. Properties showing a difference in mean and/or variance between the classes are weighted to be more influential in the model. Those showing minimal difference are down-weighted or omitted. Weighting factors, f, can be included in the calculation of intersample distance as shown below:

$$d_{ij} = \sqrt{\sum_{h=1}^{m} f_h (x_{ih} - x_{jh})^2} \quad (2)$$

Such weighting factors can be calculated in a variety of ways, e.g. Fisher weights are a ratio of "between groups" sums of squares to the "within groups" sums of squares. Forbes et al. [11] describe a variation of Fisher weights which is suitable for multiclass data and specifically designed for the analysis of mass spectral data.

4.3.2.4 QSAR Examples of kNN

An early example of the use of kNN in chemistry is given by Kowalski and Bender [6]. Much development of the kNN method has taken place in analytical chemistry, but some examples of applications are available in the QSAR literature [12–18]. In 3 of these examples, discussed below, kNN has been compared with discriminant classification methods.

Sjöström and Kowalski [16] analyzed 6 assorted chemical and biological data sets, comparing kNN with SIMCA, Linear Discriminant Analysis (LDA), the Linear Learning Machine (LLM) and the Bayesian Classification Rule. Although none of their examples was specifically related to QSAR, the authors provide a comprehensive and comparative study of these methods in related fields. Overall, kNN performed as well as the other methods. Overlapping classes were postulated by the authors for some sets when each analysis method resulted in a different misclassification. These observations show the value of using more than one classification method when analyzing a data set.

Henry and Block [17] classified a set of 51 compounds into 5 therapeutic classes (analgesics, antidepressives, antihistamines, anticholinergics and anti-Parkinson drugs). They used 4 connectivity indices and the molar refractivity of substituents at 8 different positions on the overlaid structures, giving 40 properties in total. Again, *k*NN was found to perform as well as LDA and Quadratic Discriminant Analysis. By discarding 4 of the positions and using a single connectivity index at the remaining 4 positions, they achieved approximately 75% correct classification for the training set on a leave-one-out basis. This reference provides an example of the different results that can be obtained from using different values of k.

Gombar, Jaeger and Jurs [18] reported a more recent example of a comparative study of *k*NN with Adaptive Least Squares, Bayes Linear and Bayes Quadratic Discrimination, Iterative Least Squares and the Linear Learning Machine (LLM). They analyzed the activity of a training set of 68 hypolipidemic arylproprionic acid derivatives using calculated physico-chemical descriptors. Prior to analysis, the descriptor set was reduced from 64 to 19 by excluding collinear properties and those with minimal variation. From these 19, 10 were then selected depending on their predictive power in LLM. In this study *k*NN performed poorly compared to the other methods, achieving only 75% correct classification of the training set (with $k = 1$) compared with $>89\%$ for the other methods based on cross-validation. However, the variable selection method employed may have influenced the results in favor of LLM.

Overall, *k*NN has proved reasonably successful as a classifier in QSAR even though the more sophisticated modifications to the basic method have not been widely implemented. It is less ambitious than other classification methods which base predictions of activity on some underlying model. Such methods can be more powerful and have the advantage of interpretability, but do depend on the validity of the form of model chosen.

4.3.3 Single Class Discrimination

4.3.3.1 Overview of Methods

Single Class Discrimination (SCD) [7, 8] is a collection of methods specifically developed for the analysis of embedded data. The essential feature of embedded data is that the active class is tightly clustered in property space compared to the inactive class. SCD looks for subspaces of the full property space where such a pattern is most evident. A fixed point is defined (often the mean of the active class) as the "centre of activity", and the variance about this point can be used to represent the spread of each class. The relative spread of the 2 classes, which we aim to maximize, is estimated by the ratio of these variances. SCD now proceeds in a manner similar to Principal Component Analysis (PCA) [9], but using the above variance ratio rather than total variance as the criterion to extract axes in order of importance. Thus, SCD is, principally, a dimensionality-reduction technique.

The output from SCD is also similar to PCA, consisting of scores, loadings and eigenvalues. The results can, therefore, be viewed as two- or three-dimensional scores plots and informative properties identified from the loadings. The plots should depict a clustering of the actives about the model origin and a dispersal of the inactives. The number of informative axes can be determined from a scree plot (a plot of eigenvalue against PC number), for example, and the Euclidean distance of a compound from the origin in this subspace can be calculated. This distance is known as the "Model Vector Length" (MVL) of a compound and may, under some circumstances, provide an estimate of class membership, with inactives having a greater MVL than actives.

SCD was originally developed to analyze classified activity data [7], but has since been generalized to accommodate a continuous measure of activity such as IC_{50} or ED_{50} values [8]. The generalized methods, termed Generalized Single Class Discrimination (GSCD), are also tolerant of classified or mixed activity data, where the more active compounds are expressed as an IC_{50} value while the poorly active compounds may have an activity quoted as "$>100\ \mu M$" for example. This mixed type of activity is common in pharmaceutical biological testing and such data is not readily processed by conventional methods of analysis.

A number of specific algorithms for implementing SCD were originally described, namely SCD-PCA I, II and III, SCD-CVA, GSCD-PCA I, II, and III and GSCD-CVA. From these, 4 have emerged as being of primary importance and these are SCD-PCA I, SCD-CVA, GSCD-PCA I and GSCD-CVA. The following discussion is restricted to these 4 algorithms and the methods are named after the algorithm used to identify informative axes, i.e. Principal Component Analysis (PCA) or Canonical Variate Analysis (CVA) [9]. The distinguishing features of these 4 methods are given in Table 1.

The different algorithms can be distinguished by 2 factors: a) how they deal with covariance between the descriptors in the active set, and b) whether the measure of biological activity results is classified or continuous.

With the former, the PCA-based algorithms (SCD-PCA I and GSCD-PCA I) ignore covariance between the descriptors in the active set, whilst the CVA based algorithms (SCD-CVA and GSCD-CVA) take full account of such covariance. As to which algorithm should be employed, would depend entirely on the structure of the data. The CVA algorithms tend to produce a tighter clustering of the active set than the PCA algorithms. However, because of the requirement to calculate many additional parameters (the actives covariances) a relatively large active set is needed to obtain a stable model. This is usually somewhat of a luxury in QSAR! For smaller active sets, a preliminary dimensionality reduction of the property matrix, e.g. by

Table 1. Summary of the Differences Between the 4 Main (G)SCD Algorithms.

Name	Biological activity type	Underlying algorithm	Treatment of covariance of actives	What the Axes Maximize
SCD-PCA I	Categorized	PCA	Ignored	Variance of inactives, after autoscaling actives
SCD-CVA	Categorized	CVA	Evaluated	Ratio of variance of inactives to actives
GSCD-PCA I	Continuous or categorized	PCA	Ignored	Variance of inactives, after autoscaling actives
GSCD-CVA	Continuous or categorized	CVA	Evaluated	Ratio of variance of inactives to actives

PCA, may often be advizable prior to analysis by (G)SCD in order to improve stability.

As regards to point b), it is of course always possible to analyze continuous data in a classified fashion, after drawing a line at some, possibly arbitrary, activity level. However, this often leads to loss of information, and the GSCD methods described here offer a more flexible approach. Essentially, the activity measure for each compound is used to construct an "activity weight", lying between 0 and 1, which represents the degree of activity. It can also be thought of as showing the degree of membership of the active class. Here, 1 denotes a member of the active class, 0 denotes a member of the inactive class, and in-between values are used for compounds of intermediate activity, with a foot in both camps. This idea of the "degree of class membership" is based on Fuzzy Theory [19].

In the GSCD algorithms the activity weights are used to calculate "weighted" means and standard deviations at points where, for classified activity, members of either just one class or of the other class would be used to calculate ordinary means and standard deviations. Indeed, the SCD methods can be viewed as special cases of their GSCD counterparts, as the generalized algorithms work perfectly well on classified data using weights that are either 0 or 1. However, it clarifies the ideas if they are considered separately – at least in the first instance – and so both approaches are described here.

The SCD-PCA I, GSCD-PCA I, SCD-CVA and GSCD-CVA methods are described below, with a view to a relatively simple implementation. A more complete mathematical description of the algorithms is given by Rose et al. [7, 8].

First, some definitions are given:

Training set	Compounds with known activity used to generate the model
Test set	Compounds for which an activity prediction is required
n	Total number of compounds in the training set

na	Number of active compounds
ni	Number of inactive compounds
nt	Number of compounds in the test set
p	Number of properties
X	An *n* by *p* matrix of properties for the training set
XA	An *na* by *p* matrix of properties for the active set
XI	An *ni* by *p* matrix of properties for the inactive set
XW	An *n* by *p* matrix of properties for the active weighted set
XV	An *n* by *p* matrix of properties for the inactive weighted set
Z	An *nt* by *p* matrix of properties of the test set
c	A vector of activities of length, *n*
w	A vector of active weights of length, *n*
v	A vector of inactive weights of length, *n*

Bold upper-case characters denote a matrix, bold lower-case denotes a vector and lower-case denotes scalars.

4.3.3.2 SCD-PCA I

Since the activity is classified, the property matrix, ***X***, can be divided into 2 sub-matrices, ***XA*** of dimension *na* by *p* which contains data on the active compounds, and ***XI*** of dimension *ni* by *p* which contains data on the inactive compounds. The column means and standard deviations of ***XA*** are calculated. These means, which estimate the centre of activity, are subtracted from the columns of both ***XA*** and ***XI***. ***XA*** and ***XI*** are then scaled by dividing the columns by the standard deviations of ***XA***. This results in ***XA*** becoming autoscaled, but ***XI*** will be composed of properties with different variances. Large variances in ***XI*** are due to a greater property spread in ***XI*** relative to ***XA*** (embedded data), or a difference in the means of the 2 classes (non-embedded). PCA is then carried out on the inactive matrix, ***XI***, and an appropriate number of axes are retained. (Note: this PCA should be carried out without further centering or scaling of ***XI***). ***XA*** is mapped to this space by post-multiplying it by the matrix of PC loadings.

Test compounds, held in the matrix ***Z***, are simply mapped to the SCD model by passing ***Z*** through the above process. The columns of ***Z*** are centered and scaled using the column means and standard deviations of ***XA***, and the resulting matrix is post-multiplied by the PC loadings to obtain the scores for each test compound.

4.3.3.3 GSCD-PCA I

The activity vector, ***c*** is transformed to a vector of active weights, ***w***, which lie within the range $0-1$, such that active compounds have a high value of ***w*** and inactives have a low value of w. One approach, appropriate for $-\log IC_{50}$ data for example, is to scale c to lie in the range of 0.05 to 0.95. This avoids assigning compounds to the boundaries (i.e. w_i values of 0 or 1). Thus, w can be calculated as follows:

$$w_i = 0.05 + 0.90\left(\frac{c_i - min}{max - min}\right) \qquad \text{for } i = 1 \text{ to } n \tag{3}$$

where, *min* is the activity of the least active compound and *max* is the activity of the most active compound. Compounds with activity quoted as inactive or "$>100\,\mu M$", for example can be given w values of 0. For SCD-PCA I, active compounds are given a value of 1, and inactives a value of 0.

The inactive weights, v, are calculated as:

$$v_i = 1 - w_i \qquad \text{for } i = 1 \text{ to } n \tag{4}$$

The active weighted mean, *wm*, is calculated for each property by:

$$wm_j = \frac{\sum_{i=1}^{n} (x_{ij} w_i)}{sumw} \qquad \text{for } j = 1 \text{ to } p \tag{5}$$

where, *sumw* is the sum of w.

The active weighted standard deviation, *ws*, of each property is calculated as:

$$ws_j = \sqrt{\frac{\sum_{i=1}^{n} (x_{ij} - wm_j)^2 w_i}{sumw}} \qquad \text{for } j = 1 \text{ to } p \tag{6}$$

The active weighted means are subtracted from the columns of X and the columns of X are divided by the active weighted standard deviations. Each row of X is then multiplied by the square root of its inactive weight, $\sqrt{v_i}$, to give the "inactive" matrix, XV. PCA is performed on XV to obtain a matrix of PC loadings. Finally, X is post-multiplied by the loadings to obtain the final scores for the compounds.

The scores of test compounds are determined by centering and scaling Z using the active weighted means and standard deviations of X and post-multiplying the resultant matrix by the PC loadings of XV.

In GSCD models of non-embedded, or partially embedded, data, the active weighted mean is not the best value to use for centering the matrices. In these situations, the variance ratio can, in fact, be increased by using a different value further from the mean of the inactive set to center the data. A simple method for estimating this alternative center, termed the "optimal mean", is given in [8].

4.3.3.4 SCD-CVA

X is divided into the submatrices XA and XI, and the mean of each column of XA is calculated and subtracted from XA and XI. We then solve the eigenvalue equation:

$$(XI^{T}XI - \lambda_j^2 XA^{T}XA)g_j = 0 \qquad \text{for } j = 1 \text{ to } p \tag{7}$$

where, g_j is the vector of loadings for the jth axis and λ_j is the jth eigenvalue. This can be solved either using a general latent root and vector algorithm, or a CVA algorithm where $XI^{T}XI$ takes the place of the "between groups" sums of squares matrix, whilst $XA^{T}XA$ corresponds to the "within groups" sums of squares matrix. The scores of test compounds are calculated by subtracting the means of XA from Z and post-multiplying by the CVA loadings.

4.3.3.5 GSCD-CVA

The active weighted (or "optimal") means of X, wm, are calculated as described above and are then substracted from the columns of X. Each row of X is multiplied by the square root of its active weight, $\sqrt{w_i}$, to give a matrix, XW, weighted to activity. Similarly, each row of X is multiplied by the square root of its inactive weight, $\sqrt{v_i}$, to give a matrix, XV, weighted to inactivity. We then solve the eigenvalue equation, as described previously:

$$(XV^{T}XV - \lambda_j^2 XW^{T}XW)g_j = 0 \qquad \text{for } j = 1 \text{ to } p \tag{8}$$

The scores of test compounds are calculated by subtracting the active weighted means of X from Z and then post-multiplying by the CVA loadings.

4.3.3.6 Significance Testing

Once an SCD model has been generated the significance of the axes can be determined by using random permutations to test the null hypothesis that there is no relationship between activity and the physico-chemical properties. Essentially, this involves randomizing the activity vector of the compounds, but not the property matrix, and recalculating the SCD model. This is performed a large number of times (e.g. 500) to calculate the likelihood of obtaining by chance axes with similar eigenvalues to those of the real model. Confidence limits of 95% can be obtained using this approach as outlined above for the first axis and then with adjustments for subsequent axes. The method is discussed fully by Wood et al. [20] for the classified case.

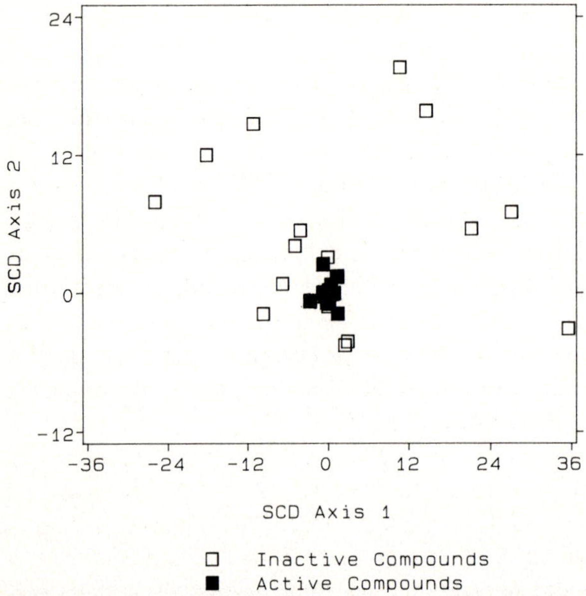

Figure 5. Results of the SCD-PCA II analysis on the Antimycin A analogs.

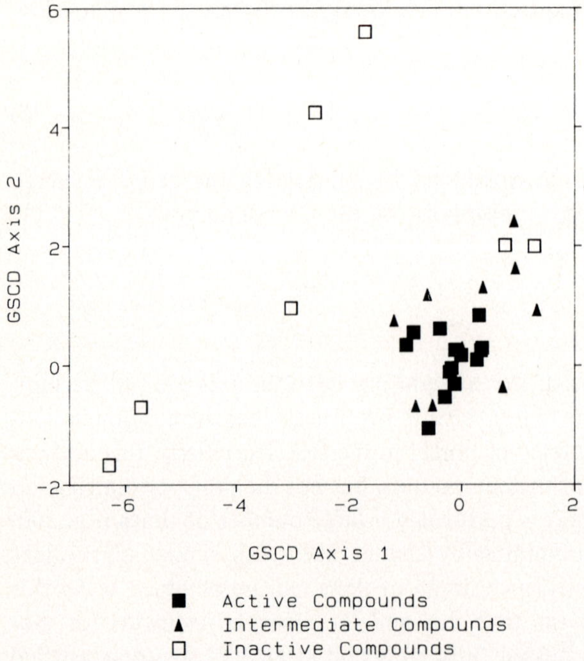

Figure 6. Results of the GSCD-PCA II analysis on the Antimycin A analogs.

4.3.3.7 QSAR Applications of SCD

As the SCD methods are relatively new, there are few examples of their application in the QSAR literature. In the original papers [7, 8], artifical data was widely used to portray the features of these methods. However, two QSAR data sets were also analyzed using both the classified (SCD) and continuous (GSCD) activity methods. The data of Selwood et al. [21], which consisted of activity data and 53 physico-chemical properties for 31 antifilarial Antimycin A analogues, was analyzed using SCD-PCA II [7] and GSCD-PCA II [8]. For the SCD-PCA II analysis, the 15 most active compounds were classified as active and the remainder as inactive. The GSCD-PCA II used the $-\log ED_{50}$ values directly. Both approaches gave a good clustering of the more active compounds and the models could be interpreted with respect to chemical structure. The results are shown in Figs. 5 and 6.

The other QSAR data set analyzed in these papers was the data of Goodford et al. [22] on the toxicity of methoxychloro analogues to houseflies. ED_{50} values and 5 substituent constants were available for 25 compounds. The 13 most toxic compounds were classified as "active" in the SCD analysis. SCD-PCA I and GSCD-PCA I successfully grouped the toxic compounds and generated chemically interpretable axes. The results of the analyses are shown in Figs. 7 and 8.

In both the above examples, the "optimal" mean was employed in preference to the "active weighted" mean in the GSCD analyses.

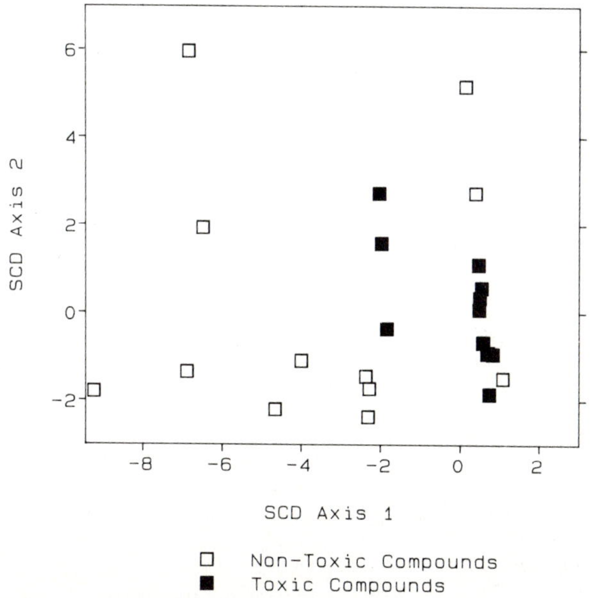

Figure 7. Results of the SCD-PCA I analysis on the methoxychloro analogs.

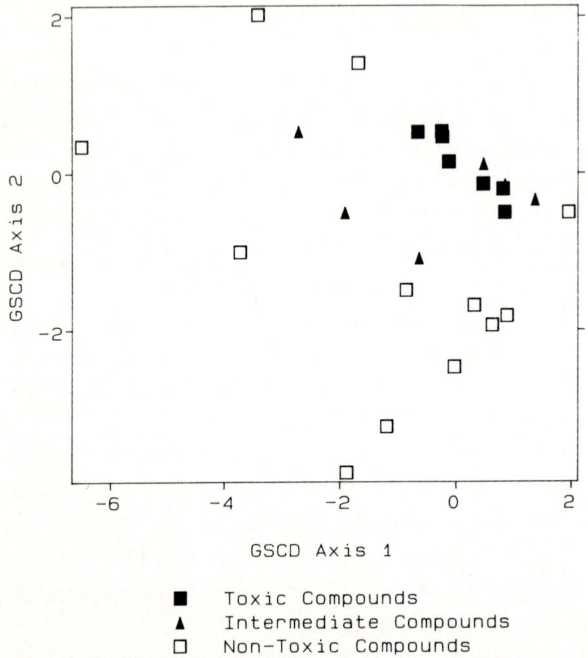

Figure 8. Results of the GSCD-PCA I analysis on the methoxychloro analogs.

The preliminary results from (G)SCD look encouraging and it is a valuable new approach for analyzing non-linear QSARs in multivariate property space.

References

[1] Magee, P.S., *Parameter Focussing – A New QSAR Technique.* In: *IUPAC Pesticide Chemistry: Human Welfare and the Environment,* Miyamtot, J. and Kearney P.C., eds., Pergamon Press, Oxford, 1983, p. 251–260
[2] McFarland, J.W. and Gans, D.J., *J. Med. Chem.* **29**, 505–514 (1986)
[3] McFarland, J.W. and Gans, D.J., *J. Med. Chem.* **30**, 46–49 (1987)
[4] Dunn III, W.J. and Wold, S., *J. Med. Chem.* **23**, 595–599 (1980)
[5] Wold, S., *Pattern Recog.* **8**, 127–139 (1976)
[6] Kowalski, B.R. and Bender, C.F., *Anal. Chem.* **44**, 1405–1411 (1972)
[7] Rose, V.S., Wood, J. and MacFie, H.J.H., *Quant, Struct.-Act. Relat.* **10**, 359–368 (1991)
[8] Rose, V.S., Wood, J. and MacFie, H.J.H., *Quant. Struct.-Act. Relat.* **11**, 492–504 (1992)
[9] Chatfield, C. and Collins, A.J., *Introduction to Multivariate Analysis*, Chapman and Hall, London, 1980
[10] Coomans, D. and Massart, D.L., *Anal. Chim. Acta* **136**, 15–27 (1982)
[11] Forbes, R.A., Tews, E.C., Freiser, B.S., Wise, M.B. and Perone, S., *J. Chem. Inf. Comput. Sci.* **26**, 93–98 (1986)

[12] Kowalski, B.R. and Bender, C.F., *J. Am. Chem. Soc.* **96**, 916–918 (1974)
[13] Chou, J.T. and Jurs, P.C., *J. Med. Chem.* **22**, 792–797 (1979)
[14] Clare, B.W., *J. Med. Chem.* **33**, 687–702 (1990)
[15] Stouch, T.R. and Jurs, P.C., *J. Med. Chem.* **29**, 2125–2136 (1986)
[16] Sjöström, M. and Kowalski, B.R., *Anal. Chim. Acta.* **112**, 11–30 (1979)
[17] Henry, D.R. and Block, J.H., *J. Med. Chem.* **22**, 465–472 (1979)
[18] Gombar, V.K., Jaeger, E.P. and Jurs, P.C., *Quant. Struct.-Act. Relat.* **7**, 225–234 (1988)
[19] Otto, M., *Chemom. Intell. Lab. Sys.* **4**, 101–120 (1988)
[20] Wood, J., Rose, V.S. and MacFie, H.J.H., *Chemom. Intell. Lab. Sys.* **23**, 205–212 (1994)
[21] Selwood, D.L., Livingstone, D.J., Comley, J.C.W., O-Dowd, A.B., Hudson, A.T., Jackson, P., Jandu, K.S., Rose, V.S. and Stables, J.N., *J. Med. Chem.* **33**, 136–142 (1990)
[22] Goodford, P.J., Hudson, A.T., Sheppey, G.C., Wootton, R., Black, M.H., Sutherland, G.J. and Wickham, J.C., *J. Med. Chem.* **19**, 1239–1247 (1976)

4.4 Quantitative Analysis of Structure-Activity-Class Relationships by (Fuzzy) Adaptive Least Squares*

Klaus-Jürgen Schaper

Abbreviations

ALS	Adaptive Least Squares
ANN	Artificial Neural Network
FALS	Fuzzy Adaptive Least Squares
*k*NN	*k*-Nearest Neighbor
LDA	Linear Discriminant Analysis
LLM	Linear Learning Machine
LOO	Leave-One-Out (cross-validation)
MLR	Multiple Linear Regression
MMG	Mean Membership Grade
NLR	Non-linear Regression
PCA	Principal Component Analysis
PCVR	Principal Component after VARIMAX Rotation
QSAR	Quantitative Structure-Activity Relationship
s.e.	standard error

Symbols

a_j	activity score of class j after scaling
$C_i^{(t)}$	correction for compound i at iteration t
ci	contribution index of a variable
δ_i	distance of calculated activity score to near cut-off point
E	error function value
FL	fuzzy level (a constant)
g	no. of groups (categories, activity classes)
m	no. of independent variables (descriptors)
m_c	calculated (fitted or predicted) mid-rank value

*Dedicated to Prof. Dr. Joachim K. Seydel, Borstel, on the occasion of his 65th birthday, with gratitude for two decades of generous support

m_o	observed mid-rank value
$M(Z)$	membership function value corresponding to Z_{calc}
n	total no. of compounds or measurements
n_j	size of group j
n_{mis}	no. of misclassified objects
r	no. of iterations without improvement of R_s or E
R_{mult}	multiple correlation coefficient
R_S	Spearman rank correlation coefficient
R_x	descriptor correlation matrix
$s.e._k$	standard error of regression coefficient k
t	total no. of ALS-iterations
t^*	no. of temporary best iteration
Z	matrix of activity ranks after ALS scaling
Z_x	matrix of scaled descriptor values

4.4.1 Introduction

The biological potency of drugs, herbicides, antifungals, etc. is often recorded in the form of activity ratings such as "inactive" (−), "weakly active" (+), "active" (+ +) and "strongly active" (+ + +). The method of choice for analyzing class data in QSAR investigations usually is the technique of linear discriminant analysis (LDA) [1, 2]. However, LDA is not considered to be appropriate for investigating ordered class data (except in a 2-group case), mainly for the following reasons: (i) LDA has been developed to deal with the problem of discriminating independent classes or categories; (ii) the assumption of multivariate normal distribution of descriptor data in each category and of equality in within-group covariance matrices is often not fulfilled in the case of g groups ($g \geq 3$); (iii) since g discriminant functions are derived, they are difficult to interpret in terms of a structure-activity relationship model.

With the aim of overcoming these unfavorable aspects of LDA, Ikuo Moriguchi and coworkers [3–16] initiated the development and application of a new discrimination method called *Adapative Least Squares* (ALS) analysis almost 20 years ago. Since 1988 *Fuzzy* Adaptive Least Squares (FALS), wich is a more powerful version of the basic ALS technique, has been investigated by the same group [17–22]. As FALS was only recently developed, until now only one paper by a different group [23] comparing the use of this method with other methods had been published, whereas ALS has been widely applied to the field of QSAR by other investigators [24–39].

The ALS method is a non-parametric classifier which has been devised to derive a single QSAR equation irrespective of the number of observed activity classes. Non-parametric in contrast to parametric statistics does not require normally distributed data. Parametric statistics involve parameters which define the population (i.e. the mean, μ, and the standard deviation, σ) whereas non-parametric statistics are independent of the distribution of data. The ordinal scale of bioactivity data is analyzed by ALS/FALS and the activity scale is ordered (i.e. with increasing activity in the classes 1, 2, 3...), but not necessarily at equal intervals. Thus, the non-parametric ALS method is a technique which accounts for the observed ranks of bioactivity in terms of the physico-chemical properties of compounds and compares calculated (i.e. fitted or predicted) with observed ranks. The degree of similarity between ranks is quantified by the non-parametric Spearman rank correlation coefficient, R_S.

4.4.2 The ALS Algorithm

The ALS method allows decisions for ordered g-group discrimination ($g \geq 2$) by using a single discriminant function. This function is obtained by iterative application of multiple linear regression (MLR) analysis with a stepwise adaptation of the dependent activity variable. The procedure is repeated for a given number of iterations, or until all substances are correctly classified, and the best discriminant function is selected. In contrast to LDA, the discriminant function is non-unique. If the objects that define the observed classes are sufficiently apart, many discriminants may be found that would equally well distinguish between the classes [28, 38].

An overview of the algorithm discussed in this section is given by the flow chart shown in Fig. 1. To the best of our knowledge, the computer program mentioned in Ref. [40] is the only one that is commercially available.

4.4.2.1 Scaling of Ranked Activity Data and Further Data Preprocessing

The observed activity ranks, in general, are numbered in ascending order, with rank 1 denoting the lowest activity class. In the first paper on ALS [3] as well as in two other papers [30, 31], no special scaling was performed. However, as the data can be very skew (different numbers of compounds in activity classes), all other papers used a specific method of scaling, taking into account the number of objects of a certain rank. As a standard numerical score for ordered categories the so-called "ridit" has been proposed [41]. The choice of "ridit", as a numerical score, is based on the assumption that only the potency order of groups is reliable, whereas quantitative differences in the potency between groups as well as between compounds within a group are uncertain. Since 1980 Moriguchi et al. [4] used a modified ridit as defined by Eq. (1):

$$a_j = 4[(n_j/2+\Sigma)/n]-2 \qquad j = 1..g \tag{1}$$

with

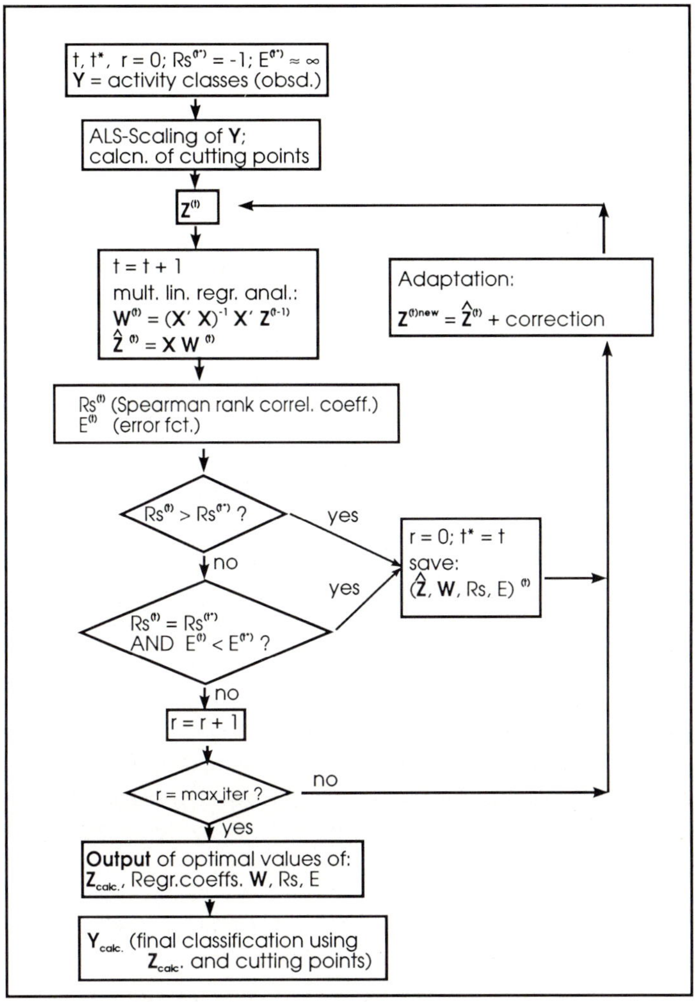

Figure 1. Flow chart of the ALS algorithm (see Sec. 4.4.2.).

$$\Sigma = 0 \qquad \text{if } j = 1$$

$$\Sigma = \sum_{i=1}^{j-1} n_i \qquad \text{if } j > 1$$

Here a_j is the scaled activity score of a compound in class j, n_j and n_i are the size of groups, j and i, respectively, and n is the total number of compounds. According to Eq. (1) the mean value of a_j over n compounds becomes zero. After the calculation of starting scores, a_j, for the members of class j ($j = 1..g$) the cut-off points*

* Called "cutting points" by Moriguchi et al. [4–22]

$b_{j,j+1}$ ($j = 1..g-1$) between classes are fixed in advance before undertaking the ALS iteration. The cut-off points $b_{j,j+1}$ are taken as the midpoints between a_j and a_{j+1}:

$$b_{j,j+1} = (a_j + a_{j+1})/2 \tag{2}$$

The next step in the preprocessing of activity data is the replacement of the a priori rank value of each compound by its corresponding a_j value, thus, giving rise to the (n*1)-matrix $Z^{(t=0)}$ (t is the number of ALS iterations). This matrix contains the starting scores of activity that are adapted in subsequent ALS iterations and are used as the dependent data set to be described by physico-chemical descriptors. Finally, before entering the ALS iteration procedure, it is useful to calculate the standard deviation of all independent variables selected for the analysis. Standard deviations are used for the calculation of contribution indexes of descriptors (see Sec. 4.4.2.3).

4.4.2.2 The ALS Iteration

The ALS algorithm is fundamentally an iterative multiple linear regression analysis [42] that uses an error corrective feedback for discriminant development. Therefore, in ALS as in MLR the same precautions must be taken, e.g. selection of a training set of compounds of low collinearity and with a high variance in molecular descriptors [43–46]. The degree of descriptor collinearity of a given set of compounds must be tested, e.g. by calculating [2] the correlation matrix, $\boldsymbol{R}_X$. However, just a mere inspection of a matrix of simple correlation coefficients is not sufficient. Often, several descriptors encode partially related information. Therefore, multicollinearities among the descriptors can also be a problem, giving rise to numerical instabilities in the analysis. As a diagnostic, the multiple correlation coefficient of the regression between one independent variable, i, and the remaining $(m-1)$ regressors can be obtained directly [43] from the ith main diagonal element, C_{ii}, of the inverted correlation matrix, $(\boldsymbol{R}_X)^{-1}$, by:

$$R_{\mathrm{mult.},i} = (1 - 1/C_{ii})^{1/2} \tag{3}$$

In the case of two or more coefficients with, e.g. $R_{\mathrm{mult.}} > 0.6$, a detailed analysis suggests descriptors that may be deleted. A better solution to the problem of multicollinearity, however, is the application of principal component analysis [47, 48] (see also Sec. 4.4.3.4). A further problem to be considered before running an MLR-type QSAR analysis is the occurrence of outliers or points of high leverage in the X-descriptor space [49, 50].

The ALS iteration is started at cycle number one (t = 1) with the activity data matrix $\boldsymbol{Z}^{(t=0)}$ obtained in the preprocessing step and the $(n*m)$ descriptor matrix $\boldsymbol{D}$ that is expanded by one column with a vector of ones to give the $[n \times (m+1)]$ regressor matrix $\boldsymbol{X}$ (see Eq. (5)). As in MLR analysis the estimate of the $[(m+1) \times 1)$ vector of weights, $\boldsymbol{W}$, (i.e. slopes, regression coefficients) is obtained by:

$$W^{(t)} = (X'X)^{-1}X'Z^{(t-1)} \quad (4)$$

In Eq. (4) the superscript, t, denotes the tth ALS iteration cycle. The next step is the computation of calculated (fitted) $\hat{Z}$-values for each compound by:

$$\hat{Z}^{(t)} = XW^{(t)} \quad (5)$$

where

$$\hat{Z} = \begin{vmatrix} z_1 \\ z_2 \\ \cdot \\ \cdot \\ z_n \end{vmatrix} \qquad X = \begin{vmatrix} 1 \ x_{11} \ldots & x_{1m} \\ 1 \ x_{21} \ldots & x_{2m} \\ \cdot \ \cdot & \cdot \\ \cdot \ \cdot & \cdot \\ 1 \ x_{n1} \ldots & x_{nm} \end{vmatrix} \qquad W = \begin{vmatrix} w_0 \\ w_1 \\ \cdot \\ \cdot \\ w_m \end{vmatrix}$$

The element x_{ij} in matrix X is the jth descriptor of compound, i. Eq. (5) can be reformulated without matrix notation as:

$$\hat{z}_i = w_0 + w_1 x_{i1} + w_2 x_{i2} + \ldots + w_m x_{im} \qquad (i = 1..n) \quad (5\,\text{a})$$

Finally, all compounds are classified (a posteriori classification) on the basis of the calculated values of $\hat{Z}$ and the previously fixed cut-off points $b_{j,j+1}$:

if $\hat{z}_i \leq b_{1,2}$ then assign substance i to class 1;

if $b_{j-1,j} < \hat{z}_i \leq b_{j,j+1}$ then assign substance i to class j;

if $b_{g-1,g} < \hat{z}_i$ then assign substance i to class g.

For the evaluation of intermediate and final classification results, two criteria are used (order of priority: a) > b)):

a) The Spearman rank correlation coefficient, R_S, i.e. the correlation between observed and calculated activity class mid-ranks;

b) the Error function value, E, (apparent variance of errors).

In ALS, the Spearman rank correlation coefficient must be calculated by the same procedure as in simple linear regression analysis [51, 52]. The only difference is the use of observed and calculated mid-ranks (m_o, m_c) instead of x- and y-values:

$$R_S = \frac{\sum_{i=1}^{n} [(m_{o,i} - m_{o,\text{mean}})(m_{c,i} - m_{c,\text{mean}})]}{\left[\sum_{i=1}^{n} (m_{o,i} - m_{o,\text{mean}})^2 \sum_{i=1}^{n} (m_{c,i} - m_{c,\text{mean}})^2\right]^{1/2}} \tag{6}$$

Mid-ranks are averaged ranks of "tied" compounds (identical class). The error function, E, is defined empirically by:

$$E = \left[\sum_{i=1}^{n} e_i^2\right] / (n - m - 1) \tag{7}$$

where

$$e_i = 0 + |z_i^{(t=0)} - \hat{z}_i^{(t)}| \quad \text{(if correctly classified)} \tag{8}$$

$$e_i = 1 + |z_i^{(t=0)} - \hat{z}_i^{(t)}| \quad \text{(if misclassified)}$$

Now, assume that the ALS analysis is started at iteration t = 0 with extremely unfavorable arbitrary values for R_s and E (e.g. $R_S^{(t=0)} = -1$ and $E^{(t=0)} \approx \infty$) which must be improved during subsequent iterations (see Fig. 1). This means at least that at iteration $t = 1$ these starting values are improved. Therefore, the corresponding matrices $\boldsymbol{W}^{(t)}$ and $\hat{\boldsymbol{Z}}^{(t)}$ are retained as well as $R_S^{(t)}$ and $E^{(t)}$. Subsequently, to complete the first iteration cycle, the adaptation step of ALS is performed. By adaptation of the elements of $\hat{Z}^{(t)}$, the deviations mainly between observed activity scores ($\hat{\boldsymbol{Z}}^{(t=0)}$) and calculated activity scores ($\hat{\boldsymbol{Z}}^{(t)}$) of misclassified objects are reduced by applying the correction term, $C_i^{(t)}$:

$$Z_i^{(t)_{\text{new}}} = \hat{Z}_i^{(t)} + C_i^{(t)} \tag{9}$$

In the early stages of development [3–6], several empirically defined expressions were investigated for C_i. Since 1984 [7] the following terms have been used:

If correctly classified:

$$C_i^{(t)} = 0 \tag{10a}$$

If misclassified *below* observed class:

$$C_i^{(t)} = +[0.1 + 0.1/(0.45 + \delta_i^{(t)})^2] \tag{10b}$$

If misclassified *above* observed class:

$$C_i^{(t)} = -[0.1 + 0.1/(0.45 + \delta_i^{(t)})^2] \tag{10c}$$

where

$$\delta_i^{(t)} = |\hat{Z}_i^{(t)} - b_{j,j+1}| \tag{11}$$

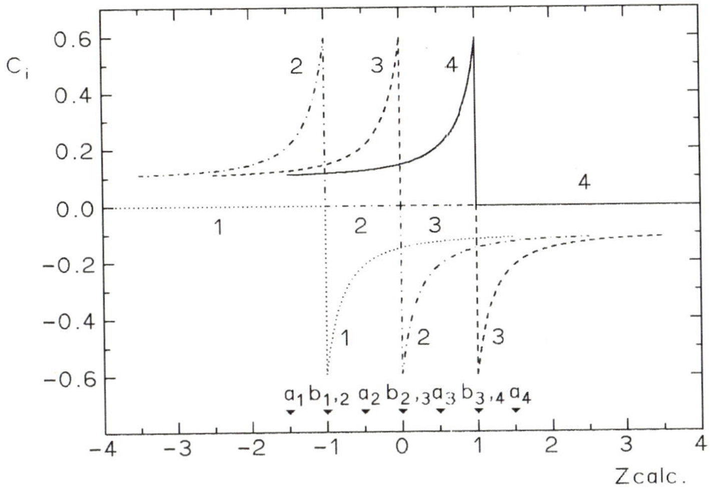

Figure 2. Graphical representation of the correction term C_i as a function of calculated activity scores in a four-class problem assuming that a compound has been observed to belong to class no. 1, 2, 3, or 4 which are all identical in size.

$\delta_i^{(t)}$ is the distance between the calculated activity score of compound i and the cut-off point of its observed class that is nearer to $\hat{Z}_i^{(t)}$ (e.g., use $b_{1,2}$ if i has been observed as class 2 but calculated as class 1; use $b_{2,3}$ if the calculated class is 3). Fig. 2 shows a graphical illustration of the correction term, C_i, as a function of calculated activity scores assuming that a compound has been observed as belonging to classes 1, 2, 3 or 4 (indicated by numbers on the graph). This graph shows that no correction was applied if the calculated activity score of a compound observed as class j is within the range of $b_{j-1,j}$ and $b_{j,j+1}$. Below this range, the score increases (positive C_i) and vice versa. Compounds with physico-chemical properties leading to calculated scores that are far from the observed score $z_i^{(t=0)}$ are shifted only by $C_i \approx 0.1$. Those near to the cut-off point are considerably more shifted to enable a correct classification.

The new $Z^{(t)}$ values obtained in the final adaptation step of this ALS iteration are used to form the dependent variable for the next iteration cycle that hopefully will lead to further improvement of R_S and E. In the case of increased R_S, the corresponding data ($\hat{Z}$, W, R_S, E retained in the hitherto obtained best iteration t^*) is again updated and the cycle continued. The same procedure is performed if $R_S^{(t)} = R_S^{(t^*)}$ AND $E^{(t)} < E^{(t^*)}$. The adaptation step is performed without updating if $R_S^{(t)} < R_S^{(t^*)}$, or if $R_S^{(t)} = R_S^{(t^*)}$ AND $E^{(t)} \geq E^{(t^*)}$.

According to Moriguchi et al. [7, 10] the best result obtained after 20 iterations is used as the final result. However, we found that an often better result is obtained with more iterations. Therefore, we modified the algorithm slightly (see Fig. 1). In the modified procedure the number, r, of iterations, without further improvement of R_S or E, is counted and at each iteration *with* improvement, the number r is reset to $r = 0$. The iteration cycle is abandoned, only if, r is equal to the predetermined number "max_iter" (e.g. max_iter = 60, for a leave-one-out cross-validation (LOO) max_iter = 50). To avoid long computation times, an improvement of E is only acceptable if $E^{(t)} < (E^{(t^*)} - 1 \times 10^{-6})$.

To show that better results were obtained by this version, we re-analyzed the anti-inflammatory data of 38 furoindoles investigated by Moriguchi et al. [10]. Using the standard version (20 iterations) we reproduced Eq. (4) reported by Moriguchi et al. [10] (here Eq. (12)), however we were unable to reproduce the LOO results:

$$\hat{Z} = \underset{(0.97)}{\underset{(6.70)}{3.057}}\,B_1(1) - \underset{(0.56)}{\underset{(3.87)}{1.778}}\,F(1) + \underset{(0.38)}{\underset{(3.09)}{0.639}}\,N_{c\beta a} - \underset{(6.72)}{4.286} \quad (12)$$

(t): (6.70) (3.87) (3.09) (6.72)
(ci): (0.97) (0.56) (0.38)

$n = 38$ $E = 1.310$ $R_S = 0.828$ $n_{mis} = 7\ (0)$
LOO cross-val.: $R_S = 0.771$ $n_{mis} = 9\ (0)$
LOO cross-val.: $R_S = 0.74$ $n_{mis} = 10\ (0)$ (Ref. [10])

Using the modified version (iteration until $r = 60$, LOO: $r = 50$) we obtained Eq. (13), which indicated a clearly improved relationship. In addition to the reduction of the number of misclassified compounds, the shift in the regression coefficients (Eq. 12→13) obtained by the modified version in Fig. 1 is indeed quite remarkable:

$$\hat{Z} = 3.802\,B_1(1) - 2.513\,F(1) + 0.814\,N_{c\beta a} - 5.435 \quad (13)$$

(t): (7.48) (4.91) (3.53) (7.65)
(ci): (1.21) (0.79) (0.48)

$n = 38$ $E = 1.310$ $R_S = 0.854$ $n_{mis} = 6\ (0)$
LOO cross-val.: $R_S = 0.771$ $n_{mis} = 9\ (0)$

In these equations, n_{mis} is the number of misclassified or LOO-mispredicted compounds; the figure in parentheses after the value of n_{mis} is the number misclassified by two grades. The first line below the equations shows the t-test value [42] of the regression coefficients (t = coefficient/standard error of coefficient) while the second line contains the "contribution indices" (ci) of the variables (see Sec. 4.4.2.3).

4.4.2.3 Validation of ALS-Discriminants

For the validation of the regression equations obtained by the ALS technique, Moriguchi et al. [4, 5, 10] proposed the contribution index of each independent variable. The contribution index (ci_k) of descriptor, k, is a measure of its contribution to the discriminant scores. It is defined as the product of the regression coefficient of the descriptor and its standard deviation:

$$ci_k = |w_k| * s_k \tag{14}$$

The ci values are identical to the slopes if scaled descriptors are used (scaled to a mean of zero and standard deviation, $s_k = 1$). Thus, this is a way of comparing the slopes of variables that are on the same scale. This index has been used by Moriguchi [4, 5, 8 – 11, 19 – 22] and others [36] to validate the importance of descriptors and, furthermore, to form a basis for their backward stepwise elimination. Descriptors with $ci < 0.1$ were tested as candidates for elimination. However the final set of descriptors was selected primarily on the basis of cross-validation results using the LOO prediction technique (therefore, even $ci < 0.03$ is possible [8, 9, 20, 22]). The discriminant function giving the best LOO prediction is the one that is finally adopted.

Obviously ci is mainly a reflection of the steepness of the gradient but not of the scatter around the regression line (hyperplane) like the t-test in conventional regression analysis. Therefore, in addition to ci, we also calculate the t-test value [42]. Since the t-statistics were developed for normally distributed continuous data, t-values cannot, strictly speaking be interpreted on face value and merely provide a rough estimation as to which descriptor should be eliminated (if $t \lesseqgtr 1.5$).

To illustrate the way in which the t-test is more sensitive compared to the ci-test, the ALS analysis published by Garcia et al. [36] was repeated to enable a comparison of ci- and t-values. Garcia determined the mutagenic potency of 10 triazino indoles in three bioassay systems and found three relationships (Eq. 4 – 6 [36]) between classified activity data (2 classes) and substituent lipophilicity expressed by π. We repeated the analysis by the same procedure (standard version of ALS with 40 iterations (!) [36]) and obtained identical results in two of the equations, whereas one of the equations differed slightly for reasons unknown:

			E	R_S	n_{mis}	
$\hat{Z} =$	$-0.859\,\pi$	$+0.158$	0.222	1	0	(15)
(t):	(5.66)	(1.03)				
(ci):	(0.89)					
$\hat{Z} =$	$-0.737\,\pi$	$+0.183$	0.148	1	0	(16)
(t):	(5.94)	(1.46)				
(ci):	(0.76)					
$\hat{Z} =$	$-0.677\,\pi$	-0.207	0.933	0.817	1	(17)
(t):	(2.59)	(0.78)				
(ci):	(0.70)					

The standard deviation of π is identical in all three equations and, therefore, the *ci*-values are exactly collinear with the regression coefficients of π and only differ slightly. In contrast the *t*-value of Eq. (17) is much lower compared to Eqs. (15) and (16), which obviously highlights the fact that at least one derivative is not adequately described by this equation.

Both *ci*- and *t*-values may give an overoptimistic picture of the importance of variables included in the ALS analysis. ALS is obviously extremely flexible and often gives rise to an excellent description of the training set data. As described previously, calculated activity scores of compounds are iteratively shifted in the direction of the observed class scores if this is in any way compatible with their physico-chemical properties and the relationships observed within the whole data set. Therefore, to ensure that no overfitting occurs, as well as to obtain a realistic feeling for the predictive capability of the model, cross-validation [53, 54] e.g. by the LOO prediction, is strongly recommended. Furthermore, the contribution of each individual descriptor to the predictive capability of a model can be assessed by the deterioration of results that occurs when that particular descriptor is eliminated [38].

4.4.3 Application of ALS

In this section, first the analysis of the activity ranking of mitomycin derivatives is discussed. Then material from several publications [14, 21, 31] in which ALS is applied to some interesting data sets is briefly mentioned. Finally, the ALS analysis of the dose- and property-dependent biological effect classes of antihypertensive acryloylpiperazinoquinazolines is presented in more detail.

4.4.3.1 Antitumor Activity of Mitomycins

The antitumor activity of mitomycins (I) was one of the earliest data sets to have been analyzed by the ALS method [3, 5]. This data set has been used several times for comparing ALS with other techniques of pattern recognition (see Sec. 4.4.4).

In the first analysis by Moriguchi and Komatsu [3], the antitumor activity against solid sarcoma in mice was categorized into five classes. Later, classes 1 and 2 were

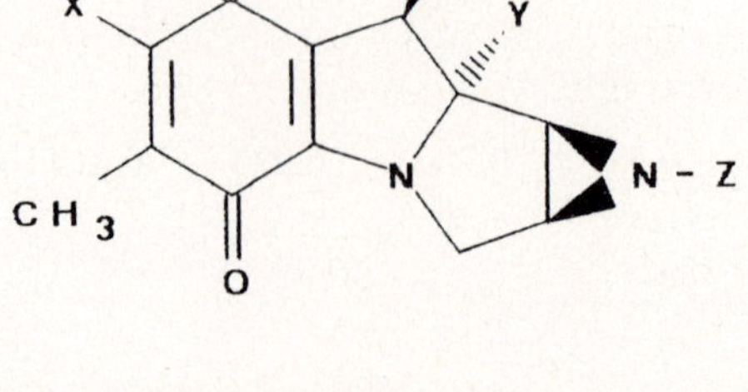

(I)

combined to class 1, and classes 4 and 5 were combined to give class 3. Furthermore, the activity against ascites sarcoma (3 classes) was investigated [5]. All activity data and corresponding descriptor values used by Moriguchi et al. [3, 5] are listed in Table 1. In the analysis of the solid sarcoma 5-class data set [3] a very early version of ALS was used which did not perform "ridit" scaling and, furthermore applied a different correction term. Nevertheless, the resulting discriminant function (Eq. (18)) is similar to the one (Eq. (19)) obtained from the flow chart outlined in Fig. 1 [55]:

$$\hat{Z} = -4.33\,\mathrm{F}_X - 2.46\,\mathrm{V}_{\mathrm{W}_X} + 2.48\,I_{\mathrm{Y=OMe}} + 2.28\,I_{\mathrm{Y=OH}} + 0.77\,E_{\mathrm{s,Z}} + 1.42 \quad (18)$$

$$a_1 = 0.5 \quad a_2 = 1.5 \quad a_3 = 2.5 \quad a_4 = 3.5 \quad a_5 = 4.5$$

$$b_{1,2} = 1 \quad b_{2,3} = 2 \quad b_{3,4} = 3 \quad b_{4,5} = 4$$

$$n = 16 \quad R_s = 0.963 \quad n_{\mathrm{mis}} = 1(1)$$

$$\hat{Z} = -4.49\,\mathrm{F}_X - 2.32\,\mathrm{V}_{\mathrm{W}_X} + 2.25\,I_{\mathrm{Y=OMe}} + 2.11\,I_{\mathrm{Y=OH}} + 0.79\,E_{\mathrm{s,Z}} - 1.11 \quad (19)$$

(t):	(2.8)	(2.0)	(4.2)	(3.1)	(3.3)	(1.8)
(ci):	(0.5)	(0.4)	(1.0)	(0.7)	(0.7)	

$$\mathrm{a}_1 = -1.625 \quad \mathrm{a}_2 = -1.125 \quad \mathrm{a}_3 = -0.375 \quad \mathrm{a}_4 = 0.875 \quad \mathrm{a}_5 = 1.75$$

$$b_{1,2} = -1.375 \quad b_{2,3} = -0.75 \quad b_{3,4} = 0.25 \quad b_{4,5} = 1.3125$$

$$n = 16 \quad E = 0.769 \quad R_{\mathrm{S}} = 0.963 \quad n_{\mathrm{mis}} = 1(1)$$

Both discriminants misclassified compound 15 as belonging to class 3, whereas class 1 was the observed class (see Table 1). According to both equations, the activity decreased with increasing bulk and field effects of substituents at the X position as well as with the steric effects of Z substituents. Favorable effects are observed if Y = H is replaced by Y = OMe or Y = OH as indicated by the positive regression coefficients of the two indicator variables which point to the presence ($I = 1$) or absence ($I = 0$) of a specific substructure.

In the second ALS analysis of the mitomycin data (3-group classification!) Moriguchi et al. [5] replaced the indicator variables of Eqs. (18) and (19) by σ^*_{Y}, a descriptor for the polarizing effects of the Y substituents and, furthermore, expressed the steric effect of the Z substituents by the STERIMOL variable $B_{1,2}$. Unfortunately, there are only three different substituents in position Y. Therefore, there is an ideal intercorrelation between σ^*_{Y} and $I_{\mathrm{Y=OMe}}$ and $I_{\mathrm{Y=OH}}$. Thus, compounds described by such indicator variables can be equally described by σ^*_{Y}. This does not mean to say that activity depends solely on this descriptor, although this could not be ruled out completely with the given data set. In this analysis Moriguchi et al. [5] did not use the correction term given in Eqs. (10b) and (10c) but the following:

$$C_i^{(t)} = \pm[\beta(\hat{Z}_i^{(t)} - Z_i^{(t=0)})^2 + 0.1/(0.45 + \delta_i^{(t)})^2] \quad (20)$$

with $\beta = 0.01$, 0.03 or 0.05.

Table 1. Property descriptors and antitumor activity of mitomycin derivatives (I).

No.	X	Y	Z	F_X	$\sigma_{m,X}$	V_{W_X}	σ^*_Y	$E_{S,Z}$	$B_{1,Z}$	$B_{4,Z}$	Activity Class									
											Solid Sarcoma						Ascites Sarcoma			
											o[a]	c[ab]	o[a]	c[ace]	p[ac]	p[ae]	o[a]	c[adf]	p[ad]	p[af]
1	NH_2	OMe	H	0.02	−0.16	0.177	1.81	1.24	1.00	1.00	5	5	3	3	3	3	3	3	3	3
2	NHEt	OMe	H	−0.11	−0.24	0.493	1.81	1.24	1.00	1.00	5	5	3	3	3	3	3	3	**2**	3
3	NH_2	OMe	Me	0.02	−0.16	0.177	1.81	0	1.52	2.04	4	4	3	3	3	3	3	3	3	3
4	NH_2	OMe	Et	0.02	−0.16	0.177	1.81	−0.07	1.52	2.97	4	4	3	3	3	3	2	2	2	2
5	NH_2	OMe	Ac	0.02	−0.16	0.177	1.81	−0.47	1.90	2.93	4	4	3	3	**2**	**2**	2	2	2	2
6	NH_2	OH	Me	0.02	−0.16	0.177	1.55	0	1.52	2.04	4	4	3	3	3	3	2	2	2	2
7	NMe_2	OMe	H	0.10	−0.15	0.441	1.81	1.24	1.00	1.00	4	4	3	3	3	3	3	3	3	3
8	NH_2	OMe	COPh-2-Cl	0.02	−0.16	0.177	1.81	−1.19	2.36	5.98	3	3	2	2	2	2				
9	NH_2	OMe	COPh-4-Cl	0.02	−0.16	0.177	1.81	−1.19	2.36	5.98	3	3	2	2	2	2				
10	NHPh	OMe	H	−0.02	−0.12	0.892	1.81	1.24	1.00	1.00	3	3	2	2	2	2	2	2	2	2
11	OMe	OMe	H	0.26	0.12	0.304	1.81	1.24	1.00	1.00	3	3	2	2	2	2	3	3	3	3
12	OMe	OMe	Me	0.26	0.12	0.304	1.81	0	1.52	2.04	3	3	2	2	2	2	2	2	**3**	2
13	OMe	OH	Me	0.26	0.12	0.304	1.55	0	1.52	2.04	2	2	1	1	**2**	1	1	**2**	**2**	**2**
14	NH_2	H	Me	0.02	−0.16	0.177	0.49	0	1.52	2.04	1	1	1	1	**2**	1	1	1	1	1
15	NH_2	OMe	SO_2Me	0.02	−0.16	0.177	1.81	−1.54	2.11	3.15	1	**3**	1	**2**	**2**	**2**	3	**2**	**2**	**2**
16	OMe	H	Me	0.26	0.12	0.304	0.49	0	1.52	2.04	1	1	1	1	1	1	1	1	1	1

[a] o = obsd., c = calc., p = pred.; [b] Eqs. (18) and (19); [c] Eq. (21); [d] Eq. (22); [e] Eq. (40); [f] Eq. (41)

For each β value twenty iterations are performed and finally the best one is selected. As values for *ci* and *E* are not given by Moriguchi et al. [5] we recalculated the equations of Moriguchi et al. In the case of the solid sarcoma data we successfully reproduced the equation published (with $\beta = 0.03$) but found different LOO predictions, whereas for the ascites sarcoma data we obtained identical LOO misclassifications but a slightly different discriminant function:

Solid Sarcoma

$$\hat{Z} = -4.33\,\sigma_{m,X} - 2.57\,V_{W_X} + 1.56\,\sigma^*_Y - 1.43\,B_{1,Z} + 0.03 \qquad (21)$$

(*t*):	(2.80)	(2.11)	(3.52)	(2.88)	(0.02)
(*ci*):	(0.56)	(0.49)	(0.70)	(0.67)	

$a_1 = -1.5 \quad a_2 = -0.375 \quad a_3 = 1.125 \quad b_{1,2} = -0.938 \quad b_{2,3} = 0.375$

$n = 16 \quad E = 0.915 \quad R_S = 0.969 \quad n_{mis} = 1(0)$

LOO: $R_S = 0.833 \quad n_{mis} = 4(0)$

Ascites Sarcoma

$$\hat{Z} = -2.10\,V_{W_X} + 1.75\,\sigma^*_Y - 0.53\,B_{4,Z} - 1.12 \qquad (22)$$

(*t*):	(1.53)	(3.75)	(1.53)	(0.94)
(*ci*):	(0.42)	(0.83)	(0.41)	

$a_1 = -1.571 \quad a_2 = -0.429 \quad a_3 = 1.143 \quad b_{1,2} = -1 \quad b_{2,3} = 0.357$

$n = 14 \quad E = 1.333 R_S = 0.876 \quad n_{mis} = 2(0)$

LOO: $R_S = 0.706 \quad n_{mis} = 4(0)$

It was interesting to find that in contrast to the solid sarcoma test system the X substituents appeared to have no electronic influence on the ascites test results and that, furthermore, the steric effects of the Z substituents in this data set is best described by the largest width, B_4. Compounds which have been misclassified in the training sets and mispredicted in the LOO cross-validation are shown in Table 1. According to Moriguchi et al. [5, 19] and Tetko et al. [23] compounds ***5, 13, 15*** and ***16*** are mispredicted in a LOO run of Eq. (21) whereas we found mispredictions for the derivatives ***5*** and ***13, 14,*** and ***15*** in all runs with $\beta = 0.01$, 0.03 and 0.05, or when applying the procedure outlined in Fig. 1. Probably two values have been inadvertently interchanged in the analysis undertaken by Moriguchi et al. [5].

4.4.3.2 Inhibition of Calmodulin Activated Phosphodiesterase

This next section deals with chloropromazine-type inhibitors of calmodulin activated phosphodiesterase [14, 21]. Although activity data of 53 compounds were available, the mode of inhibition was not completely understood. However, it was known that

24 of the analogues could be categorized into three different types (I, II, III) of calmodulin inhibitors, whereas 29 compounds could not be categorized as such. The 24 compounds for which the inhibitory mode of action was known were analyzed by LDA to clarify the structural characteristics of the three groups of inhibitors and this information was used to classify the 29 analogs. Finally, the combined set of 31 type I compounds (6 derived from the first group of 24, and 25 obtained from the second group of 29) was investigated by ALS to identify properties responsible for their assignment to the three classes of activity. In this investigation physico-chemical characteristics of stable and semi-stable "candidate conformers" of flexible derivatives were considered (22 compounds with one conformer, 7 with two conformers and 2 with three conformers). Using conformation-dependent descriptors in the QSAR analysis, a *simultaneous* selection of the best set of conformers and the best subset of descriptors was performed [56!]. (A similar analysis with simultaneous selection of sets has also been performed by Yoshii et al. [15]). The technique of simultaneously selecting the best set of descriptors and conformers is certainly prone to giving chance correlations. Nevertheless, this approach of, first, determining a group of isomechanistic compounds by LDA, and then investigating QSARs for the activity ranking within the homogeneous group, is very interesting. Moriguchi et al. [5] showed that the discrimination between compounds with different types of activity can also be performed by ALS when only two types are considered or one type is compared with several others.

4.4.3.3 Fungicidal Methyl *N*-Phenylcarbamates

A few years ago Takahashi and Kirino et al. [31] determined the fungicidal activity of methyl *N*-phenylcarbamates against gray mold of cucumber caused by *Botrytis cinerea* resistant to benzimidazole fungicides. Two test systems were investigated:

a) the preventive activity (2 classes) after foliar application in pot tests that can be viewed as a type of "in vivo" test, and
b) an in vitro test for the determination of the concentration required for the 50% inhibition of mycelial growth (pIC_{50}).

Interestingly, the activity classes observed for 19 derivatives in test a) could be quantitatively described in a linear ALS equation by pIC_{50} values and by substituent lipophilicity ($n_{mis,LOO} = 0$). Obviously, these variables are descriptors of specific activity and transport properties. Furthermore, pIC_{50} values of 69 compounds showed a significant correlation with the physico-chemical substituent effects. Similar relationships between quantitative in vivo and in vitro test results have often been found, e.g. in medicinal chemistry. This type of intercorrelation is, however, not frequently described for quantitative and semi-quantitative data. Clearly the two relationships mentioned above can be used to predict the in vivo classification of homologous compounds.

A successful prediction of the activity rating of (morpholinocarbonyl)furoindoles with analgesic and anti-inflammatory activities has been described by Kawashima and Moriguchi et al. [10, 11]. After analyzing the semi-quantitative data (3 classes) of 38 derivatives by the ALS technique these researchers synthesized 15 additional compounds to confirm the correlations obtained. Fortunately, in both test sytems 13 of the additional derivatives showed class 3 activity.

4.4.3.4 Antihypertensive Acryloylpiperazinoquinazolines

In this latter part of Sec. 4.4.3 dose- and property-dependent activity classes of antihypertensive acryloylpiperazinoquinazolines (II) are analyzed by the ALS technique [55].

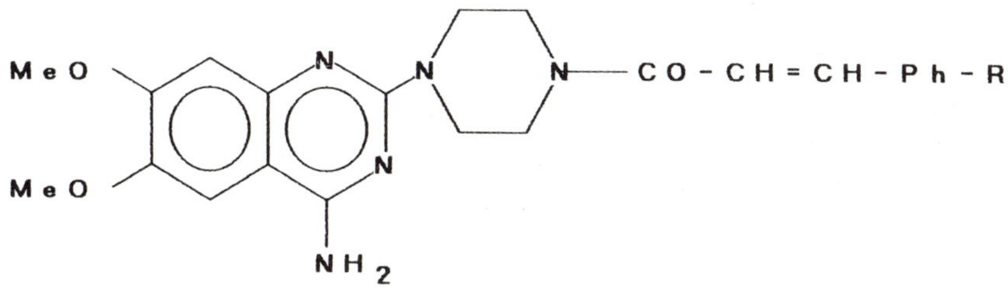

(II)

A few years ago Schaper [57] showed that single or all measurements from several dose-response curves can be included in a correlation of logit(% effect) with physico-chemical descriptors ***and*** log(concentration):

$$\text{logit}(\%) = \log(\%/(100-\%)) = b^*\log 1/ED_{50} + b^*\log C \quad (23)$$

$$\log 1/ED_{50} = a_0 + a_1X_1 + a_2X_2 + \ldots \quad (24)$$

$$\text{logit}(\%) = b_0 + b_1X_1 + b_2X_2 + \ldots + b^*\log C \quad (25)$$

Using ALS, Schaper and Saxena [34] later showed that this same approach could be applied to activity classes, thus, taking into consideration experimental uncertainties in the data as well as allowing for the inclusion of 0% and 100% effect data for which the logit transformation is not defined. This method is demonstrated here with the example of the antihypertensives (II) investigated by Sekiya et al. [58]. This data set was analyzed previously using ALS [59] and other methods [22, 32, 35, 60] without taking into consideration the dose-dependence. To make the analysis easier we restricted ourselves to aryl acryloyl derivatives (II, aryl = substituted phenyl)

Table 2. The observed, calculated and predicted (LOO-cross-validated) activity data for a given dose of 2-(4-Acryloylpiperazino)-4-amino-6,7-dimethoxyquinazolines (II)

R	Dose log M/kg	Activity Class						Activity Score Z	
		Obsd.		Calc.		Pred.		Calc.	Pred.
		3 h	6 h	3 h[a]	6 h[b]	3 h[a]	6 h[b]	3 h[a]	3 h[a]
H	−5.182	2	2	2	2	2	2	0.194	0.214
	−5.659	2	2	2	2	2	2	−0.180	−0.200
	−6.182	1	2	1	2	1	2	−0.589	−0.712
2-Me	−4.646	3	3	**2**	**2**	**2**	**2**	0.547	0.521
	−5.169	2	2	2	2	2	2	0.137	0.135
	−5.646	1	1	1	1	**2**	**2**	−0.236	−0.238
3-Me	−4.655	2	2	2	2	**3**	**3**	1.253	1.447
4-Me	−5.227	2	1	2	**2**	2	**2**	1.044	1.075
4-*i*-Pr	−4.664	1	1	1	1	**2**	1	−0.243	0.073
2-MeO	−4.699	1	1	1	1	1	**2**	−0.518	−0.614
3-MeO	−4.710	2	2	2	2	2	2	0.249	0.272
4-MeO	−4.653	3	3	3	3	3	3	1.382	1.415
	−5.176	2	3	2	3	2	**2**	0.972	1.003
	−5.653	2	2	2	2	2	2	0.599	0.624
2-EtO	−4.714	1	–	1	–	1	–	−0.534	−0.549
4-EtO	−5.189	2	3	2	**2**	2	**2**	0.212	0.200
	−5.666	2	3	2	**2**	2	**2**	−0.162	−0.158
	−6.189	1	1	1	1	1	1	−0.571	−0.624
4-*i*-PrO	−4.726	2	2	2	2	2	2	0.526	0.512
3,5-$(MeO)_2$	−4.709	1	1	1	1	1	1	−2.202	−2.454
2,3,4-$(MeO)_3$	−4.707	2	2	2	2	2	**3**	0.860	0.918
4-Cl	−4.702	1	2	1	2	**2**	2	−0.235	−0.026
3,4-Cl_2	−4.689	1	1	1	1	1	1	−1.373	−1.436
4-Br	−4.750	2	2	2	2	2	2	−0.103	−0.103
3-NO_2	−4.715	1	1	1	1	1	1	−1.686	−1.814
3-CF_3	−4.696	1	1	1	1	1	1	−0.681	−0.757

[a] calc. by Eq. (29), see cut-off points in Table 6.
[b] calc. by Eq. (30), see cut-off points in Table 6.

whereas previously published analyses investigated a more heterogeneous set of compounds (aryl = substituted phenyl, furyl, thienyl). The molar doses and activity ratings observed three and six hours after oral administration of drugs to rats are listed in Table 2. A graphical illustration of the dose-dependent activity classes 3 h after administration is given in Fig. 3. Table 3 presents a list of descriptor values used in the analysis. Values for the resonance effect r^*R and the field effect f^*F were obtained from Ref. [61] and the remaining data from Ref. [62].

Preliminary ALS analyses with different sets of descriptors showed inconclusive results because of the numerical instability of some regression coefficients. Indeed,

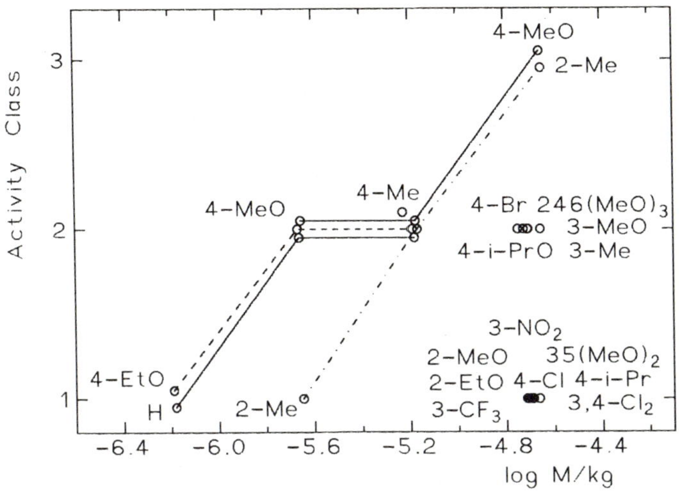

Figure 3. Classes of antihypertensive activity of 18 cinnamoyl-piperazinoquinazolines (II) observed 3 hours after oral administration of different doses to rats. The R groups are shown with the position of substitution on the phenyl ring indicated. (Class 1: no significant blood pressure reduction (BPR) (BPR < 10%); class 2: 10% ≤ BPR < 20%; class 3: 20% ≤ BPR < 30%).

this effect could be traced back to intercorrelations among descriptors. In such a situation, principal component analysis (PCA) [47, 48] may be useful for obtaining uncorrelated parameters. Table 4 shows the correlation matrix of the continuous variables considered in this analysis. The multiple correlation coefficients $R_{mult.}$ of the regression between every one descriptor and the remaining regressors is given in the bottom row. An inspection of the correlation matrix suggests that there is no serious intercorrelation among descriptors. However, the $R_{mult.}$ values clearly show the opposite and underline the necessity for a more detailed analysis. (It is unusual though to use *F*, *R* and σ simultaneously in one regression analysis, but here they are considered in conjunction in order to obtain a better description of the electronic effects). Performing a PCA with subsequent VARIMAX rotation [47] based on the correlation matrix in Table 4 results in the loadings of descriptors onto four VARIMAX-rotated principal components (*PCVR*s), listed in Table 5, and onto the PCVR scores of the measurements, listed in Table 3. The high loading values in Table 5 show that *PCVR* 1 is mainly an expression of the resonance effect, whereas *PCVR* 2, *PCVR* 3 and *PCVR* 4 represent the field effect, the bulk effect of *meta/para* positions and the overall substituent lipophilicity, respectively. As *PCVR* 3 and *PCVR* 4 represent almost pure substituent effects, they may also be used in the form of $(PCVR3)^2$ and $(PCVR4)^2$ to investigate whether non-linear relationships between the activity ranking and these substituent effects do exist. In previous analyses [22, 32, 35, 59, 60] performed on one section of the data given by Sekiya

Table 3. Physico-chemical property data of 2-(4-acryloylpiperazino)-4-amino-6,7-dimethoxyquinazolines (II); ($\Sigma o,m,p$-position for columns 1 to 4, $\sigma_o = \sigma_p$; $\Sigma MR = \Sigma MR_{m,m',p}$; $MR_o = MR_{ortho}$; $PCVR1$ to $PCVR4$ = scaled and VARIMAX-rotated principal components derived from the values given in columns 1 to 5).

R	ΣfF	ΣrR	$\Sigma\sigma$	$\Sigma\pi$	ΣMR	MR_o	$PCVR1$	$PCVR2$	$PCVR3$	$PCVR4$
H	0.000	0.000	0.00	0.00	3.09	1.03	0.587	−0.855	−0.753	−0.831
	0.000	0.000	0.00	0.00	3.09	1.03	0.587	−0.855	−0.753	−0.831
	0.000	0.000	0.00	0.00	3.09	1.03	0.587	−0.855	−0.753	−0.831
2-Me	−0.065	−0.122	−0.17	0.56	3.09	5.65	−0.221	−0.735	−1.197	0.504
	−0.065	−0.122	−0.17	0.56	3.09	5.65	−0.221	−0.735	−1.197	0.504
	−0.065	−0.122	−0.17	0.56	3.09	5.65	−0.221	−0.735	−1.197	0.504
3-Me	−0.051	−0.049	−0.07	0.56	7.71	1.03	0.413	−1.147	0.036	0.263
4-Me	−0.052	−0.141	−0.17	0.56	7.71	1.03	−0.004	−1.074	−0.078	0.320
4-*i*-Pr	−0.080	−0.120	−0.15	1.53	17.04	1.03	0.292	−1.634	1.712	2.102
2-MeO	0.515	−0.432	−0.27	−0.02	3.09	7.87	−1.060	0.744	−1.611	−0.537
3-MeO	0.405	−0.174	0.12	−0.02	9.93	1.03	0.753	−0.217	0.546	−1.043
4-MeO	0.413	−0.500	−0.27	−0.02	9.93	1.03	−0.800	0.044	0.116	−0.828
	0.413	−0.500	−0.27	−0.02	9.93	1.03	−0.800	0.044	0.116	−0.828
	0.413	−0.500	−0.27	−0.02	9.93	1.03	−0.800	0.044	0.116	−0.828
2-EtO	0.356	−0.383	−0.24	0.38	3.09	12.47	−0.988	0.478	−1.664	0.302
4-EtO	0.363	−0.444	−0.24	0.38	14.53	1.03	−0.474	−0.394	1.118	−0.155
	0.363	−0.444	−0.24	0.38	14.53	1.03	−0.474	−0.394	1.118	−0.155
	0.363	−0.444	−0.24	0.38	14.53	1.03	−0.474	−0.394	1.118	−0.155
4-*i*-PrO	0.488	−0.724	−0.45	1.05	19.12	1.03	−1.540	−0.000	1.454	1.344
3,5-$(MeO)_2$	0.809	−0.347	0.24	−0.04	16.77	1.03	0.921	0.418	1.846	−1.256
2,3,4-$(MeO)_3$	1.333	−1.105	−0.42	−0.06	16.77	7.87	−2.278	2.279	0.558	−0.747
4-Cl	0.690	−0.161	0.23	0.71	8.09	1.03	0.593	0.855	−0.511	0.718
3,4-Cl_2	1.366	−0.217	0.60	1.42	13.09	1.03	1.133	2.466	−0.115	2.190
4-Br	0.727	−0.176	0.23	0.86	10.94	1.03	0.662	0.763	0.065	0.951
3-NO_2	1.087	0.054	0.71	−0.28	9.42	1.03	2.409	1.263	0.453	−1.680
3-CF_3	0.618	0.065	0.43	0.88	7.08	1.03	1.421	0.625	−0.541	0.999

Table 4. Correlation matrix and multiple correlation coefficients of the continuous descriptors of Table 3 ($n = 26$).

	$\Sigma fF_{o,m,p}$	$\Sigma rR_{o,m,p}$	$\Sigma\sigma_{p,m,o=p}$	$\Sigma\pi_{o,m,p}$	$\Sigma MR_{m,m',p}$
$\Sigma fF_{o,m,p}$	1.000				
$\Sigma rR_{o,m,p}$	−0.422	1.000			
$\Sigma\sigma_{p,m,o=p}$	0.439	0.620	1.000		
$\Sigma\pi_{o,m,p}$	−0.036	0.155	0.123	1.000	
$\Sigma MR_{m,m',p}$	0.490	−0.574	−0.075	0.270	1.000
$R_{mult.}$:	0.994	0.997	0.996	0.624	0.894

Table 5. Loadings of five variables (columns 1 to 5 of Table 3) onto VARIMAX-rotated principal components PCVR 1 to 4 and corresponding eigenvalues of *PCVR*s.

	PCVR 1	*PCVR* 2	PCVR 3	*PCVR* 4
ΣfF	0.075	**0.957**	0.274	−0.044
ΣrR	**0.836**	−0.400	−0.357	0.112
$\Sigma\sigma_{p,\mathrm{m,o=p}}$	**0.915**	0.397	−0.022	0.057
$\Sigma\pi_{p,m,o}$	0.093	−0.038	0.134	**0.986**
$\Sigma MR_{m,m',p}$	−0.186	0.268	**0.929**	0.176
eigenvalue	1.586	1.307	1.084	1.021

[59], the indicator variable $I_{o-\mathrm{OR}}$ was found to be significant. Therefore, we have also included this descriptor as well as the indicator $I_{o-\mathrm{R}}$ and MR_o (= MR_{ortho}). Finally, of course, the dose (log M/kg) is also added to the descriptor set. Using log M/kg, the *PCVR*s and one of the *ortho*-effect descriptors the ALS analysis could be performed without encountering any problems. The stepwise regression analysis of the 3 h data leading to the most significant relationship (Eq. (29)) is illustrated in Table 6. The final equation for the 6 h data (Eq. (30)) was obtained analogously and is also presented in Table 6.

Using $I_{o-\mathrm{OR}}$ or $I_{o-\mathrm{R}}$ instead of MR_o did not lead to improved equations. This is in contrast to the ALS equation by Sekiya [59] (referred to by Moriguchi et al. [60]) which shows that *o*-OR groups are unfavorable for activity. Unfortunately, the paper by Sekiya is not available in this laboratory. According to Moriguchi et al. [60] the more heterogeneous data set given by Sekiya [58, 59] is, furthermore, described mostly in terms of lipophilicity and electronic effects, both having a negative effect on activity. The same dependence on electronic effects is indicated by Eqs. (29) and (30), whereas lipophilicity reaches the border of significance. Interestingly, both equations show a non-linear decrease of activity with steric bulk at the *meta* and *para* position ($\approx$ *PCVR* 3). This effect was possibly not tested by Sekiya [59, 60]. In Table 2 the observed activity classes are compared with the ratings calculated using Eqs. 29 and 30 and with those predicted by the LOO cross-validation. Calculated and predicted activity scores $Z_{\mathrm{calc.}}$ obtained from Eq. 29 are also listed in Table 2. Fig. 4 shows the dose-dependence of $Z_{\mathrm{calc.}}$ as calculated from Eq. (29) for the 18 compounds (with $n = 26$ doses) of the 3 h data set. The only misclassification occurs at the highest dose in regression line 8 (R = 2−Me) with a $Z_{\mathrm{calc.}}$ value almost in the center of the class 2 range (class 3 had been observed). For this measurement, the LOO procedure (mis)predicts almost the same value ($Z_{\mathrm{pred.}} \approx Z_{\mathrm{calc.}}$, see Table 2) whereas all other mispredicted values have $Z_{\mathrm{pred.}}$ values near to the cut-off points.

Table 6. Stepwise ALS regression analysis of 3 h activity ratings (Eqs. (26) – (29)) and of 6 h data (Eq. (30)) using the activity data in Table 2 and descriptor values in Table 3[a,b].

log M/kg	$PCVR1$ ($\approx R$[c])	$PCVR2$ ($\approx F$[d])	$PCVR3$ ($\approx \Sigma MR_{mm'p}$)	$(PCVR3)^2$	$PCVR4$ ($\approx \Sigma\pi$)	$(PCVR4)^2$	MR_o	intercept	n_{mis}	R_S	LOO n_{mis}	LOO R_S	Eq.
1.093	−0.66	−0.55	−0.02	−0.52	−0.23	0.005	−0.04	5.81	1(0)	0.977	7(0)	0.641	(26)
(2.82)[e]	(2.89)	(2.68)	(0.06)	(2.30)	(1.30)	(0.03)	(0.38)	(2.98)					
(0.51)[f]	(0.66)	(0.55)	(0.02)	(0.54)	(0.23)	(0.01)	(0.14)						
1.02	−0.66	−0.54		−0.53	−0.25		−0.06	5.78	1(0)	0.977	7(0)	0.641	(27)
(3.05)	(4.05)	(3.29)		(3.32)	(1.71)		(0.94)	(3.31)					
(0.50)	(0.66)	(0.54)		(0.55)	(0.25)		(0.17)						
0.91	−0.59	−0.55		−0.57	−0.21			5.14	1(0)	0.977	7(0)	0.641	(28)
(2.71)	(4.08)	(3.43)		(3.93)	(1.44)			(2.97)					
(0.45)	(0.59)	(0.55)		(0.59)	(0.21)								
0.78	−0.68	−0.58		−0.63				4.51	1(0)	0.977	4(0)	0.773	(29)
(2.32)	(4.49)	(3.49)		(4.19)				(2.58)					
(0.39)	(0.68)	(0.58)		(0.66)									
0.62	−0.52	−0.47		−0.58	−0.32			3.57	4(0)	0.866	9(0)	0.644	(30)
(1.36)	(2.58)	(2.12)		(2.73)	(1.56)			(1.50)					
(0.31)	(0.52)	(0.48)		(0.57)	(0.32)								

[a] 3 h data ($n = 26$): $a_1 = -1.154$ $a_2 = 0.692$ $a_3 = 1.846$ $b_{1,2} = -0.231$ $b_{2,3} = 1.269$
[b] 6 h data ($n = 25$): $a_1 = -1.280$ $a_2 = 0.320$ $a_3 = 1.600$ $b_{1,2} = -0.480$ $b_{2,3} = 0.960$
[c] R = resonance effect [d] F = field effect [e] (t) = t-test value [f] (ci) = contribution index

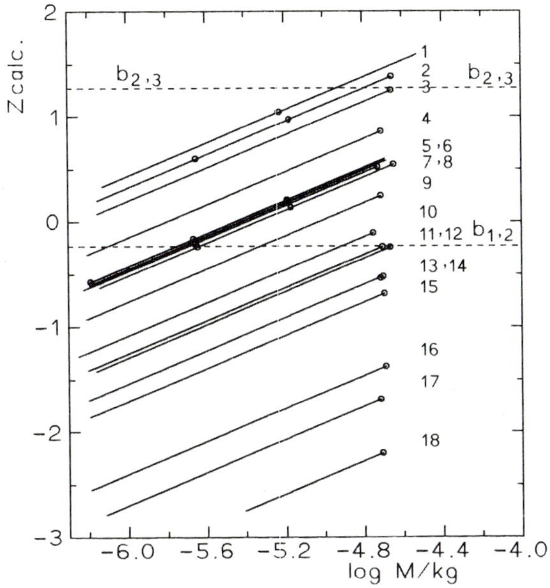

Figure 4. Illustration of the dose-dependence of 3 h activity scores $Z_{calc.}$ obtained from Eq. (29) for the 18 cinnamoylpiperazinoquinazolines (II). The following are the R groups substituted on the phenyl group with positions indicated. **1** 4-Me; **2** 4-MeO; **3** 3-Me; **4** 2,3,4-$(MeO)_3$; **5** 4-EtO; **6** H; **7** 4-*i*-PrO; **8** 2-Me; **9** 3-MeO; **10** 4-Br; **11** 4-Cl; **12** 4-*i*-Pr; **13** 2-MeO; **14** 2-EtO; **15** 3-CF_3; **16** 3,4-Cl_2; **17** 3-NO_2; **18** 3,5-$(MeO)_2$.

4.4.4 Comparison of ALS with Other Methods

ALS has been extensively compared with other methods for classification and pattern recognition such as with for instance LDA [2–6, 9, 24, 27], *k*NN [2, 4, 6, 9, 24, 26, 28, 29, 38], LLM [24, 29, 38], FALS [19, 22, 23, 60], Artificial Neural Networks (ANN) [22, 23, 32, 35, 39], Funclink [22, 60], SIMCA [9], MLR [3], Bayes statistics [29], Iterative Least Squares [29] and Non-linear Regression (NLR) [39]. Generally, ALS is a good contender compared with other methods (except for FALS and Funclink). Frequently, ALS provides a better description of the training set data, whereas its predictive power is approximately similar to other techniques. Specific results obtained by different methods for the previously discussed 3-class mitomycin data sets in Table 1 are compared in Table 7. The favorable results obtained by Moriguchi et al. [60] with ANN for the training set in contrast to the unfavorable results obtained from the LOO cross-validation are obviously an indication of overfitting by estimating too many ANN weights (memory effect).

Wiese and Schaper [39] recently compared ALS with ANN and NLR in the analysis of dose- and property-dependent % effect data [63] of acaricidal chloromethanesulfonamides ($ClCH_2SO_2NR^1R^2$). They also analyzed the effect of

Table 7. Comparison of the classification results obtained by different methods for the 3-class data sets of Table 1; (a) training set data, (b) LOO cross-validation.

Method		Solid Sarcoma		Ascites Sarcoma		Ref.
		n_{mis}	R_S	n_{mis}	R_S	
ALS	(a)	1(0)	0.969	2(0)	0.878	[5, 60]
	(b)	4(0)	0.833	4(0)	0.706	[5, 60]
LDA	(a)	–	0.946	–	0.795	[5]
	(b)	–	0.601	–	0.667	[5]
FALS	(a)	1(0)	0.969	2(0)	0.876	[19, 60]
	(b)	2(0)	0.901	2(0)	0.876	[19, 60]
Funclink	(a)	1(0)	0.969	1(0)	0.931	[60]
	(b)	2(0)	0.901	1(0)	0.931	[60]
ANN	(a)	0(0)	1	0(0)	1	[60]
	(b)	5(2)	0.503	7(0)	0.548	[60]
	(b)	4(0)	0.854	–	–	[23]

Table 8. Comparison of the classification results for acaricidal chloromethanesulfonamides obtained by different methods with different class limits[a]; (a) training set data, (b) LOO cross-validation; $n = 43$.

Method		I[a]		II[a]		III[a]	
		n_{mis}	R_S[b]	n_{mis}	R_S[b]	n_{mis}	R_S[b]
ALS	(a)	5(0)	0.964	9(0)	0.906	12(1)	0.886
	(b)	12(0)	0.885	17(2)	0.822	18(2)	0.810
ANN	(a)	10(0)	0.900	11(0)	0.913	11(0)	0.917
	(b)	12(1)	0.878	15(2)	0.847	17(2)	0.826
NLR	(a)	11(0)	0.911	14(0)	0.883	15(0)	0.887
	(b)	13(3)	0.852	15(2)	0.847	17(2)	0.826

[a]

	class 1	class 2	class 3	class 4
class limits (I):	<60%,	60 to <80%	80 to 90%	>90%
class limits (II):	<35%,	35 to <65%,	65 to 85%,	>85%
class limits (III):	<30%,	30 to <70%,	70 to 85%,	>85%

[b] Ref. [55]

different a priori classification schemes on the predictive power, i.e. unequally spaced class limits in the test data chosen by the authors [63] were changed to more evenly spaced % effect ranges. The class limits and classification results are listed in Table 8. In this investigation, unclassified % effect data were analyzed by ANN and NLR with a posteriori classification according to class limits. The results showed that the classification capability of all three methods was dependent on the chosen class limits and that the class ranges (I) appeared to be optimal. Furthermore, it was found that ALS could describe the observed activity classes better than NLR and ANN. On comparing LOO predictions all three methods gave similar results.

4.4.5 Non-linear ALS Analysis

With the exception of one paper [33], only linear regression type ALS analysis has until now been described to our knowledge. However, in the last decade it has become obvious that increasingly non-linear functions are required to obtain a satisfactory description of biological activity data. Therefore, non-linear relationships should also be considered for the analysis of classified data. Within the framework of the ALS method this means that the MLR procedure must be replaced by non-linear regression (NLR) analysis [42]. Unfortunately, NLR is not as straightforward as MLR. Within each ALS cycle an iterative optimization of NLR coefficients must be performed, and furthermore, a specific (non-linear) function must be provided.

In the field of QSAR "bilinear" relationships are often determined between activity and $\log P$ [62, 64]. Therefore, an artificial data set based on Eq. (31) has been generated to show a non-linear ALS analysis. In Sec. 4.4.5.1 experimental data are analyzed.

$$Y = 1.5 \log P - 2 \log (10^{-1}P + 1) + 5 \tag{31}$$

Unclassified activities Y calculated using Eq. (31) and the corresponding "observed" activity classes are listed in Table 9. Usually it is possible to approximately fit bilinear data to a parabolic relationship. If this approach is applied to this data set, the following equation is obtained by linear (!) ALS analysis:

$$\hat{Z} = 0.451 \log P - 0.102 (\log P)^2 + 0.385 \tag{32}$$

$$\begin{array}{llll} (t): & (5.22) & (6.26) & (1.66) \\ (ci): & (1.45) & (1.74) & \end{array}$$

$$a_1 = -1.375 \quad a_2 = -0.125 \quad a_3 = 1.125 \quad b_{1,2} = -0.75 \quad b_{2,3} = 0.5625$$

$$n = 16 \quad E = 0.804 \quad R_S = 0.963 \quad n_{mis} = 1(0)$$

$$\text{LOO:} \quad R_S = 0.702 \quad n_{mis} = 6(0)$$

Obviously, the description of the total (training) set by this equation is quite good. However, the LOO results clearly show that this model is not really suitable. Of course the bilinear fit is definitely superior:

$$\hat{Z} = 0.738 \log P - 1.189 \log (10^{-1.574} P + 1) + 0.339 \tag{33}$$

$$\begin{array}{lllll} (t): & (5.79) & (7.65) & (2.76) & (1.35) \end{array}$$

Table 9. Artificial activity data Y generated by a bilinear relationship (Eq. 31) and the corresponding "observed" activity classes analyzed by linear (parabolic) and non-linear (bilinear) ALS technique (Eqs. (32) and (33)).

No.	log P	Y^{a}	Activity Class					$\hat{Z}$	$\hat{Z}$
			"obsd"[b]	calc.[c]	pred.[c,e]	calc.[d]	pred.[d,e]	calc.[d]	pred.[d,e]
1	−4	−1.000	1	1	1	1	1	−2.615	−2.960
2	−3	0.500	1	1	1	1	1	−1.877	−2.052
3	−2	1.999	1	1	**2**	1	1	−1.138	−1.212
4	−1.5	2.747	1	**2**	**2**	1	**2**	−0.769	−0.624
5	−1	3.491	2	2	2	2	2	−0.401	−0.452
6	−0.5	4.223	2	2	2	2	2	−0.035	−0.028
7	0	4.917	2	2	2	2	2	0.325	0.277
8	0.5	5.511	3	3	**2**	3	**2**	0.666	0.518
9	1	5.898	3	3	**2**	3	3	0.955	0.986
10	1.5	6.011	3	3	3	3	3	1.130	1.256
11	2	5.917	3	3	3	3	3	1.144	1.317
12	2.5	5.723	3	3	3	3	3	1.025	1.240
13	3	5.491	3	3	3	3	3	0.839	1.000
14	4	4.999	2	2	**3**	2	**3**	0.405	0.733
15	6	4.000	2	2	**1**	2	2	−0.494	−0.495
16	8	3.000	1	1	1	1	1	−1.395	−1.708

[a] Eq. (31); [b] class 1: $Y<3.25$; class 2: $3.25 \le Y \le 5.25$; class 3: $5.25 < Y$; [c] Eq. (32); [d] Eq. (33); [e] LOO cross-validation

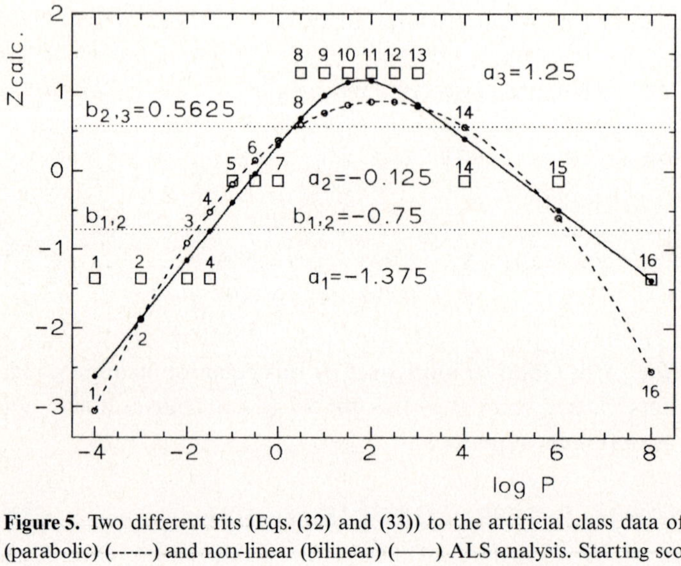

Figure 5. Two different fits (Eqs. (32) and (33)) to the artificial class data of Table 9 obtained by linear (parabolic) (------) and non-linear (bilinear) (——) ALS analysis. Starting scores are indicated by squares

$$n = 16 \quad E = 0.299 \quad R_S = 1 \qquad n_{mis} = 0$$
$$\text{LOO:} \qquad R_S = 0.853 \quad n_{mis} = 3(0)$$

The optimal $\log P$ values as defined by Eqs. (31), (32) and (33) are 1.48, 2.21 and 1.79, respectively. Fig. 5 shows the starting scores (a_i values) of the compounds as well as the curves calculated by Eqs. (32) and (33).

4.4.5.1 Non-linear ALS Analysis of Activity Data of Enantiomeric Mixtures

Recently Schaper [33, 65] showed that the biological activity of combined drugs that are mutually exclusive in their binding to a common receptor site can be described by the non-linear relationship (34). This relationship holds true also for mixtures of stereoisomers/enantiomers.

$$Y_{1..j} = \log \left[\sum_{i=1}^{j} (f_i{*}10^{Y_i}) \right] \tag{34}$$

where

$Y_{1..j}$ = activity ($\log 1/EC_{50}$) of the mixture of drugs/isomers $1..j$;
Y_i = activity of the pure isomer/enantiomer i;
f_i = mole fraction of isomer i in the mixture.

The activities of pure enantiomers are dependent on their physico-chemical properties. This dependence may also be recognized by QSAR analysis of activity data of incompletely resolved enantiomeric mixtures.

This approach has, indeed, been applied by Schaper [33] to the pED_{50} data given by Manabe et al. [66] who investigated the fungicidal effect of chiral *N*-acylimidazoles (acyl = $4{-}X{-}Ph{-}CH_2{-}C{*}H(t{-}Bu){-}CO{-}$) against powdery mildew on barley. Ten compounds were tested as racemates and four of these racemates were also tested after being resolved into the (+)- and (−)-enantiomers. However, all purified isomers were still contaminated with the other optical isomer by up to 6.5%. This contamination of course is accounted for in terms of the mole fractions in Eqs. (34) and (35). The absolute configuration of the isomers is unknown, but as all (+)-enantiomers were found to be more active than the (−)-enantiomers it may be safe to assume that all (+)-isomers have the same configuration. For two enantiomers ((−)-H, (−)-OMe, see Table 10) the activity was determined as $pED_{50} < 3.7$ [66]. Schaper [33] employed an arbitrary value of $pED_{50} = 3.4$ in the NLR analysis although assuming that this approach might not be correct. Therefore we classified all compounds into four activity classes (Table 10) which classify the two inactives into the lowest activity class 1. Using the same equation as in the NLR analysis [33], an excellent description of the class data by the field

Table 10. Fungicidal activity of chiral 2-(4-X-Benzyl)-2-*t*-butylacetylimidazoles towards powdery mildew (*Erysiphe graminis*) on barley; comparison of the observed pED_{50} values and activity classes with classes obtained by non-linear ALS analysis (Eq. (35)).

No.	X	Opt. $f_{(+)}$	Purity $f_{(-)}$	*E. gram.* pED_{50}	Activity Class obsd.[a]	calc.[b,d]	pred.[c,d]	$Z_{calc.}$[d]	F	B_1
1	(+)-H	0.999	0.001	4.33	2	2	2	−1.102	0.00	1.00
2	(−)-H	0.003	0.997	<3.70	1	1	1	−2.404	0.00	1.00
3	(±)-H	0.500	0.500	4.34	2	2	**1**	−1.382	0.00	1.00
4	(+)-OMe	0.980	0.020	5.15	3	3	3	−0.043	0.26	1.35
5	(−)-OMe	0.011	0.989	<3.70	1	1	**2**	−1.612	0.26	1.35
6	(±)-OMe	0.500	0.500	5.15	3	3	3	−0.329	0.26	1.35
7	(+)-Cl	0.950	0.050	6.35	4	4	4	0.898	0.41	1.80
8	(−)-Cl	0.037	0.963	4.82	2	**3**	**3**	−0.471	0.41	1.80
9	(±)-Cl	0.500	0.500	6.35	4	4	**3**	0.621	0.41	1.80
10	(+)-Br	0.935	0.065	6.29	4	4	4	1.164	0.44	1.95
11	(−)-Br	0.049	0.951	5.22	3	3	3	−0.097	0.44	1.95
12	(±)-Br	0.500	0.500	6.06	4	4	4	0.893	0.44	1.95
13	(±)-I	0.500	0.500	5.89	4	4	4	1.075	0.40	2.15
14	(±)-NO_2	0.500	0.500	5.92	4	4	4	1.074	0.67	1.70
15	(±)-CF_3	0.500	0.500	5.96	4	4	4	0.798	0.38	1.98
16	(±)-CN	0.500	0.500	6.34	4	4	4	0.576	0.51	1.60
17	(±)-Me	0.500	0.500	4.66	2	2	**3**	−0.779	−0.04	1.52
18	(±)-SMe	0.500	0.500	5.14	3	3	**2**	0.009	0.20	1.70

[a] class 1: pED_{50}<4.1; class 2: 4.1≤pED_{50}≤4.9; class 3: 4.9<pED_{50}≤5.7; class 4: 5.7<pED_{50}
[b] $a_1 = -1.778$, $a_2 = -1.111$, $a_3 = -0.222$, $a_4 = 1.111$; $b_{1,2} = -1.444$, $b_{2,3} = -0.667$, $b_{3,4} = 0.444$
[c] LOO-cross-validation, [d] calc. by Eq. (35)

effect F and STERIMOL B_1 [62] was obtained with the training set (see Table 10) and satisfactory results were also obtained using a LOO procedure:

$$\hat{Z} = \log\left(f_{(+)} * 10^{Z_{(+)}} + f_{(-)} * 10^{Z_{(-)}}\right) \tag{35a}$$

with

$$Z_{(+)} = -2.467 + 2.265\,F + 1.366\,B_1 \tag{35b}$$
$$(t)\text{:} \quad (4.75) \quad (3.58) \quad (3.61)$$

$$Z_{(-)} = -2.429 + 2.265\,F \tag{35c}$$
$$(t)\text{:} \quad (7.27) \quad (3.58)$$

$$n = 18 \quad E = 0.295 \quad R_S = 0.984 \quad n_{mis} = 1(0)$$
$$\text{LOO:} \quad R_S = 0.879 \quad n_{mis} = 6(0)$$

4.4.5.2 Analysis of Embedded Data

ALS is based on the technique of MLR analysis. Therefore, in principle, ALS can discriminate only between compound groups and/or activity classes which are separated by (linear) hyperplanes in multidimensional space. However, in Fig. 5 it has been shown that in the case of a one-dimensional descriptor space, the group of actives which are embedded in inactives can be separated from the latter, if a non-linear discriminant function is fitted to the data. Usually, the unknown non-linear function is approximated by a parabolic relationship which can be analyzed by linear regression techniques. Certainly, this approach can also be applied to a multi-dimensional descriptor space. For instance, in a 2D space with actives being surrounded by inactives, Z values may be described by X_1, X_1^2 and X_2, X_2^2. In this case, concentric circles or ellipses are formed by the curves which separate actives from inactives. However, the axes of these ellipses are necessarily parallel to the descriptor axes. To allow discrimination of inactives from actives forming elliptic planes inclined against the descriptor coordinate axes, descriptor cross-terms (e.g. $X_1{}^*X_2$) must be included.

The artificial embedded data of Fig. 6 has been analyzed by this approach and a complete separation of actives from inactives was possible using Eq. (36):

$$\hat{Z} = 0.667X_1 - 0.120X_1^2 + 0.867X_2 - 0.119X_2^2 + 0.101X_1X_2 - 3.748 \qquad (36)$$

$$a_1 = -1 \quad a_2 = 1 \quad b_{1,2} = 0$$

$$n = 16 \quad E = 0.799 \quad R_S = 1 \quad n_{mis} = 0$$

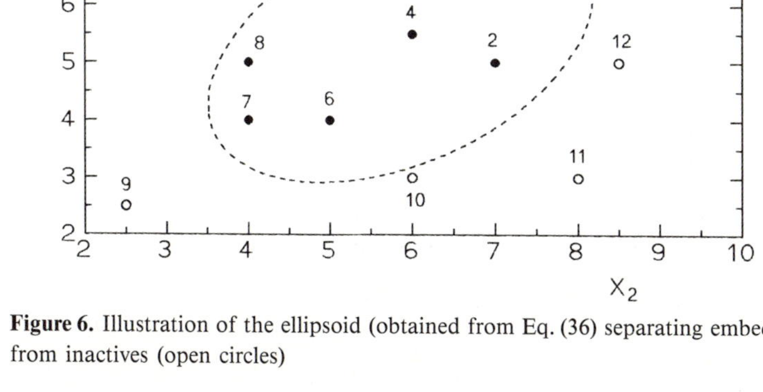

Figure 6. Illustration of the ellipsoid (obtained from Eq. (36) separating embedded actives (filled circles) from inactives (open circles)

The transition of actives to inactives is observed at all those points in X space where $\hat{Z} = b_{j,j+1}$. Therefore, to obtain the equation for the transition curve (ellipsoid) shown in Fig. 6, this expression must be substituted into Eq. (36) and the equation solved for X_1 or X_2.

This approach to the analysis of embedded data is only feasible in the case of a low dimensional descriptor space. In the case of a multi-dimensional space, many different cross-terms must be analyzed. Therefore, in this situation, more complex methods such as Cluster Significance Analysis [67] or Single Class Discriminant Analysis [68] are preferred.

4.4.6 Fuzzy Adaptive Least Squares (FALS)

Between 1988 and 1990 Moriguchi et al. [17–19] introduced an advanced version of the ALS method, named Fuzzy Adaptive Least Squares (FALS) and this technique is still undergoing further development [20–22]. We shall be discussing only the latest developments in this section.

A novel feature of FALS lies in the classification of objects using the concepts of fuzzy theory. Ordered categories contain not only statistical uncertainties, which stem from inaccuracies of measurement, but also a "built-in vagueness", as a result of the subjective criteria employed for the classification. Such uncertainties can be conceptualized by introducing a membership function that indicates the extent to which an object belongs to an activity class. The theory of fuzzy sets [69–71] enables the representation and handling of vague statements and the uncertainties of classifications in a data set. Instead of assigning a sample to a single class, samples may belong to more than one class with different degrees of membership in each class. With the theory of fuzzy sets decisions can be made on the basis of a characteristic function that increases or decreases monotonically with the variable of interest (in this case $Z_{\text{calc.}}$). This function is the membership function that represents a fuzzy set over $Z_{\text{calc.}}$. In contrast to fuzzy set theory, a conventional or crisp subset A of a given universe of elements is usually defined by specifying for every element of the universe if it is a member of A or not. Mathematically, this can be expressed by the membership function that assigns a value of $M(Z) = 1$ to every element that is a member of the subset A and a value of $M(Z) = 0$ to elements that are not members of A. Fuzzy sets are obtained by a generalization of the concept of a membership function to allow for membership values between 0 and 1. To illustrate this, for example, the membership in a certain activity class would be ascertained for an object by determining all those values of Z that have a membership value $M(Z)$ which is greater than, say, 0.5. The membership function chosen for solving a problem by fuzzy set theory depends on the classification problem at hand.

FALS involves the following changes and extensions of ALS:

1. FALS is performed in two steps. Step 1 is a complete ALS analysis with 20 iterations. The best result is used as a starting point in Step 2 (the actual FALS).

2. For each object the membership value $M(Z)$ is calculated as a function of $Z_{\text{calc.}}$.

3. The ALS correction term C_1 depends on $M(Z)$.

4. The optimization criterion is the product of R_S and the mean membership grade: $R_S{*}MMG$.

A flow chart of FALS is shown in Fig. 7. In FALS the a posteriori classification is performed by the same algorithm as in ALS (comparison of $Z_{\text{calc.}}$ with cut-off points, (refer to text following Eq. (5)). The main difference between ALS and FALS is the adaptation step, which in the case of FALS is dependent on the $M(Z)$ values. Moriguchi et al. [17–22] defined the following membership function for a compound with observed class 1 (Eq. (37a)), class j ($j = 1..g$, Eqs. (37b)–(37d) or class g (Eq. (37e)):

$$M(Z) = 1 \quad \text{if } j = 1 \text{ AND } Z \leq (b_{1,2} - FL_{1,2}) \tag{37a}$$

$$M(Z) = \frac{1}{1 + [(Z - b_{j-1,j})/FL_{j-1,j} - 1]^4} \quad \text{if j} > 1 \text{ AND } Z \leq (\text{b}_{j-1,j} + FL_{j-1,j}) \tag{37b}$$

$$M(Z) = 1 \quad \text{if } 1 < j < g \text{ AND } (b_{j-1,j} + FL_{j-1,j}) < Z \leq (b_{j,j+1} - \text{FL}_{j,j+1}) \tag{37c}$$

$$M(Z) = \frac{1}{1 + [(b_{j,\text{j}+1} - Z)/FL_{j,\text{j}+1} - 1]^4} \quad \text{if } j < g \text{ AND } (b_{j,\text{j}+1} - FL_{j,j+1}) < Z \tag{37d}$$

$$M(Z)_g = 1 \quad \text{if } j = g \text{ AND } (b_{g-1,g} + FL_{g-1,g}) < Z \tag{37e}$$

where $Z = \hat{Z} = Z_{\text{calc.}}$ is the Z-value calculated by MLR, and $FL_{j,j+1}$ is the fuzzy level, a constant quantifying the "fuzziness" or indistinct nature of the boundary between class j and class $j+1$. FL determines the gradient of the ($M(Z)$ vs $Z_{\text{calc.}}$)-curves in the vicinity of the class boundaries. The gradient decreases with increasing FL values

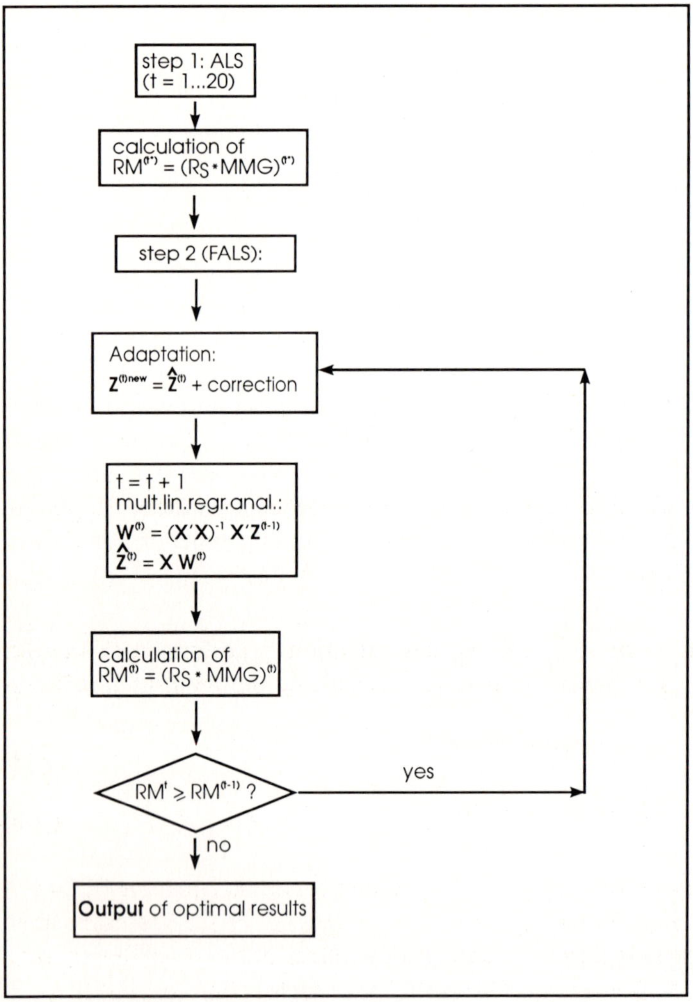

Figure 7. Flow chart of FALS

or increasing fuzziness of the boundary between classes. For a three-class problem Eqs. (37 a) – (37 e) describe three curves shown in Fig. 8 which were calculated with $FL = 0.2$ for all class boundaries [72!]. It becomes clear when comparing Fig. 8 with Eqs. (37 a) – (37 e) that the upwardly sloping gradient of a membership curve is described by Eq. (37 b), whereas the downwardly sloping gradient is described by Eq. (37 d). Furthermore, it is found that $M(Z) = 0.5$ whenever

$$Z_{\text{calc.}} = b_{j-1,j} \text{ or } Z_{\text{calc.}} = b_{j,j+1} \, .$$

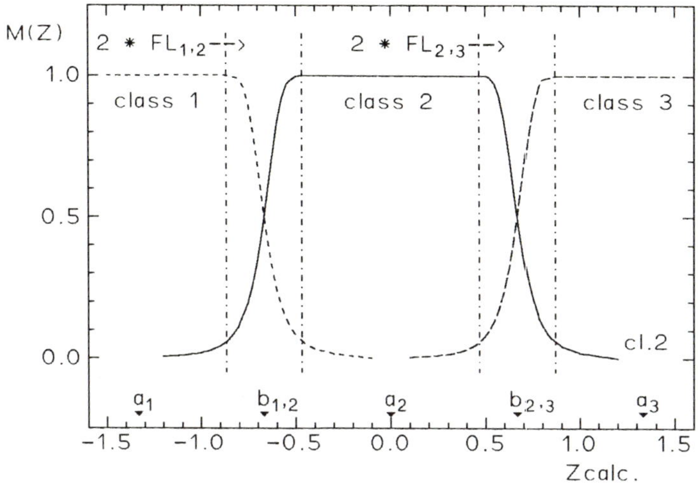

Figure 8. Membership function curves calculated for a three-class classification problem with identical class sizes using fuzzy level $FL = 0.2$ at all class boundaries

In FALS, the adaptation of $Z_{\text{calc.}}$ for the next iteration is performed as in ALS (compare Figs. 1 and 7), the correction term is calculated, however, by a different equation:

$$C_i^{(t)} = \alpha^*\{[1 - M(\hat{Z}_i^{(t)})]^* FL_{j-1}\}^{1/2} \quad \text{if } \hat{Z}_i^{(t)} \leq a_j \tag{38a}$$

$$C_i^{(t)} = -\alpha^*\{[1 - M(\hat{Z}_i^{(t)})]^* FL_j\}^{1/2} \quad \text{if } a_j < \hat{Z}_i^{(t)} \tag{38b}$$

In Eq. (38) α is a constant (usually $\alpha = 0.5$). According to this equation $C_i^{(t)} = 0$ if $Z_{\text{calc.}}$ is within the range of $(b_{j-1,j} + FL_{j-1,j})$ to $(b_{j,j+1} - FL_{j,j+1})$ because then $M(Z) = 1$. The correction term is added (Eq. (38a)) if $Z_{\text{calc.}}$ is lower than $(b_{j-1,j} + FL_{j-1,j})$, it is subtracted (Eq. (38b)) if $Z_{\text{calc.}}$ is higher than $(b_{j,j+1} - FL_{j,j+1})$. The correction profiles, corresponding to the membership functions of Fig. 8, are shown in Fig. 9.

After adaptation of $Z_{\text{calc.}}$ and subsequent MLR analysis the results obtained are evaluated by calculating R_S as well as the mean membership grade *MMG*,

$$MMG = \left[\sum_{i=1}^{n} M(\hat{Z}_i)\right] / n \tag{39}$$

The product of R_S and *MMG* is used as the criterion for the best discrimination in Step 2 of FALS. *MMG* provides a measure of the accuracy of the calculated classification, whereas R_S supplements the information concerning the misclassification of more than one rating. Thus, the iterative least squares calculation

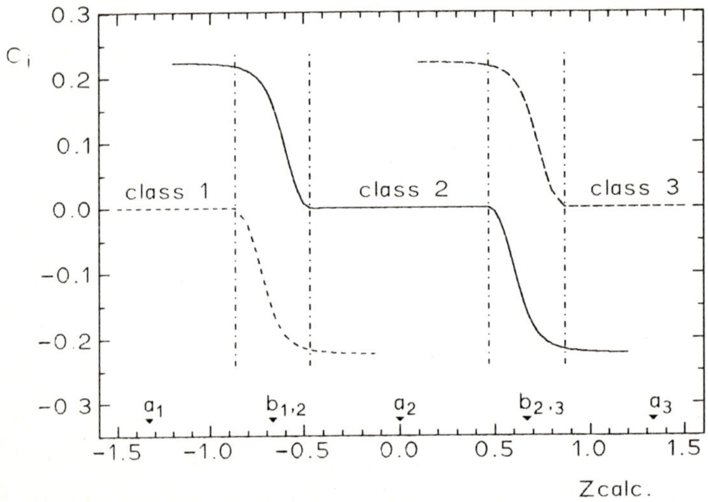

Figure 9. Correction profiles corresponding to Figure 8 calculated for the objects observed in classes 1, 2, or 3 with identical fuzzy levels ($FL = 0.2$) at all class boundaries

is carried out to maximize $\Sigma M(Z)$ or to minimize ΣC_i. Iterations in Step 2 are continued until no further improvement of R_S*MMG is observed (with a maximum of at least 20 iterations). Clearly also, the results of FALS are validated by the LOO prediction. The discriminant function with a scientifically reasonable subset of descriptors which gives the best LOO prediction is finally adopted.

Because of the recent development of FALS, until now only a few papers comparing the use of this method with other techniques have been published [19, 21–23, 60]. Generally, FALS compares very favorably with other methods (see Tables 1 and 7). By applying FALS to the mitomycin data of Table 1 the following equations were obtained by Moriguchi et al. [19]:

Solid Sarcoma

$$\hat{Z} = -5.65\,\sigma_{m,X} - 3.20\,V_{W_X} + 1.66\,\sigma^*_Y - 1.64\,B_{1,Z} + 0.58 \tag{40}$$

(*ci*): (0.70) (0.59) (0.72) (0.74)

$n = 16$ $MMG = 0.925$ $R_S = 0.969$ $n_{mis} = 1(0)$ (*FL* values not given)
LOO: $MMG = 0.883$ $R_S = 0.901$ $n_{mis} = 2(0)$

Ascites Sarcoma

$$\hat{Z} = -1.91\,V_{W_X} + 1.73\,\sigma^*_Y - 0.55\,B_{4,Z} - 0.88 \tag{41}$$

(*ci*): (0.37) (0.79) (0.41)

$n = 14$ $MMG = 0.859$ $R_S = 0.876$ $n_{mis} = 2(0)$ (*FL* values not given)
LOO: $MMG = 0.857$ $R_S = 0.876$ $n_{mis} = 2(0)$

4.4.7 Advantages and Disadvantages of (F)ALS

The advantages are:

- only *one* QSAR equation is obtained which can be easily interpreted;
- only a small number of misclassifications is obtained in the training set; cross-validation results are at least comparable with other techniques;
- the approximate significance of regression coefficients can be tested by a *t*-test;
- non-linear dependences of activity class rankings on physico-chemical descriptors may be analyzed by non-linear (F)ALS.

Furthermore (not shown):

- calculated activity scores ($Z_{calc.}$) obtained by ALS analysis of classified artificial data are highly collinear with unclassified artificial data;
- (F)ALS regression coefficients obtained from classified artificial data are proportional to the coefficients used to generate the unclassified artificial data;

The disadvantages:

- because of the inherent extreme adaptability of (F)ALS the possibility of overfitting must be considered and cross-validation is absolutely necessary;
- for iterative techniques such as (F)ALS, cross-validation requires long computation times (especially in non-linear (F)ALS);
- the basic (F)ALS method provides meaningful results only for linearly separable data;
- if data with more than two activity classes is analyzed by (F)ALS then different rankings of only one type of activity are allowed;
- unequivocal statistical tests on the significance of regression coefficients are not yet available.

Acknowledgement

The author is grateful to Dr. K. Visser and Dr. M. Wiese (Borstel) for the useful criticisms on reading the manuscript for this text.

References

[1] van de Waterbeemd, H., in volume 2 of this series

[2] Thorndike, R. M., *Correlational Procedures for Research*, Gardner Press, New York, 1978

[3] Moriguchi, I. and Komatsu, K., *Chem. Pharm. Bull.* **25**, 2800–2802, 3440 (errata) (1977)

[4] Moriguchi, I., Komatsu, K. and Matsushita, Y., *J. Med. Chem.* **23**, 20–26 (1980)

[5] Moriguchi, I., Komatsu, K. and Matsushita, Y., *Anal. Chim. Acta* **133**, 625–636 (1981)

[6] Moriguchi, I. and Komatsu, K., *Eur. J. Med. Chem.* **16**, 19–23 (1981)

[7] Akahane, K., Momose, D.-I., Iizuka, K., Miyamoto, T., Hayashi, H., Iwase, K. and Moriguchi I., *Eur. J. Med. Chem.* **19**, 85–88 (1984)

[8] Komatsu, K., Hirono, S. and Moriguchi, I., *Chem. Pharm. Bull.* **33**, 4081–4084 (1985)

[9] Moriguchi, I., *Kagaku Zokan (Kyoto)* **107**, 103–116 (1986)

[10] Kawashima, Y., Amanuma, F., Sato, M., Okuyama, S., Nakashima, Y., Sota, K. and Moriguchi, I., *J. Med. Chem.* **29**, 2284–2290 (1986)

[11] Kawashima, Y., Okuyama, S., Sato, M., Hatada, Y., Amanuma, F., Nakashima, Y., Sota, K. and Moriguchi, I., *Chem. Pharm. Bull.* **35**, 402–408 (1987)

[12] Moriguchi, I., Hirono, S. and Liu, Q., *16th Symp. Structure-Activity Relationships,* Kyoto, Oct. 1988, p. 300–303

[13] Liu, Q., Hirono, S., Matsushita, Y., Nakagawa, T. and Moriguchi, I., *17th Symp. Structure-Activity Relationships,* Osaka, Nov. 1989, p. 224–227

[14] Liu, Q., Hirono, S. and Moriguchi, I., *Chem. Pharm. Bull* **38**, 2184–2189 (1990)

[15] Yoshii, F., Liu, Q., Hirono, S. and Moriguchi, I., *Chem. Senses* **16**, 319–328 (1991)

[16] Kawashima, Y., Kameo, K., Kato, M., Hasegawa, M., Tomisawa, K., Hatayama, K., Hirono, S. and Moriguchi, I., *Chem. Pharm. Bull.* **40**, 774–777 (1992)

[17] Moriguchi, I., Hirono, S. and Liu, Q., *Abstracts of Papers, 16th Symp. on Structure-Activity Relationships, Kyoto, Oct. 1988*, p. 300–303

[18] Liu, Q., Hirono, S., Matsushita, Y., Nakagawa, T. and Moriguchi I., *Abstracts of Papers, 17th Symp. on Structure-Activity Relationships, Osaka, Nov. 1989*, p. 224–227

[19] Moriguchi, I., Hirono, S., Liu, Q., Matsushita, Y., and Nakagawa, T., *Chem. Pharm. Bull.* **38**, 3373–3379 (1990)

[20] Liu, Q., Hirono, S., Matsushita, Y. and Moriguchi, I., *Environ. Toxicol. Chem.* **11**, 953–959 (1992)

[21] Moriguchi, I., Hirono, S., Liu, Q. and Nakagome, I., *Quant. Struct.-Act. Relat.* **11**, 325–331 (1992)

[22] Moriguchi, I., Hirono, S., Matsushita, Y., Liu, Q. and Nakagome, I., *Chem. Pharm. Bull.* **40**, 930–934 (1992)

[23] Tetko, I. V., Luik, A. I. and Poda, G. I., *J. Med. Chem.* **36**, 811–814 (1993)

[24] Rose, S. L. and Jurs, P. C., *J. Med. Chem.* **25**, 769–776 (1982)

[25] Sekiya, T., Hata, S. and Yamada, S., *Chem. Pharm. Bull.* **31**, 2432–2437 (1983)

[26] Jurs, P. C., Stouch, T. R., Czerwinski, M. and Narvaez, J. N., *J. Chem. Inf. Comput. Sci.* **25**, 296–308 (1985)

[27] Gombar, V. K., *Arzneim.-Forsch./Drug Res.* **35**, 1633–1636 (1985)

[28] Stouch, T. R. and Jurs, P. C., *J. Med. Chem.* **29**, 2125–2135 (1986)

[29] Gombar, V. K., Jaeger, E. P. and Jurs, P. C., *Quant. Struct.-Act. Relat.* **7**, 225–234 (1988)

[30] Kirino, O., Takayama, C., Yoshida, M., Inoue, S. and Yoshida, R., *Agric. Biol. Chem.* **52**, 561–568 (1988)

[31] Takahashi, J., Kirino, O., Takayama, C., Nakamura, S., Noguchi, H., Kato, T. and Kamoshita K., *Pestic. Biochem. Physiol.* **30**, 262–271 (1988)

[32] Aoyama, T., Suzuki, Y. and Ichikawa, H., *J. Med. Chem.* **33**, 905–908 (1990)

[33] Schaper, K.-J., *Pharmacochem. Libr.* **16**, 25–32 (1991)

[34] Schaper, K.-J. and Saxena, A. K., *Pharmacochem. Libr.* **16**, 45–48 (1991)

[35] Aoyama, T. and Ichikawa, H., *Chem. Pharm. Bull.* **39**, 1222–1228 (1991)

[36] Garcia, E., Lopez-de Cerain, A., Martinez-Merino, V. and Monge, A., *Mutation Res.* **268**, 1–9 (1992)
[37] Lavine, B.K., *Signal Processing and Data Analysis.* In: *Practical Guide to Chemometrics,* Haswell, S.J. ed., Marcel Dekker, New York, 1992, p. 211–238
[38] Jaeger, E.P., Jurs, P.C. and Stouch, T.R., *Eur. J. Med. Chem.* **28**, 275–290 (1993)
[39] Wiese, M. and Schaper, K.-J., *SAR and QSAR Environm. Res.* **1**, 137–152 (1993)
[40] *Data Desk® Professional 2.0* (1988), Odesta Corporation, Northbrook, IL, program derived from ALS81
[41] Bross, I.D.J., *Biometrics* **14**, 18–38 (1958)
[42] Draper, N. and Smith, H., *Applied Regression Analysis.* 2nd edn., J. Wiley, New York, 1981
[43] Schaper, K.-J., *Quant. Struct.-Activ. Relat.* **2**, 111–120 (1983)
[44] Hellberg, S., Sjöström, M., Skagerberg, B., Wikström, C. and Wold, S., *Acta Pharm. Jugosl.* **37**, 53–65 (1987)
[45] Pleiss, M.A. and Unger, S.H., *The Design of Test Series and the Significance of QSAR Relationships.* (Comprehensive Medicinal Chemistry, Vol. **IV**) Hansch, C., Sammes, P.G., Taylor, J.B. and Ramsden, C.A., eds., Pergamon Press, Oxford, 1990, p. 561–587
[46] Morgan, E., *Chemometrics: Experimental Design,* J. Wiley, Chichester, New York, 1991
[47] Schaper, K.-J. and Kaliszan, R., *Applications of Statistical Methods to Drug Design.* In: *Trends in Medicinal Chemistry,* Mutschler, E. and Winterfeldt, E., eds., VCH, Weinheim, 1987, p. 125–139
[48] Franke, R. and Gruska, A., *Principal Component and Factor Analysis.* In: *Chemometric Methods in Molecular Design* (Methods and Principles in Medicinal Chemistry, Vol. II), van de Waterbeemd, H., ed., VCH, Weinheim, 1995
[49] Mager, H., *Quant. Struct.-Act. Relat.* **3**, 147–153 (1984)
[50] Mager, P.P., Coburn, R.A., Solo, A.J., Triggle, D.J. and Rothe, H., *Drug Design & Discovery* **8**, 273–289 (1992)
[51] Johnson, N.L. and Leone, F.C., *Statistic and Experimental Design in Engineering and the Physical Sciences.* Vol. **I**, J. Wiley, New York, 1977, p. 319–325
[52] Press, W.H., Flannery, B.P., Teukolsky, S.A., and Vetterling, W.T., *Numerical Recipes; The Art of Scientific Computing,* Cambridge University Press, Cambridge, 1986, p. 489–491
[53] Cramer, R.D., Bunce, J.D., Patterson, D.E. and Frank, I.E., *Quant. Struct.-Act. Relat.* **7**, 18–25 (1988)
[54] Wold, S., *Quant. Struct.-Act. Relat.* **10**, 191–193 (1991)
[55] Schaper, K.-J., this contribution
[56] Unfortunately only the ALS analysis in [21] could be reproduced. Using the property data of "best conformers" in [14] quite a different QSAR equation is obtained. These values do not coincide with data of Moriguchi [21]. Obviously in both analyses different data have been used.
[57] Schaper, K.-J., *Pharmacochem. Library* **10**, 58–60 (1987)
[58] Sekiya, T., Hiranuma, H., Hata, S., Mizogamo, S., Hanazuka, M. and Yamada, S., *J. Med. Chem.* **26**, 411–416 (1983)
[59] Sekiya, T., In: *Structure-Activity Relationship and Drug-Design,* Fujita, T. ed., Kagakudojin, Kyoto, 1986, p. 129–135, cited in Refs. [22, 32, 35, 60]
[60] Liu, Q., Hirono, S. and Moriguchi, I., *Quant. Struct.-Act. Relat.* **11**, 318–324 (1992)
[61] Williams, S.G. and Norrington, F.E., *J. Amer. Chem. Soc.* **98**, 508–516 (1976)
[62] Seydel, J.K. and Schaper, K.-J., *Chemische Struktur und Biologische Aktivität von Wirkstoffen; Methoden der Quantitativen Struktur-Wirkung-Analyse,* Verlag Chemie, Weinheim, 1979
[63] Tamaru, M., Ogawa, H., Nishimura, T., Takahashi, Y. and Sasaki, S.-I., *J. Pesticide Sci.* **13**, 1–6 (1988)
[64] Kubinyi, H., *Arzneim.-Forsch./Drug Res.* **26**, 1991–1997 (1976)
[65] Schaper, K.-J., *Progr. Clin. Biol. Res.* **291**, 41–44 (1989)

[66] Manabe, A., Kirino, O., Furuzawa, K., Takano, H., Hisada, Y. and Tanaka, S., *Agric. Biol. Chem.* **7**, 1959–1965 (1987)

[67] McFarland, J. W. and Gans, D. J., *Cluster Significance Analysis.* In: *Chemometric Methods in Molecular Design* (Methods and Principles in Medicinal Chemistry, Vol. II), van de Waterbeemd, H., ed., VCH, Weinheim, 1995

[68] Rose, V. S., Wood, J. and Macfie, H. J. H., Chap. 4.3 of this volume

[69] Zadeh, L. A., *Inf. Control* **8**, 338–353 (1965)

[70] Novak, V., *Fuzzy Sets and their Applications,* Adam Hilger, Bristol, 1989

[71] Otto, M., *Chemometrics Intelligent Lab. Systems* **4**, 101–120 (1988)

[72] In the earlier versions of FALS [17–20] different *FL* values were tested at the class boundaries. Usually two types of gradient were employed, i.e., "steep" with $FL = 0.1$ and "gentle" with $FL = 0.5$. This means that in case of a g class problem (with $g-1$ class boundaries) and of $f = 2$ different fuzzy levels the FALS analysis has to be performed with $(g-1)*f$ different combinations of fuzzy levels. Those fuzzy levels giving the best LOO prediction are selected for the final result. In the last version of FALS [21, 22] a single *FL* value is used ($FL = 0.1$) to ensure that maximum *MMG* value is obtained.

4.5 Alternating Conditional Expectations in QSAR

Brian W. Clare

Abbreviations and Symbols

ACE	Alternating Conditional Expectations
C	molar concentration of triazene causing 30 mutations above background in 10^8 bacteria
K_i	apparent inhibition constant for dihydrofolate reductase
MR	molar refractivity
n	number of points (compounds)
π	hydrophobic substituent constant
p	probability (i.e. statistical significance level)
P	octanol-water partition coefficient
QSAR	Quantitative Structure-Activity Relationships
q_{HOMO}	density of highest occupied molecular orbital on triazene nitrogen 1
r	multiple correlation coefficient
r^2_{cv}	cross-validated r^2
s	standard error of estimate
SS_r	residual sum of squares
SS_t	sum of squares of deviations from the mean.

4.5.1 Introduction: Non-Linearity and ACE

In any quantitative and empirical field of research, the technique of multiple linear regression is likely to be useful in discovering a relationship between a variable of interest (the response variable) and a set of predictor variables. There is, however, no a priori reason to expect relationships to be linear. Indeed, in the field of quantitative structure-activity relationships (QSAR) the classic equation of Hansch contains a term quadratic in log *P*. Until recently, there was no easy method of exploring non-linearity in data except by extremely laborious and quite unreliable trial-and-error procedures, involving the trial of non-linear functions of the predictor variables as additional predictor variables. In addition, there was no systematic procedure to determine optimal non-linear transformations. However, the situation has now changed following the introduction of the Alternating Conditional Expectations (ACE) method by Breiman and Friedman [1]. This method was first applied to QSAR by Franke and Lanteri [2].

For a dependent variable y and a set of independent variables x_i, the ACE method involves fitting a set of functions in the form of:

$$f(y) = \Sigma_i g_i(x_i) + \varepsilon \tag{1}$$

where f and the g_i are univariate arbitrary non-linear functions, and ε is a residual error term. The functions f and g_i are chosen such that the fit is optimized (i.e. the sum of squares of the residuals), being subject only to the condition that they are smooth functions.

In order to remove indeterminacy, it is assumed that $f(y)$ and the $g_i(x_i)$ have an expectation value of zero, and that $f(y)$ has unit variance. No particular form is assumed for these functions and they are obtained iteratively, starting with a linear transformation and then alternatingly improving $f(y)$ and the $g_i(x_i)$, and hence the name, ACE. Each of the functions is obtained as table, with a pair of numbers representing each point. These functions may then be plotted or subjected to conventional curve-fitting procedures. It should be noted that ACE is not a general non-linear data fitting method and is restricted to fitting sums of univariate non-linear functions.

Within the program it is possible to selectively force one or more of the functions to be monotonic or linear, or to be excluded. A smoothing algorithm is applied, which requires the user to supply a parameter, SPAN. If SPAN is between zero and one, then it represents the fraction of the range of y or x_i which is averaged in the smoothing process. If SPAN is zero, then the range of the smoother is adaptive, and is generated by the program. This should only be undertaken when the number of points is large (more than 30). Since an optimal fit is then obtained, it is expected that this would usually allow for a more parsimonious description of a system than would be expected from linear methods, using trial-and-error transformations.

The value of ACE lies in its ability to give an insight into the form of the optimal transforms. While one can often find effective transforms by trial-and-error, the amount of work involved escalates rapidly as the number of variables to be tried and the number of transformations increases. If m transformations are to be tried on each of n variables, the number of regressions to be tried in an exhaustive search is n^m. The ACE procedure places no restrictions on the number of transformations; all possible transformations are tried at once.

While it would be unusual to solve a regression problem in a single application of ACE, the method does furnish a solution in much less time than would be expected in an exhaustive search, and at the same time ensures that no unexpected or unusual transformation has been missed. At the same time, it must be recognized that ACE can give misleading results, especially with small samples or when there are large errors in the data. Therefore, the robustness of the transformations should be checked by comparison with ACE runs on the same data but with constraints imposed on the transformations, and by applying of linear or non-linear regression with the transformations have been explicitly imposed. Particular care should be taken when the dependent variables is transformed, as the properties of this method have not been fully understood and is still under investigation.

The variance of each transformed variable should be calculated, and provides a measure of the relative importance of that variable in the resulting QSAR. Those variables with low variance are candidates for constraint to linearity, or for deletion. This also applies to variables which give physically implausible transformation plots.

Trials with simulated data show that ACE performs extremely well when the error (either random or lack-of-fit) is very small, but that it is undermined when the error is larger, especially when the number of points is small, or the number of variables large. Under these circumstances, the plots of the transformations become degraded in form, and it becomes harder to judge the analytic form of the appropriate transformation from the plots. In particular, with completely random data, apparent good correlations can often be obtained under these conditions. Cross-validation serves to control this tendency.

4.5.2 Cross-Validation with ACE

The use of any statistical technique requires testing the significance of its findings. ACE is a complex technique for which no formal statistical tests are available and confidence is placed in the non-parametric technique of cross-validation. By analogy with conventional regression, an r^2 may be defined as $(SS_t - SS_r)/SS_t$, where SS_t is the sum of squares of deviations from the mean of the response variable, and SS_r is the sum of squares of the residuals. A cross-validated r^2_{cv} may be defined analogously, by leaving out groups of points, then running ACE on the points which remain, and finally, predicting the response variable values for the points which had been left out. This is repeated until every point has been left out once, and only once. The sum of squares of the predicted residuals calculated in this way is used instead of SS_r mentioned above to give r^2_{cv}. The r^2 and r^2_{cv} values calculated as described above, provide a measure of the goodness of fit and the robustness of the ACE regression, respectively.

It has usually been found that r^2 is an excessively optimistic estimate, in that on applying standard multiple regression to the data set, with the transformations suggested by the ACE plots, the r^2 obtained is not as large as that given by ACE. On the other hand, the r^2_{cv} is usually smaller than that indicated by conventional regression. This occurs because of large errors in the predicted residuals when one or more of the predictor variables is extremal. Trials with synthetic data [3] showed that this tendency of cross-validation to give misleadingly pessimistic results is was further exacerbated when either the number of variables was large, the number of points small, or when there was considerable random error present in the data.

An interesting observation of ACE in QSAR work is the frequent occurrence of piecewise linear transformations [2, 3]. This can sometimes be explained by physical models such as compartment distribution as derived by Kubinyi [4], but in other cases, the causes are more obscure [3].

4.5.3 The Randomization Test

Topliss and Edwards [5] have studied the phenomenon of chance correlation in the context of multiple linear regression and have published the results of computer simulation. This work has been widely interpreted to mean that no useful results can be obtained from selecting variables by multiple regression, despite a statement to the contrary by these authors. Their results represented the worst case, in which the independent variables were uncorrelated. In the case of where there is correlation, the effective number of variables is smaller than the actual number, but an extra problem now arises, namely multicollinearity. The problem of chance correlation may be dealt with by a method suggested by Topliss and Edwards [5] with a modification by Giles (Giles, D. E., Murdoch University, personal communication, 1990).

In this method, the values for the dependent variables are randomly reassigned to the drugs, keeping the independent variables constant. This procedure is repeated a number of times. If the fit obtained with the real data is consistently better than that with the reassigned data, the correlation obtained with the real data can be confidently assumed not to be due to chance. Random reassignment of dependent variables in conjunction with ACE effectively demonstrates the relevance of the unselected set of variables to the problem in hand, and can also be employed in a stepwise linear regression.

4.5.4 Stepwise Regression with ACE

Stepwise regression, both forward and backward in the context of multiple linear regression has been employed for a long time in statistics and in QSAR analysis. Stepwise regression has been relatively unexplored in the context of ACE, but experience suggests that with small data sets at least, the equations obtained are not robust, and that the transformations derived are unacceptably complex, representing overfitting. The validation technique described in Sec. 4.5.3 can be applied. First, a maximum acceptable number of independent variables must be decided in advance, and any run giving more than that number must be discarded. The stepwise selection process is then carried out with the randomized dependent variable, just as in the case of unmodified data, and is repeated a number of times. The resulting correlation coefficient data may then be normalized with the Fisher transformation (Afifi and Azen [6]) and a significance test may then be applied. In most real cases, this would require excessive demands on computer time and forward stepwise regression with ACE should probably be avoided for the above reasons.

4.5.5 Examples

Two examples of ACE applied to data taken from the literature are provided. These are not necessarily the "best" examples of QSARs in the papers from which they were taken, but merely serve as examples of how ACE can illuminate certain aspects of the data, and provide transformations which were not apparent in the original treatment.

4.5.5.1 DHFR Inhibitors

Selassie et al. [7] have presented a set of QSARs for the inhibition of *Lactobacillus casei* dihydrofolate reductase (DHFR) by a series of 68 benzyldiaminopyrimidines of the general structure:

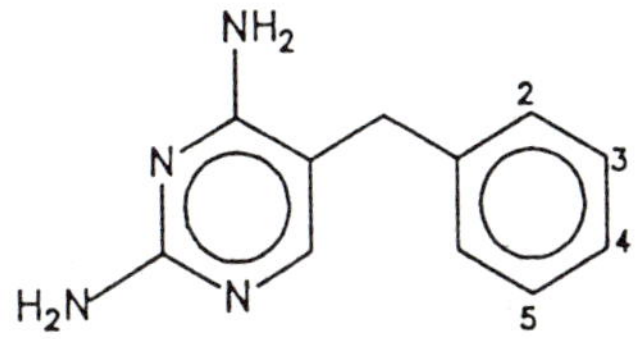

The best equation is as follows:

$$\begin{aligned}\log(1/K_i) = {} & 1.13(\pm 0.23)MR'_4 + 0.47(\pm 0.30)MR'_3 + 0.53(\pm 0.51)MR_5 \\ & -0.19(\pm 0.29)MR_5^2 + 0.34(\pm 0.30)\pi_3 + 0.25(\pm 0.18)\pi_4 \\ & -0.62(\pm 0.42)\log[\beta_3(10^{\pi_3})+1] \\ & -0.78(\pm 0.34)\log[\beta_4(10^{\pi_4})+1] + 5.43(\pm 0.19) \qquad (2)\end{aligned}$$

$$n = 68,\ r^2 = 0.726,\ s = 0.283$$

where $\log \beta_3 = -1.02$ and $\log \beta_4 = -0.98$, the π terms are hydrophobic substituent constants, the terms in parentheses are 95% confidence estimates and the MR terms are molar refractivities.

Reanalysis of the data of Selassie et al. [7] by ACE, using the adaptive smoother span, resulted in the transformation plots shown in Fig. 1, with an r^2 of 0.760, r^2_{cv} of 0.477, and with the variance 0.142 for MR'_3, 0.035 for π_3, 0.165 for π_4, 0.386 for MR'_4, and 0.069 for MR_5. The variances for π_3 and MR_5 were small, suggesting that

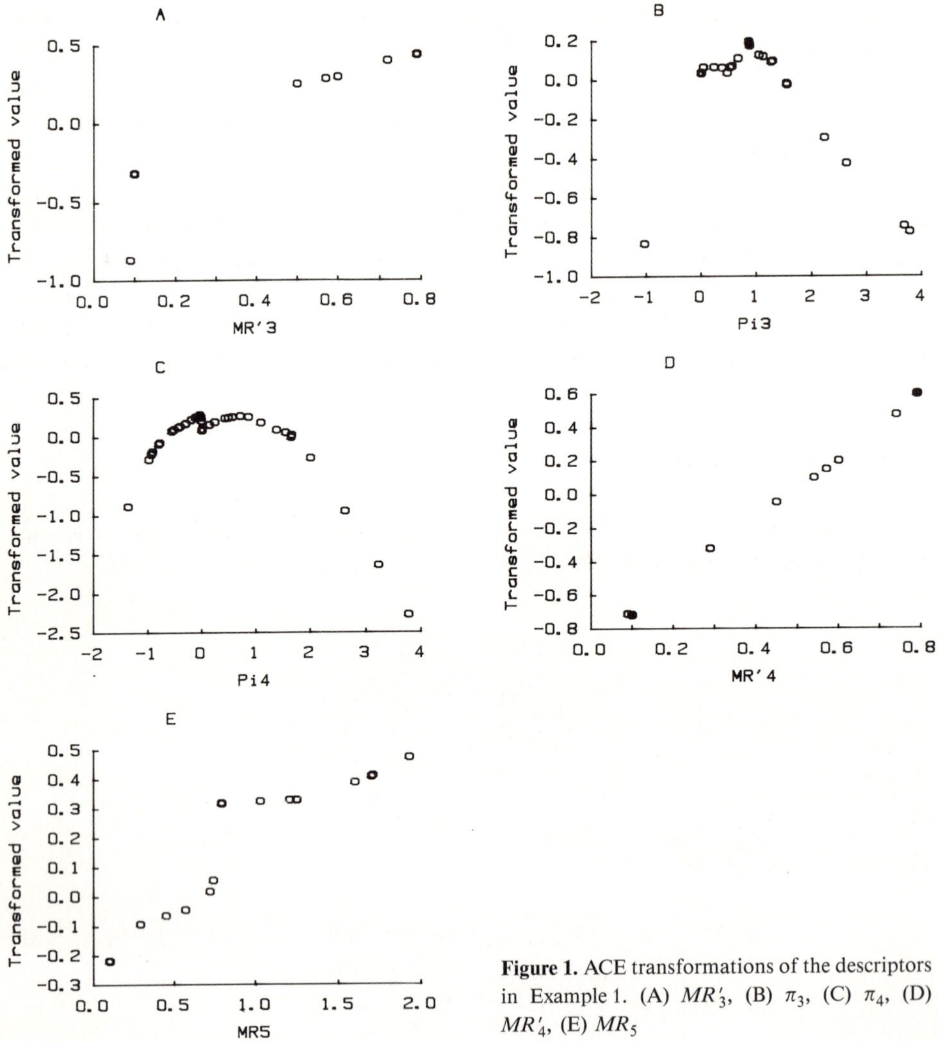

Figure 1. ACE transformations of the descriptors in Example 1. (A) MR'_3, (B) π_3, (C) π_4, (D) MR'_4, (E) MR_5

these variables were of minor importance. Constraining π_4 to linearity reduced r^2 to 0.628, which is an unacceptable loss in the goodness of fit. Constraint on any other single variable resulted in little loss. The plot for MR'_4 is almost perfectly linear, and the plot for MR'_3 deviates from linearity at one point only, which corresponds to the only 3-fluoro compound in the data set. The transformation plot for MR_5 is not feasible, and should be tested for linearity. Constraining MR'_3, MR'_4 and MR_5 to linearity gave an r^2 of 0.740 and an r^2_{cv} of 0.535. If, in addition, either π_3 or π_4 was constrained to linearity, an r^2 of 0.705 and 0.613 was obtained, respectively, resulting in an appreciable loss in the goodness of fit in the first case, and unacceptable loss in the goodness of fit in the second case. The transformations of π_3 and π_4 point considerably towards a bilinear plot of the kind described by Kubinyi [4], albeit in the case of the former, with one highly influential point. This was not the same point as in the case of MR'_3, but was one of only two compounds that contained an alcohol substituent (CH_2OH). Thus the left-hand side of the bilinear transformation of π_3 seems doubtful. A similar QSAR to that of the previous mentioned authors has been obtained with this method, except that the MR_5 transformation is linear, and the term linear in π_3 is doubtful. Examination of the equation published (Eq. (2)) showed that the MR_5^2 term is statistically not significant, and when this term was omitted, r^2 decreased only by 0.005. On omitting the term linear in π_3, r^2 was further decreased by only 0.005. This bordered on statistical significance.

4.5.5.2 Triazene Mutagenicity

Shusterman et al. [8] studied the mutagenicities of 17 N-1 and phenyl-substituted 1-methyl-3-phenyl triazenes, of the following structure:

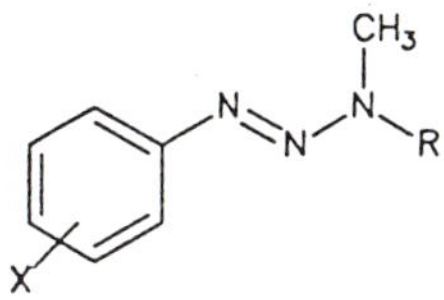

The substituents and values of the descriptors are given in Table 1.
Their equation was as follows:

$$\log 1/C = 0.92\,(\pm 0.36)\log P - 6.98\,(\pm 3.97)\,q_{HOMO} + 5.72 \qquad (3)$$

$$n = 17,\ r^2 = 0.789,\ s = 0.638$$

When ACE was applied to their data with SPAN 0.3, the plots shown in Fig. 2 were obtained, with $r^2 = 0.890$, $r^2_{cv} = 0.800$ and variances of 0.493 for log P and 0.234 for q_{HOMO}. The transformation on q_{HOMO} was almost linear, and when the ACE run was repeated with this transformation constrained to linearity, the statistics obtained were $r^2 = 0.886$ and $r^2_{cv} = 0.806$. The transformation for log P can be viewed as being sigmoidal, pointing to a function such as $\tan^{-1} x$, $\tanh x$, or the logistic function, $e^x/(e^x + 1)$. While none of these functions was derived from a particular physical model, they all point to the same conclusion: biological activity increases with increasing log P between the log P values 2 to 3, but below 2 and above 3, the effect tends asymptotically to a constant value. That is, below a log P value of 2, further decreases of log P are ineffective in reducing activity further, and similarly,

Table 1. Properties of phenyltrianzenes.

No.	R	X	log *P*	q_{HOMO}	log 1/*C*
1	CH_3	3,5-CN	2.18	0.483	3.46
2	CH_3	4-SO_2NH_2	0.98	0.507	3.49
3	CH_3	3-$CONH_2$	1.21	0.444	3.51
4	CH_3	4-$CONH_2$	1.20	0.417	4.04
5	Allyl[a]	4-$CONH_2$	2.09	0.426	4.16
6	CH_3	3-$NHCONH_2$	1.29	0.274	4.19
7	CH_3	4-CN	2.39	0.410	4.43
8	CH_3	4-$COCH_3$	2.27	0.409	4.47
9	CH_3	H	2.59	0.386	5.32
10	*n*-Butyl[b]	4-$CONH_2$	2.46	0.425	5.41
11	CH_3	4-$NHCONH_2$	1.25	0.260	5.59
12	CH_3	4-$NHCOCH_3$	1.54	0.257	5.83
13	CH_3	4-CF_3	3.70	0.482	5.99
14	CH_3	3-CH_3	2.85	0.380	6.44
15	CH_3	4-Cl	3.33	0.382	6.48
16	CH_3	4-CH_3	2.93	0.362	7.00
17	CH_3	4-C_6H_5	4.40	0.202	7.67

[a] Propen-3-yl
[b] Butan-1-yl

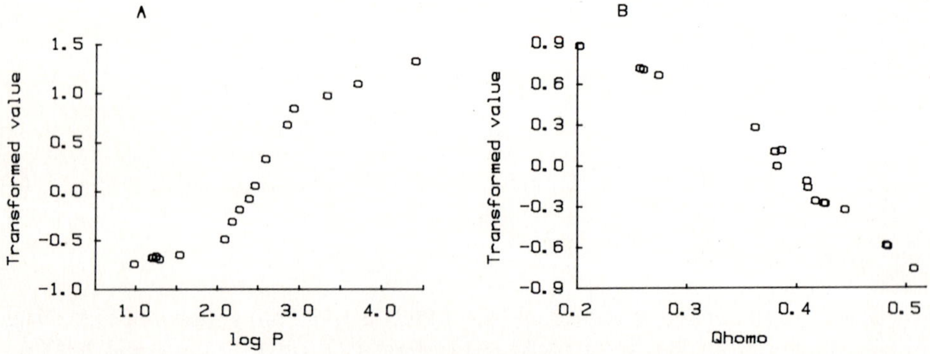

Figure 2. ACE transformations of the descriptors in Example 2. (A) log *P*, (B) q_{HOMO}

above 3, increases in log *P* are ineffective in increasing activity. Such a phenomenon has not been reported previously, and requires further investigation, if it is to be confirmed by studies using alternative models and more compounds.

Adopting the logistic function in the form of:

$$\log 1/C = a\ q_{HOMO} + b\ e^{w(x-o)}/(1+e^{w(x-o)}) + c \tag{4}$$

where $x = \log P$, o is the position of the inflection point and w is a width parameter. Non-linear regression using the algorithm of Marquardt (Bevington [9]) yielded the following results: $a = -7.246$, $b = 2.327$, $w = 7.540$, $o = 2.522$, and $c = 7.043$, with $r^2 = 0.902$ and $s = 0.436$. It can by seen that the position of the inflection point given by non-linear regression corresponds well with that obtained by ACE, and that the fit is a marked improvement over Equation 2.

A demonstration of the validity of the ACE regression in this last example is given by the randomization procedure described in Sec. 4.5.3. If the dependent variable is randomly reassigned, the values of r^2 and r^2_{cv} shown in Table 2 were obtained. It will be seen that r^2_{cv} is always negative, and that r^2 implies, perhaps surprisingly, an appreciable fit in some cases. This illustrates the tendency of ACE to overfit. If however the Fisher transformation is applied to the r^2 values followed by a test, based on the normal distribution, the r^2 from the original data is greater than that from the randomized data at a very high level of significance ($p = 2.1 \times 10^{-10}$).

Table 2. Series of ACE runs using the data of Table 1.

r^2_{cv}	r^2	$v = \frac{1}{2} \ln \frac{1+r}{1-r}$
−0.243	0.244	0.541
−0.919	0.124	0.367
−0.415	0.281	0.591
−0.565	0.414	0.764
−0.721	0.009	0.097
−0.547	0.175	0.446
−0.693	0.098	0.323
−0.202	0.369	0.750
−0.633	0.394	0.737
−0.850	0.294	0.607
−1.111	0.111	0.346
−0.935	0.112	0.348
−0.620	0.434	0.790
−0.810	0.027	0.165
−1.870	0.261	0.564
−0.370	0.125	0.370
		Mean 0.488
		S.D. 0.214

For non-randomized data: $r^2 = 0.890$, $v = 1.768$, $p = 2.1 \times 10^{-10}$. The Fisher transformation involves transforming the correlation coefficient r, which has the range -1 to 1, into the statistic v, which has the range $-\infty$ to ∞, and is approximately normally distributed [7].

4.5.6 Conclusion

The ACE method is a very useful technique for determining univariate non-linear transformations in multiple linear regression, but some caution in its application is necessary. In particular, the application of ACE in a forward stepwise manner to reduce a large pool of predictor variables leads to overfitting of data (i.e. random correlation) and, consequently, unsatisfactory transformations, unless heavy smoothing is applied. Heavy smoothing in turn leads to a loss of resolution in the fitted function, defeating the purpose of the ACE analysis. Thus, this application of ACE, at least with small samples, cannot be recommended. A backwards stepwise approach, starting with a data set, which has already been partially selected by other methods, seems more satisfactory. Transformation of the dependent variable should also be undertaken with due caution.

The only technique for assessing the quality of the fit is cross-validation. While the cross-validated r^2 decisively rejects data in which the dependent variable has been randomized, this leads to an unduly pessimistic estimate of the likely error when applied to data, in which a genuine correlation does exist. The ACE technique should, therefore, be supplemented with either classical multiple regression, linear or non-linear methods to obtain a realistic appraisal of the results.

The routine use of ACE should be considered whenever transformations are sought after in multiple regression analysis. Large amounts of random error result in poor transformation curves, and tend to disguise the form of the transformations. Provided that random errors are not dominant, ACE suggests feasible transformations at a considerable speed, and indicates when it is not possible to obtain a useful set of additive non-linear transformations.

In cross-validation, the calculation of the residual for a point, which is extremal in one or more variables, involves extrapolation of the transformation tables. The large residuals for extremal points are a warning that the ACE transformations should not be extrapolated, and also that the cross-validated r^2 is unduly pessimistic. As already remarked by Franke and Lanteri [2], the conventional r^2 is excessively optimistic, so perhaps the function of ACE should be to suggest transformations, which can then be verified and tested for statistical significance using normal linear methods for which the standard F and t tests are available.

A physical interpretation of the results from an ACE analysis should be based not on the transformation plots as such, but on the resulting conventional linear or non-linear regression, but should even then be made with due caution. Because of collinearities in the data, such interpretations are necessarily speculative. While it is usually not possible to decide, on statistical grounds, which of a group of correlated variables is most relevant, it is often found that the members of these groups of variables are all indicators of a common physical factor. It must be remembered, however, that in the selection of transformations there has been some effective loss of degrees of freedom.

ACE can, at best, give an indication of the form of an optimal transformation. This could be interpreted physically, as with the bilinear transformtion in the first example, or it could be quite empirical, as with the logistic function in the second example. It is usually possible to assign many different functional forms to such a transformation, which from a purely statistical point of view, are indistinguishable in the data set at hand. When one of these functions has theoretical significance, the ACE transformation can be said to support the physical model underlying that function. When, as in the second example, no such physical model is obvious, it is first necessary to confirm the form of the transformation. This may be accomplished by investigating other methods of selecting descriptors, to determine whether the form of the transformation is stable, and by collecting data on more cases, in this case drugs. If the transformation is stable, this strongly suggests that some non-linear functional relationship is operating, and one then has the task of discovering the nature and physical significance of such a relationship. This task is beyond the scope of statistical methods, but the form of the ACE transformation curves should provide some clues.

4.5.7 Availability

The ACE method is implemented in the statistics package S-Plus, which is available from the CSIRO Division of Mathematics and Statistics, Locked Bag 17, North Ryde, NSW 2113, Australia. The ACE subroutines are also available by email, free of charge, as "ace" from statlib over the *internet* network as follows:
ftp lib.stat.cmu.edu
Name statlib
Password (your email address)
cd general
get ace

This file should be named **ace.for**. The sub-routines are written in standard FORTRAN. A set of supporting and driver routines and their documentation are available at no charge through *internet* as follows:
ftp csuvax1.murdoch.edu.au
Name anonymous
Password (your email address)
cd pub/chem/martha
get (filename)

The required files are:
edit.for, aceprog.for, multlr.for, hpplot.for, exam_1.mar, exam_2.mar, exam_1.out, exam_2.out, martha.doc., and **readme.txt**. The .FOR files are written in Microsoft FORTRAN for the PC. They include an interactive editor, a graphics package and a conventional multiple linear regression program. Graphics output is

to a HPGL file. In addition to the above, you will need a FORTRAN compiler, preferably Microsoft, and some means of transfering the HPGL graphics output to a hardcopy or screen. It may be necessary to remove a few characters inserted by the system from the beginning and end of each file.

Acknowledgements

I am indebted to Jerome H. Friedman of Stanford University Department of Statistics for supplying me with a list of his ACE sub-routines. I also thank Paula J. McLay of Murdoch University for criticism of the manuscript for this text.

References

[1] Breiman, L. and Friedman, J. H., *J. Amer. Statist. Assoc.* **80**, 580–619 (1985)

[2] Franke, I. E. and Lanteri, S., *Chemom. Intell. Lab. Syst.* **3**, 301–313 (1988)

[3] Clare, B. W., *Chemom. Intell. Lab. Syst.* **18**, 71–93 (1993)

[4] Kubinyi, H., *J. Med. Chem.* **20**, 625–629 (1977)

[5] Topliss, J. G. and Edwards, R. J., *J. Med. Chem.* **22**, 1238–1244 (1979)

[6] Afifi, A. A. and Azen, S. P., *Statistical Analysis: A Computer Oriented Approach,* 2nd edn., Academic Press, New York, 1979, pp. 140

[7] Selassie, C. D., Fang, Z.-X., Li, R., Hansch, C., Debnath, G., Klein, T. E., Langridge, R. and Kaufman, B. T., *J. Med. Chem.* **32**, 1895–1905 (1989)

[8] Shusterman, A. J., Johnson, A. S. and Hansch, C., *Int. J. Quantum Chem.* **36**, 19–33 (1989)

[9] Bevington, P. R., *Data Reduction and Error Analysis for the Physical Sciences,* McGraw-Hill, New York, 1969

5 Neural Networks and Expert Systems in Molecular Design

5.1 Neural Networks – A Tool for Drug Design

David T. Manallack and David J. Livingstone

Abbreviations

NN	Neural Network
PC	Personal Computer
QSAR	Quantitative Structure-Activity Relationships
BP	Back Propagation
MLR	Multiple Linear Regression
ReNDeR	Reversible Non-linear Dimension Reduction
CoMFA	Comparative Molecular Field Analysis
LOO	Leave-One-Out.

5.1.1 Introduction

In the last few years there has been an explosion of interest in the field of artificial intelligence known as neural networks [1, 2]. This has not only involved research into the techniques themselves but has included the practical application of these methods to a wide variety of existing problems in our society. Neural networks are employed inter alia in the recognition of handwriting for cheque verification, underground train platform management, the forecasting of trends in stock market movements, the recognition of faces in a security system and even the control of a nuclear reactor. Along side these demonstrations of utility there has also been much "hype". The following quotes from a book devoted to PC implementations of neural networks highlights this [3]:

"Neural networks... are being touted as one of the greatest computational tools ever developed. Much of the excitement is due to the apparent ability of neural networks to imitate the brain's ability to make decisions and draw conclusions when

presented with complex, noisy, irrelevant, and/or partial information.... It is hard, especially for a person unfamiliar with the subject, to separate the substance from the hype....

Neural networks really do offer solutions to some problems that can't be solved in any other way known... It is a myth that neural networks can leap tall buildings in a single bound and that they can solve problems single handedly...."

Without doubt, neural networks (NN) have attracted considerable attention from the chemical community as can be seen from the increase in the number of papers using neural networks for chemical applications (Fig. 1) since 1987. Applications have included, QSAR data analysis, prediction of protein secondary structure, process control, analysis of spectra, prediction of chemical reactivity, etc. Networks have found applications in most areas in chemistry, not only performing established computing tasks, but tackling complex problem solving using their ability for pattern recognition. The future for networks in chemistry will be exciting as further applications are discovered. Networks are already improving productivity in chemical processes, but a considerable amount of work is still to be done to fully understand their "behavior" and some caution will need to be exercised before their full potential is realized.

The appeal of new techniques in any branch of science is their potential to perform traditional tasks more effectively and more efficiently. With increasing pressures on chemists to improve productivity, novel technologies are sometimes implemented prior to establishing the soundness of the method. To some extent, neural networks fall into this category and some "catching up" is taking place in chemistry with a re-examination of various applications in the light of neural networks. Some problems associated with networks have already been encountered by those who work exclusively in that area, but the sharing of this information across scientific fields has not always taken place. Recently, enthusiasm for the appealing results obtained with networks has come under criticism and strategies have been drawn up to dimin-

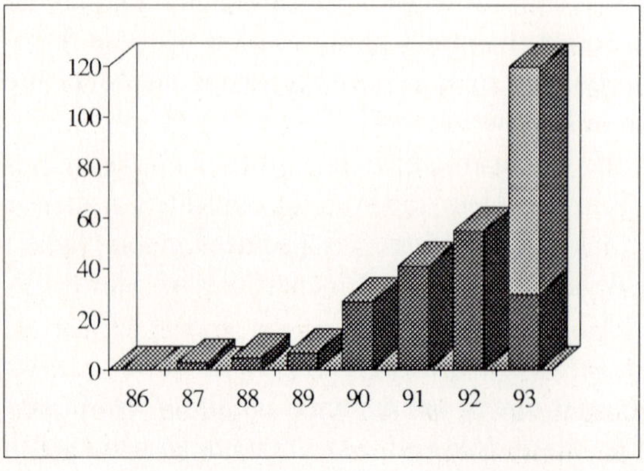

Figure 1. Frequency of publications mentioning neural networks and chemistry from the Chemical Abstracts database. The projected number of publications for 1993 has been illustrated in light shading. This was not an exhaustive search but should serve to show the trends.

ish problems such as overfitting. Criticisms have been leveled before at so-called "black box" methods where a complete understanding of the method has been overlooked because of the remarkable results which emerged. Fortunately, a more analytical approach is now taking place with a healthy respect for potential problems.

This chapter focuses on the use of neural networks to carry out the chemometric procedures involved in quantitative drug design (QSAR). Comparison is made with the standard techniques which are used for these tasks and we attempt to shed light on potential problems and suggest the necessary steps to minimize them.

5.1.1.1 Neural Network Theory

Most artificial intelligence methods seek to emulate intelligence by reproducing the decision-making functions of the brain. There is, generally, no attempt to simulate the way in which these decisions are reached, they aim simply to reproduce "what we do" rather than "how we do it". Neural networks, on the other hand, try, in a limited sense, to mimic "what we do" by copying the way in which the brain "does it". An artificial neural network consists of a number of processing units connected together, usually, but not always, in a number of distinct layers. Numerous texts have been published which describe the theory of neural networks and so in this chapter we will simply give a brief description of how one of the more popular types of network (back propagation) operates.

Each processing unit, or neuron using the brain analogy, performs 3 functions. This contrasts to the hundreds of functions carried out by biological neurons, but this approximation is adequate for the simple models used at present. The 3 functions sum the inputs that the processing unit receives, apply a transfer function to the summed inputs and produce an output (Fig. 2). The transfer function shown in the figure is a sigmoid although several other transfer functions may be used; the purpose of a transfer function is to mimic the operation of biological neurons which send out an impulse if the inputs received are above a threshold value. A number of different network-learning processes are available, but the most widely used for chemometrics is feed-forward back propagation (BP). Typically, this type of network uses processing units placed in three types of layers, input, hidden, and output, see Fig. 3, for example, and this has also been termed a multi-layer perceptron. Each unit in a layer is connected to units in adjacent layers with an associated weight (connection strength); it is the adjustment of these weights which is undertaken during network training. The output value of each unit in the final layer is compared to a target value. If we take a simple QSAR multiple linear regression case we may have, for example, a table of n compounds with y physico-chemical descriptors and the associated biological activity for the n compounds. A BP network would be set up containing y units in the input layer and one unit in the output layer (i.e. biological activ-

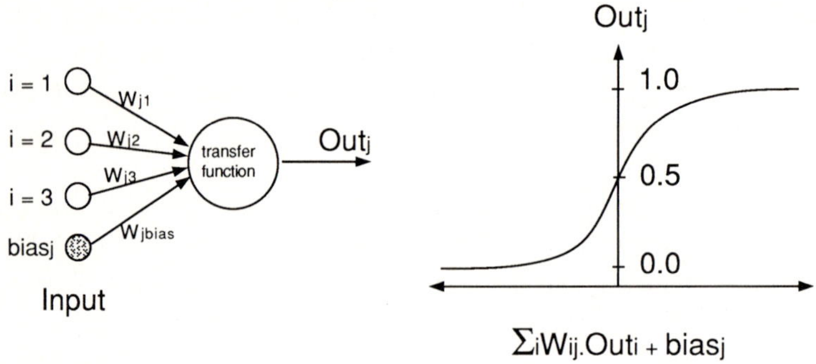

Figure 2. A. Diagram illustrating the 3 functions performed by a neural network processing element. The bias unit can be considered to be a scaling term which can be treated as the connection weight from a neuron with a constant output of 1.0 B. Sigmoidal transfer function used in back propagation neural networks (N. B., the transfer function is continuous).

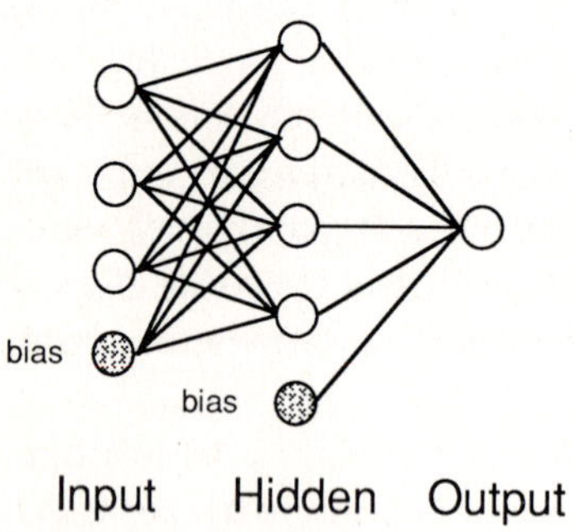

Figure 3. Example of a 3 layer neural network showing input units and two layers of active processing units.

ity) with usually one layer of hidden units between the input and output layers (Fig. 3).

Input to the network in this example would be the values of each of the *y* physicochemical properties, to each of the *y* input neurons, for each compound in turn. The target value for each compound is the biological activity. Since the data enters the network at the input layer and produces a signal at the output layer (one neuron in this case) this type of network is known as feed-forward. Network training involves iteratively changing the weights between neurons until the output signal matches the target output within a desired error limit. There are various ways in which the network weights may be altered, but one of the most commonly used methods is known as the delta rule [4] as described below.

In the delta rule, the change $\Delta_k W_{ji}$, in the weight for the connection between neuron j and neuron i the *k*th iteration is given by

$$\Delta_k W_{ji} = \beta \delta_{pj} \cdot \mathrm{out}_{pi} + \alpha \Delta_{k-1} W_{ji} \qquad (1)$$

where β is the learning rate and α a momentum term, two parameters which may be adjusted to control the speed at which the network assimilates information. The error signal δ_{pj} of the jth neuron on presentation of the pth pattern is determined as follows:
if neuron j belongs to the output layer

$$\delta_{pj} = (t_{pj} - \text{out}_{pj})\, f'(\text{inp}_{pj}) \tag{2}$$

and if neuron j belongs to the hidden layer

$$\delta_{pj} = f'(\text{inp}_{pj})\, \Sigma_z\, \delta_{pz}\, W_{zj} \tag{3}$$

where z represents the neurons to which neuron j sends its output and where $f'(\text{inp}_{pj})$ is the derivative with respect to inp_{pj} of the transfer function. This use of a derivative of the transfer function in the delta rule imposes a restriction on the type of transfer functions which may be used in this kind of network training, obviously they must be capable of differentiation.

At the beginning of training the connection weights are set to random values. The data for all of the compounds (physico-chemical descriptors) are passed through the network (i.e., feed forward) and the output responses are compared to the target data (biological activity) to give an error value. The weights are then adjusted for the second pass of the data through the network in order to reduce the above error value. Since the delta rule requires the calculation of an error at the output neuron in order to calculate errors for other neurons in the network, this is known as "back propagation of errors", hence the name back propagation (BP). Such networks are sometimes given the rather unwieldy title of "feed-forward back propagation" which we will abbreviate to BP. The entire procedure is repeated in an iterative manner until the error value reaches a minimum or other specific criteria are met (e.g., a preset number of cycles or a specific error cut-off). Finally, a regression coefficient may be calculated between the observed biological activity and the network predicted values.

5.1.1.2 Implementation (Hardware/Software)

Neural network computing systems can be implemented using either specialist hardware or software, or a combination of the two. As an imitation of the brain, which can be regarded as a huge parallel computing device, they lend themselves particularly well to parallel computers, that is to say computers which have multiple processors. Using such systems it is possible to assign a single processor to carry out the calculations for a single neuron, or group of neurons, and to connect the input and output signals of the neurons in any desired architecture (BP networks are just one such type of network). An alternative is to construct the architecture of a particular network directly into silicon so that the processing elements and their connections are "hard-wired" together. The advantage of this latter approach is the speed with which networks can be trained, since the computer does not have to control the routes which is taken by the information through the network, a disadvantage being of course that the network architecture cannot be re-organized once the chip is constructed. The use of highly parallel computers or dedicated network chips is required when very large networks (millions of connections) are constructed.

Most chemical problems, with the exception of protein structure prediction (see later), do not involve large numbers of connections and are thus generally implemented in software running on serial computers. Modern PCs have sufficient computing power to allow network training to be completed in a few hours or less and, of course, once trained, a network is able to make "predictions" very quickly. In certain circumstances, networks may need to be re-trained many times, for example, when carrying out cross-validation, in which case it may be necessary to use a more powerful computer such as a UNIX workstation. There is a large variety of software available commercially and in the public domain for the construction of BP networks (and other architectures) as outlined in Appendix 1. These programs run on most commonly available hardware platforms such as IBM PC, Macintosh, and UNIX workstations from SUN, Silicon Graphics, IBM, etc.

5.1.1.3 Chemical Applications

Some of the earliest applications of neural networks to chemistry involved the prediction of protein secondary structure from amino acid sequences [5–7]. The input layer of the networks in these examples consisted of a number of groups of neurons with each group containing 21 neurons. As the linear sequence of amino acids is presented to the network, one neuron in each group is activated according to the identity of the amino acid at that position. Twenty of the neurons in each group correspond to the naturally occurring amino acids, the extra neuron is activated to indicate the termination of a protein chain. Network training is carried out to predict the secondary structure of a central residue. For example, in the work reported by Qian and Sejnowski [5], the input layer consisted of 13 groups of neurons so a prediction was made for a residue on the basis of 6 neighbors on either side. The output layer may consist of two (helix, sheet) or three neurons (helix, coil, sheet) and various numbers of neurons have been used in the hidden layer. Variants of these experiments have involved prediction of β-turns [8] and prediction of the disulphide bonding state of cysteine residues [9]. Comparison of these network predictions with standard techniques showed that the networks performed as well, or in some cases slightly better. However, perhaps not surprisingly, it is clear that the amino acid sequence alone does not contain sufficient information for accurate prediction of the secondary structure. More recent applications of networks to protein structure problems have involved the prediction of water binding sites on proteins [10] and the pK_a value of a catalytic histidine residue [11].

Neural networks have been employed in the estimation of aqueous solubility of organic compounds [12], the interpretation of infra-red spectra [13], the prediction and classification of ^{13}C chemical shifts [14] and the prediction of physical organic substituent constants such as σ values [15] and log P [16]. Applications of networks to QSAR problems are described in the next section and references [17–19] give some recent reviews on the chemical applications of networks.

5.1.2 Applications to QSAR

One of the first reports of the application of neural networks to a QSAR problem involved the discriminant analysis of a set of anticarcinogenic mitomycin derivatives [20]. These compounds were classified into 5 activity categories and the trained network was able to correctly classify all 16 compounds. The physico-chemical data used to describe these compounds consisted of two indicator variables and four substituent constants. Thus, for a training set of 16 compounds, chance effects would not be expected to be a problem [21]. The network architecture employed in this example, however, involved a hidden layer of 12 neurons with at least 132 connections in the network. Concern that chance effects may have dominated these results led us to examine the performance of discriminant networks using random numbers [22] as is summarized in the next section. The presence of too many connections in a network may not only allow chance correlations to occur, but may also result in over fitting. That is to say, an over elaborate surface may be fitted to the training data, leading to apparently good performance in fitting, but not prediction. This problem was recognized by Andrea [23] who proposed a parameter, ρ, which could be used to characterize the relationship between data points and connections:

$$\rho = \frac{\text{No. of data points}}{\text{No. of connections}} \tag{4}$$

In the particular example reported [23] it was recommended that ρ should be greater than 1.8, to avoid memorizing the data, and less than 2.2, so that the network would have sufficient connections to fit the data. The next section discusses in greater detail the question of optimum values for ρ. The networks reported in reference [20] employed a ρ value of less than 0.5 and could, therefore, have resulted in an overfitting. Network prediction performance for this example was tested by splitting the data into two sets, a training set of 11 compounds and a test set of five compounds. After training, the network was still able to predict the training set correctly, as expected, but test set performance was poor suggesting that the network was indeed overfitted.

Others have recognized the problem of overfitting and the difficulties involved in assessing the performance of trained networks and have reanalyzed the mitomycin data set [24]. In these experiments the number of neurons in the hidden layer was reduced to a minimum which would correctly classify all of the training set molecules. Fifty random starting weight matrices were used for training, and 50 predictions were made on the test compounds. A probability was assigned to the predictions to assess the significance of the results. This was determined using a sign criterion defined in Eq. (5). For ranks of activity H_0 and H_1 after n trials, the molecule was predicted m times as H_0 and $n-m$ $(m<n-m)$ times as H_1. Then at the level p of significance:

$$\rho < \sum_{r=0}^{m+1} \frac{m!}{(m-r)!r!} 2^{-n} \tag{5}$$

the molecule is assigned rank H_1. This method has clear advantages over a single training and prediction cycle for networks which have a ρ value which indicates that over-training is a potential problem. In particular, this technique is advantageous when examining data sets with a limited number of compounds as it may be difficult to construct a network with a ρ value greater than a recommended value.

Early examples of QSAR analyses usually only considered a handful of physicochemical descriptors (e.g., log P, σ, π, E_s etc.). The general trend in more recent years has been to generate and collect as many descriptors as possible. The problems of using such "wide" data sets with multiple linear regression are well known [21], one solution to these problems would be to use a selection strategy which would reduce dimensionality [25, 26]. Networks, of course, add further complications if you take into account overfitting and over-training. To circumvent these problems Wikel and Dow [27] used neural networks to identify those properties most relevant to the dependent variable. In one of their examples, a data set of 31 compounds and 53 descriptors was used. A neural network was set up to perform multiple linear regression (MLR) using a cross-validation procedure. Training was not fully completed to convergence (i.e., the global minimum) as the ability of the network to generalize at the global minimum was found to be poor judging from the cross-validation prediction results. After training, the values of the weights were visualized to highlight those descriptors which were important in obtaining a solution. Each of the "local minima" from these analyses consistently showed the importance of the same subset of the original descriptors. These selected properties were then used to generate MLR equations. Two of these properties which had high weights were ultimately incorporated into the final equation.

This technique of using the weights to identify properties of interest appears to be very useful. Unfortunately, cross-validation is not a panacea for the problems of chance effects, and it is still possible that apparent important descriptors will be chosen by chance. Wikel and Dow [27], however, introduced two important suggestions

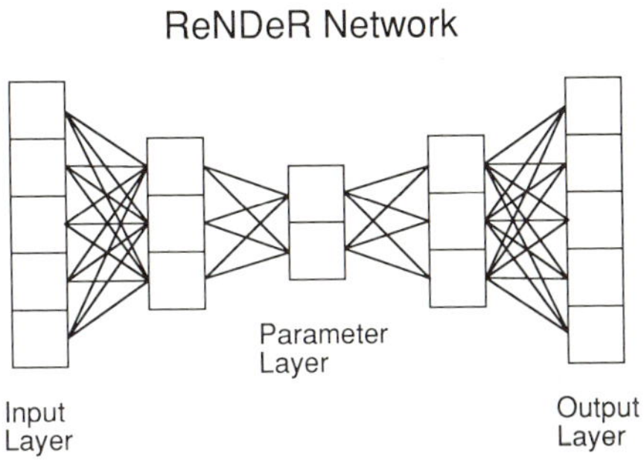

Figure 4. Example of the architecture required for a ReNDeR neural network.

for network training which might help to minimize over-training and, thus, the memorization of input data. The first of these, is to use cross-validation to monitor predictive performance and the second, is to not train to convergence.

Another early report on the use of neural networks for QSAR described a method for the display of a large number of physico-chemical properties in a small number of dimensions; this technique is known as ReNDeR (*Re*versible *N*on-linear *D*imension *R*eduction) Multivariate display techniques are of considerable importance to QSAR since they are able to reveal the patterns hidden in quite complex data [28] and yet, since they are an unsupervised learning method, they should not suffer from chance effects [29]. Figure 4 illustrates the network architecture for ReNDeR, although more units can be used in the parameter layer (e.g., 3 for 3D display). The network simply trains to reproduce at the output layer the same data that are presented to the input layer [30]. Once trained, the network effectively squeezes the data through a two-dimensional bottle-neck (the central hidden layer) so that when each compound is presented to the network two values will appear at the neurons in the central hidden layer. These values may be used as the X and Y coordinates for the construction of a two-dimensional plot. The term "unsupervised learning" refers to the fact that this method does not use any information concerning the dependent data (i.e., biological activity) for training. This technique overcomes many of the problems of using several physicochemical properties, although any clustering of activity classes due to chance cannot be completely ruled out [31]. The examples presented by Livingstone et al. [30] were encouraging and complemented the results obtained using the dimension reduction techniques, non-linear mapping and principal components analysis. In one case the ReNDeR plot was superior and appeared to be less affected by noisy data.

Richards and co-workers [32], employed a ReNDeR approach to examine a series of steroids. In this study, a large number of descriptors were generated describing the steric and electrostatic properties of the molecules. To simplify the large data set generated, each molecule was compared with each other to provide an $N \times N$ table of molecular similarities [33]. Similarities were calculated using either shape or electrostatic potential, or both together. This information was used as input to a

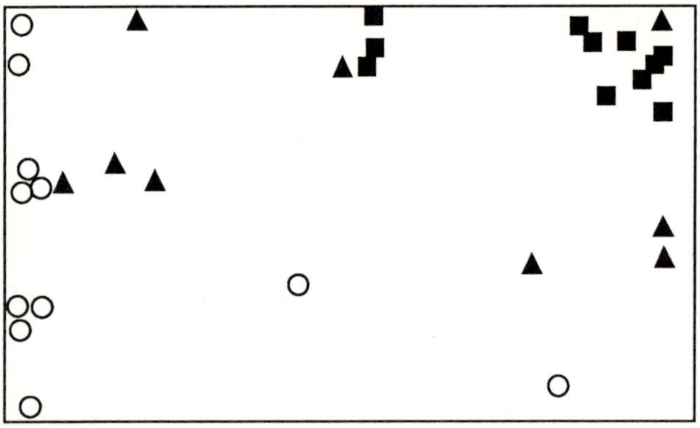

Figure 5. ReNDeR neural network plot from an electrostatic potential similarity matrix for a series of steroids. The binding affinity of each compound for the corticosteroid binding globulin was classified into 3 categories; high affinity (■); intermediate affinity (▲); low affinity (○). Diagram reproduced with kind permission from Good et al. [32].

ReNDeR network. The neuron plot using the electrostatic potential similarity matrix demonstrated good clustering of the high activity class (Fig. 5). Each similarity matrix was also examined using the CoMFA routine within SYBYL (Tripos Associates) since the steroids reported here have been previously analyzed using the CoMFA technique [34]. The results showed that the similarity approach gave comparable models to those of the first reported study, and that it can be helpful to visualize the problem using a two-dimensional display method. The testing of new compounds would require the similarity matrix and the ReNDeR network to be recalculated. Visual inspection of the plot could then be used to make activity predictions for the untested compounds.

An interesting application of the use of networks in QSAR was recently reported by Weinstein and co-workers who examined the classification of anticancer agents [35]. This study employed the screening data of 134 drugs tested for their ability to inhibit the growth of a panel of 60 different human tumor cell lines. Compounds were categorized into 6 classes according to their mechanism of action. The network architecture employed for this study used 60 neurons in the input layer, 6 neurons in the output layer and a variable number (3 to 9) of neurons in the hidden layer. There were, thus, more connections in the networks than compounds studied, although the use of dose-response information to characterize the compounds meant that the input data exceeded the number of connections. In order to test the predictive ability of these networks, a cross-validation scheme was included in the algorithm. Rather than perform the often used leave-one-out (LOO) method, Weinstein et al. chose to divide the data set into ten approximately equal subsets. The network was trained on 9/10ths of the data and a prediction was made on the remainder. This process was repeated 10 times until each group had been left out for prediction. In this case the network model performed better than linear discriminant analysis.

One of the advantages of using neural networks for regression and discriminant analysis is their ability to develop complex non-linear and cross product terms without these terms having to be specifically defined [4, 36]. Following their use of networks in discriminant analysis Aoyama and co-workers [20] used a neural network to perform MLR, and in order to simulate the way that MLR constructs linear equations, employed a linear transfer function for the final output neuron [37]. Although this might be expected to simulate the regression process, it appeared that this simply resulted in a poorer fit and required a greater number of connections to achive a comparable fit to a network using sigmoidal transfer functions for all neurons [36]. Another regression approach using a hybrid neural network system, known as FUNCLINK [38], has recently been described by Liu and co-workers [39–41]. This progam generates an expanded list of new parameters derived from the original properties using a chosen list of non-linear and cross product functions. The expanded list of parameters is then utilized for multiple linear regression (and discriminant analysis) using a two layer neural network. In a series of experiments comparing FUNCLINK to 3 layer networks, they appeared to provide superior predictive ability. However, although the performance of FUNCLINK compared well with MLR, it does not have the ability and natural advantage of neural networks to generate appropriate non-linear and cross product terms as needed. Moreover, the possibility of chance effects increases as increasingly more new parameters are considered for the data analysis component of this method [21].

5.1.3 Networks vs Statistics

Our own interest in the use of neural networks for QSAR was initiated by reports claiming superior results over traditional statistical methods. On closer inspection, it was noted that these networks often used as many connections as there were compounds under consideration. If connections could be regarded as the equivalence of independent variables, then these analyses may suffer from chance effects. Indeed, the phenomenon of overfitting is well known in the network community but appears to have been overlooked by early QSAR/network researchers. This was probably due to the exciting results that were being obtained. An exception, however, is the work reported by Andrea [23] who proposed the ρ parameter (Eq. (4)) to describe the relationship between connections and compounds.

5.1.3.1 Discriminant Analysis

We have concentrated on determining experimental guidelines to help minimize overtraining (memorization) and, thus, avoid, to a large extent, the potential problem of

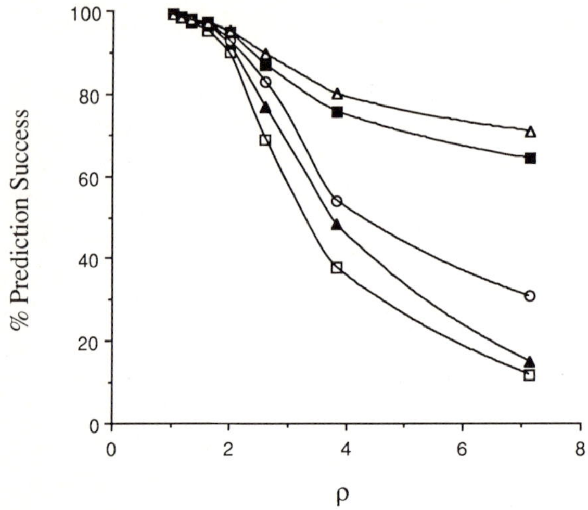

Figure 6. Plot of percentage cumulative prediction success against ρ. Results represent the percentage of output unit patterns differing from their respective target patterns by 10% (□), 20% (▲), 30% (○), 40% (■) and 50% (△).

chance effects. As a first approach to this problem we investigated how well neural networks would perform linear discriminant analysis using random numbers [22]. These experiments simulated a QSAR study involving 50 compounds classified into two categories (25 in each). The 4 input variables were created using random numbers generated by employing the RS1 data analysis package (BBN Software, Staines, U.K.). Two output units were used and training was aimed at activating the appropriate unit for the category of that compound. The number of hidden layer units was varied from 1 to 7. A plot of % prediction success against ρ (Fig. 6) demonstrates that as the number of connections approaches the number of compounds under consideration (i.e., $\rho = 1$) the percentage prediction success is close to 100%. Clearly, at ρ values less than 2.0, memorization of the input data is occurring. It is necessary, therefore, to use networks with a ρ value in excess of 3 to keep apparent successful predictions (within 0.2 of the target) at a rate lower than 50%.

The network architecture needed to perform discriminant analysis does not necessarily require as many output units as there are activity categories. A single output unit may be used by employing target values set to either end of the output range (e.g., 0.0 and 1.0) for each category. The above experiments were repeated using this protocol, in which the identical random number data sets as generated previously were employed. Fig. 7 shows that the total RMS error (a global measure of error for the differences between each of the output and target values) is lower for the $4-n-1$ discriminant analysis networks. It follows that, at the same ρ values, the $4-n-1$ network performs better than a $4-n-2$ network which suggests that it is more capable of memorizing the data.

One criticism of the above experiments is the use of random numbers to simulate physico-chemical input parameters. The structure of real and random data is differ-

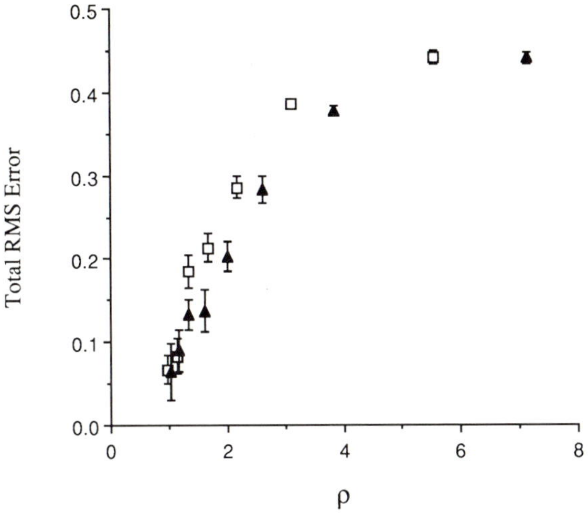

Figure 7. Plot of total RMS error vs ρ for 4−n−1 (▲) and 4−n−2 (□) networks.

ent and their behavior in the above paradigm may also differ significantly. An additional problem is that these networks cannot be tested for predictive capability, since using random numbers one would obviously expect prediction to be poor. To address these points, a previously reported discriminant analysis involving a series of antineoplastic napthoquinones [42] was used for NN training. In this study 3 physicochemical properties ($\Sigma\pi$, ΣMR and $\Sigma\pi^2$) were shown to classify 22 of the 27 compounds correctly. In our work, the squared term was not employed in the training since networks can implicitly account for non-linear relationships [4, 36]. In these experiments the number of hidden layer units was varied and the number of incor-

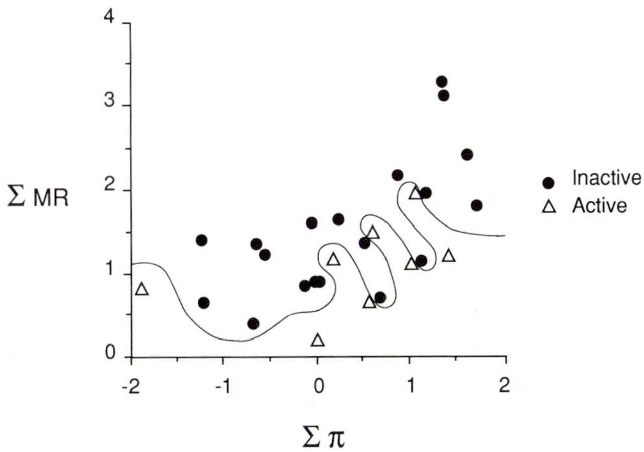

Figure 8. Plot of ΣMR vs $\Sigma\pi$ for a series of 27 antineoplastic napthoquinones [42]. A hypothetical function has been drawn to separate the active (△) and inactive (●) compounds.

rectly classified compounds was monitored. This number decreased from 7, for a hidden layer of 1 unit, to 0 (all classified correctly) with 7 hidden layer units. These results mirror those observed for random number data. To test this network system for predictive ability, 9 compounds were chosen at random and removed from the training set. The remaining compounds were used to train networks with varying numbers of hidden layer units. Each test compound was then presented to the trained network for classification. In all cases, at least 5 of the 9 were incorrectly classified demonstrating the poor predictive ability of the networks. A plot of ΣMR against $\Sigma\pi$ (Fig. 8) shows that the active and inactive compounds are in close proximity to each other which serves to highlight the inadequacy of this information for the discrimination of the 2 activity classes. An over-trained network may be able to separate each class by fitting a complex function between the classes (as shown in Fig. 8), however, this would be of no use for the purpose of prediction. This particular QSAR analysis may have been compromized by unsuitable physico-chemical data which inadequately described each compound.

5.1.3.2 Regression Analysis

In contrast to discriminant analysis which requires a yes/no decision for classification, regression analysis networks are trained toward a continuous target variable.

Table 1. Effect of varying network architecture on regression performance.

Network[a] Architecture	Connections[b]	ρ^c	$R^2 \pm SEM$
4,1,1	7	7.14	0.214 ± 0.022
4,2,1	13	3.85	0.434 ± 0.035
4,3,1	19	2.63	0.542 ± 0.041
4,4,1	25	2.0	0.743 ± 0.025
4,5,1	31	1.61	0.852 ± 0.023
4,6,1	37	1.35	0.915 ± 0.012
4,7,1	43	1.16	0.977 ± 0.005
4,8,1	49	1.02	0.985 ± 0.007
4,4,1	25 (15 cases)	0.6	0.996 ± 0.001
4,4,1	25 (45 cases)	1.8	0.770 ± 0.025
4,4,1	25 (55 cases)	2.2	0.672 ± 0.023
4,4,1	25 (135 cases)	5.4	0.318 ± 0.015

[a] Network architecture, the three numbers indicate the number of units in the input, hidden and output layers, respectively.

[b] The number of connections in the network. Results are the average of 10 experiments for each network architecture. New sets of random numbers were generated for each experiment.

[c] Ratio of the number of cases (50, unless otherwise stated) to the number of connections.

A similar series of experiments to those described above for discriminant analysis have been carried out with regression analysis networks using random numbers as input. Table 1 gives the results of network training expressed as an average correlation coefficient for the 10 data sets used for each architecture, and it can be seen that at ρ values of less than 2.0 the correlation coefficient using random numbers is above 0.74. A previously reported QSAR study using networks recommended that a ρ value between 1.8 and 2.2 was optimal for network performance, as measured by predictive capability [23]. Clearly at this ρ value the networks shown in Table 1 have memorized a significant quantity of data. The two studies, however, cannot be compared in too great a detail as the results shown in the table are based on random numbers, whereas the analysis reported by Andrea and Kalayeh [23] used real data. The structure of real data, in terms of correlations between variables, is clearly different to that of random number sets. Another difference between these two sets of results is that the performance of the networks using real data was tested by examining their predictive capability. Networks which have been trained to associate sets of random numbers might be expected to fit, but not to predict the data.

5.1.3.3 Real Examples of QSAR

Since the main purpose of fitting a linear (or other) model to a QSAR data set is to make predictions for unknown compounds, it is necessary to examine real QSAR data sets, with differing in-built structure, in order to assess how well networks might perform in prediction. Four previously published QSAR examples were selected to exemplify a number of different and commonly encountered QSAR models. Table 2 lists each of the four data sets along with the originally published equations and cross-validated correlation coefficients calculated using a leave-one-out (LOO) procedure. Each of these data sets has been analyzed using MLR neural networks with varying numbers of hidden layer units. One known problem with neural networks is that they can fall into so-called "local minima". In other words, the best solution is missed. If we imagine that the solution to the problem involves an error surface of "hills and valleys" and that the aim is to find the lowest point on the surface by going down hill, it is possible, from different starting points, to end up in different minimum positions, which may not necessarily be the lowest. Attempts to overcome this problem include perturbing the weights connecting the units in the network, followed by further training, or merely choosing another starting point and training again. A simple check on the total error will indicate which minimum is the lowest. In our own training procedures we employed a system of perturbing the weights and restarting the networks to try and locate the global minimum.

Neural networks were trained using the BIOPROP software [46] which incorporates a command language to enable control of network training (input/output/initialization/saving etc.) by use of script files. Each neural network MLR analy-

Table 2. QSAR equations. Eqs. (6), (7), (9) and (11) represent our own work repeating the MLR analyses previously reported. In these examples small differences were found in the coefficients and constants of the original equations to our own calculations. These frequent small differences may be due to computer rounding errors or to typographical errors in the original paper (data tables were carefully checked against the originals to avoid errors).

No.	Equation	n	R^2	s	F	Cross-validated R^2
	Linear example [43]					
	$\log P_{expt} = 0.40\alpha_{calc} - 0.46\mu + 0.33E_{(HOMO)} - 6.06$	37	0.826	0.6956	52.10	
(6)	$\log P_{expt} = 0.412\alpha_{calc} - 0.359\mu + 0.384E_{(HOMO)} - 7.115$	37	0.847	0.609	60.8	0.784
	Linear with indicator [44]					
	$\log 1/C = 0.45\pi + 1.05I - 0.48MR_Y$ [a]	38	0.929	0.264	17.1	
(7)	$\log 1/C = 0.424\pi + 1.090I - 0.495MR_Y + 3.374$	38	0.931	0.259	151.9	0.916
(8)	$\log 1/C = 0.424\pi + 0.165MR_Y + 3.494$	38	0.835	0.393	88.87	
	Quadratic [45]					
	$\log(1/D_{40}) = -1.40R_m^2 - 0.42R_m + 0.71\,pKa + 0.39$	50	0.828	0.25	67.9	
(9)	$\log(1/D_{40}) = -1.42R_m^2 - 0.43R_m + 0.70pKa + 0.36$	50	0.821	0.252	70.2	0.789
(10)	$\log(1/D_{40}) = -1.06R_m + 0.73\,pKa + 0.02$	50	0.709	0.317	57.23	
	Quadratic with Indicator [44]					
	$\log 1/C = 0.82\pi'_3 - 0.11\pi'^2_3 - 0.97MR_Y + 0.91I + 4.47$	34	0.878	0.343	13.3	
(11)	$\log 1/C = 0.84\pi'_3 - 0.11\pi'^2_3 - 0.97MR_Y + 0.96I + 4.40$	34	0.837	0.420	37.1	0.781
(12)	$\log 1/C = 0.34\pi'_3 - 0.03MR_Y + 4.47$	34	0.430	0.759	11.7	

[a] The constant value was not documented; presumably an omission.

sis was conducted using data scaled between the ranges of 0.0 to 1.0, and in one example between 0.2 to 0.8. The latter experiment was conducted to allow the network to extrapolate beyond the scaled range. Output values from the BIOPROP package fall in the range 0.0 to 1.0. Cross-validation was also conducted using both LOO and leave-*N*-examples out procedures (N = approximately 10% of the data set). The choice of grouping of compounds for the leave-*N*-examples out procedure was based on hierarchical clustering of the compounds as described by the physico-chemical parameters. A similarity level was chosen which divided the compounds into 4 groups and representatives from each were selected for testing. Training and testing were continued until all compounds had been left out (once) for test purposes. In those cases where the original equation used an indicator variable or a squared term, these parameters were not included for training. The removal of these properties was aimed at allowing the network to exploit its ability to develop "non-linear" relationships without these being specifically stated.

The four QSAR examples examined (Eqs. (6), (7), (9) and (11); Table 2) contain various structured data sets ranging from linear to a quadratic equation including an indicator variable. The intention was to explore their behavior in network training as the relationship between biological activity and physico-chemical properties increased in complexity going from a linear to a quadratic model. Our other aim was to determine whether networks were capable of outperforming traditional statistical methods.

Training

Fig. 9 to 12 illustrate the changes in correlation coefficient as ρ is altered. In each case, R^2 improves as the number of connections in the network is increased. At a ρ value below 2.0, the R^2 value tended to level out toward its maximum value. The linear example reached a value close to unity for training, however, the remaining "non-linear" examples failed to reach this level. One explanation for this behavior, which contrasts to the results obtained with random numbers, is that the inherent structure in the data prevents the network achieving a perfect fit. This is reasonable, for example, if 2 compounds have the same values for their physico-chemical input data but differ in their biological response, then it will not be possible to accurately predict the activity of both compounds.

In addition to the neural network results, Fig. 9 to 12 show the results obtained using regression analysis. The networks all performed well, and provided R^2 values exceeding those obtained using traditional methods at ρ values less than 3 (in 2 cases

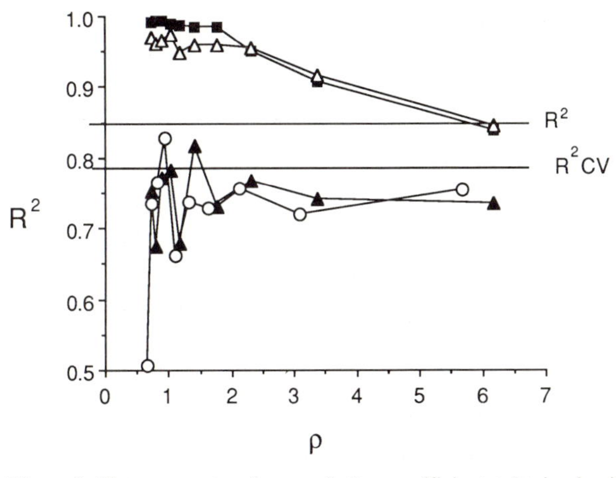

Figure 9. Plot comparing the correlation coefficients obtained using traditional statistics and neural networks for the "Linear" example listed in Table 2 [43]. For each curve, the correlation coefficient is plotted against ρ. The top two curves represent training of the data using networks employing a $3-n-1$ architecture with the data scaled between 0.0–1.0 (■) and 0.2–0.8 (△). The cross-validation curves employed the LOO (▲) and leave-N-out (○) procedures, respectively; in this case N is approximately 10% of the data set. The horizontal lines represent the results obtained using traditonal statistics which are detailed in Table 2. The cross-validation result using traditional statistics used a LOO procedure.

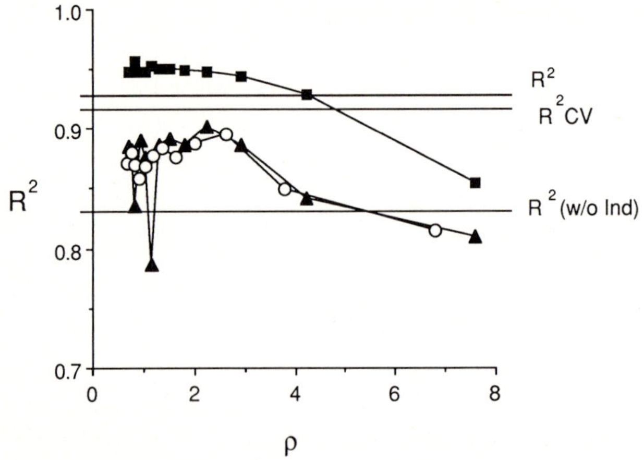

Figure 10. Plot comparing the correlation coefficients obtained using traditional statistics and neural networks for the "Linear with Indicator" example listed in Table 2 [44]. For each curve, the correlation coefficient is plotted against ρ. The top curve represents training of the data using networks employing a $2-n-1$ architecture with the data scaled between 0.0–1.0 (■). The cross-validation curves employed the LOO (▲) and leave-N-out (○) procedures, respectively; in this case N is approximately 10% of the data set. The horizontal lines represent the results obtained using traditional statistics which are detailed in Table 2. In addition, a line has been drawn showing the correlation coefficient in which the indicator variable was omitted. The cross-validation result using traditional statistics used a LOO procedure.

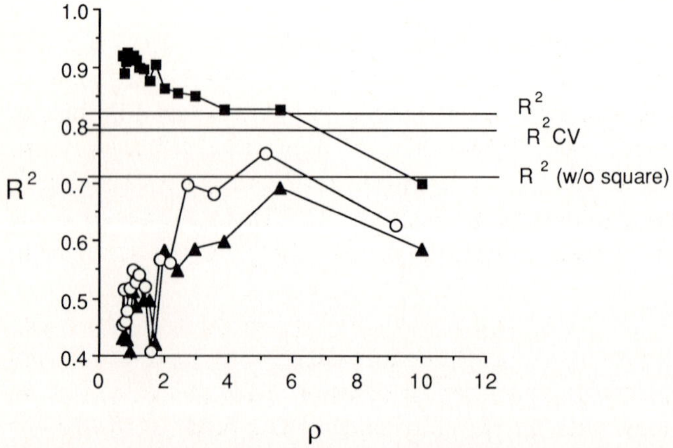

Figure 11. Plot comparing the correlation coefficients obtained using traditional statistics and neural networks for the "Quadratic" example listed in Table 2 [45]. For each curve, the correlation coefficient is plotted against ρ. The top curve represents training of the data using networks employing a $2-n-1$ architecture with the data scaled between 0.0–1.0 (■). The cross-validation curves employed the LOO (▲) and leave-N-out (○) procedures, respectively; in this case N is approximately 10% of the data set. The horizontal lines represent the results obtained using traditional statistics which are detailed in Table 2. In addition, a line has been drawn showing the correlation coefficient in which the quadratic variable was omitted. The cross-validation result using traditional statistics used a LOO procedure.

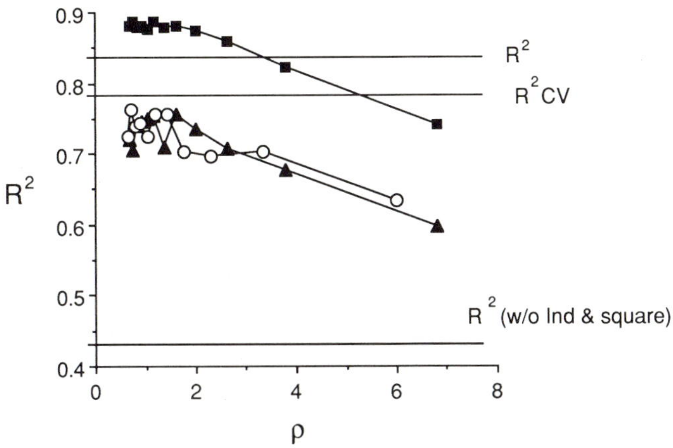

Figure 12. Plot comparing the correlation coefficients obtained using traditional statistics and neural networks for the "Quadratic with Indicator" example listed in Table 2 [44]. For each curve, the correlation coefficient is plotted against ρ. The top curve represents training of the data using networks employing a $2-n-1$ architecture with the data scaled between $0.0-1.0$ (■). The cross-validation curves employed the LOO (▲) and leave-N-out (○) procedures, respectively; in this case N is approximately 10% of the data set. The horizontal lines represent the results obtained using traditional statistics which are detailed in Table 2. In addition, a line has been drawn showing the correlation coefficient in which the quadratic and indicator variables were omitted. The cross-validation result using traditional statistics used a LOO procedure.

ρ was less than 6). It should be kept in mind of, course, that in our training procedures the squared terms and indicator variables were not used as input for network training. The networks have, therefore, discovered, the "non-linear" relationships of the input properties to the target data. To illustrate this further, the MLR equations were also derived without the square or indicator terms Eqns. (8), (10) and (12); Table 2) and their values have been shown on Fig. 10 to 12. Clearly, the networks are performing well at ρ values less than 8.0 and far exceed the fit using traditional methods.

One concern with the training "predictions" was the fact that the data were scaled between 0.0 and 1.0. As the network operates in this range, then it is unable to extrapolate beyond these limits. Such constraints may give artificially higher R^2 values. To test this hypothesis, the linear data set was also trained using data scaled between $0.2-0.8$. The result (Fig. 9) shows that the correlation coefficient is reduced, and in future experiments we may employ this smaller scaling range to avoid any potential problems.

Cross-Validation

Although the training results are impressive, this does not give any indication of how well the network will perform predictively. This can be measured to some extent by using cross-validation techniques. Cross-validation provides an idea of robustness but is normally used to give an indication of how well a model predicts when *n* cases are left out of the analysis. In our study we used both LOO and leave-*N*-out procedures and the results are shown in Fig. 9 to 12 (as expected, the leave-*N*-out procedure takes less time to calculate). Both methods give similar results with the exception of the quadratic example where the leave-*N*-out gave higher R^2 values at lower ρ values. Although the cross-validated results were variable at low ρ values, the behavior of each example was somewhat similar. Once again, the quadratic example appears to behave differently giving a maximum cross-validated R^2 at $\rho \approx 5.5$.

Interestingly, the cross-validated R^2, in all cases except for a few points on the linear example, fell below the results obtained with traditional statistical methods. Thus, although the networks appear to fit these data sets very well, they perform poorly in prediction as measured by cross-validation. It may be argued that this result is a consequence of over-training and highlights a disadvantage of networks which perform statistical procedures. On the other hand, the networks are able to fit the more complex data sets without having to specify indicators or non-linear transforms and appear to predict more accurately than the regression models without these terms. A more effective test of the performance of both neural networks and standard statistics is the use of a train/test set procedure. This involves the introduction of a test set after training has completed in order to determine predictive capability. Unfortunately, the generation of test and training sets requires careful consideration, unlike a LOO procedure in which every compound in the set is left out once. The major advantage of a training/test set approach, is that those compounds left out should be representative of the entire data set, if chosen correctly. Unless a data set is particularly "well-behaved", in terms of the disposition of compounds in parameter space, the LOO procedure will select some highly unsuitable (i.e. outlier) compounds since every compound in the set is considered.

5.1.4 Conclusions

Neural networks have found applications in many areas of chemistry including several different approaches to the generation of quantitative structure-activity relationships. They appear to offer some advantage over the fitting of traditional statistical models, such as regression and discriminant functions in that it is not necessary to specify the particular functional form of the relationship. This means, for example, that non-linear and cross-product terms of the input variables are identified without having to explicity state these terms. Similarly, indicator variables are also not required for network training.

There are also disadvantages in the use of neural networks to fit such models. Interpretation of the contribution of individual variables is difficult, in addition to assessing the "significance" of network fitting, since there are no equivalent terms in the statistical methods used to judge regression and discriminant analysis. An appropriate architecture is essential to avoid over-fitting as demonstrated by the near perfect results obtained using random numbers. Our experiments with these random number sets suggest guidelines for the value of ρ when constructing networks for both discriminant and regression analysis. Judging from the analysis of real data sets, it appears that networks perform well in fitting with the correlation coefficient for the network fit being greater than that of the regression model. The performance of networks in prediction, however, as measured by cross-validation, was considerably worse than that of the regression equations.

The apparent disadvantages of analyzing data in this way with neural networks may seem to outweigh the advantages. This may be true, and it is clear that caution must be exercized in the interpretation of network models, particulary when they are used for prediction. They may be useful, however, as "idea generators". A neural network may be able to fit a model to a set of data where statistical methods have failed to do so. Such models, of course, will need careful inspection, for example, by examination of cross-validated predictions.

One area in which networks may offer something new is in data reduction. The ReNDeR network is a novel approach to the display of multivariate data which has advantages over existing methods. One of these advantages is the ability to move between the two-dimensional display produced by the network and the starting data [30]. This is particularly important when the primary aim of the data set display is to identify clusters of interesting samples, so that further samples might be predicted.

We hope that this chapter has demonstrated some interesting applications of neural networks and that it will encourage others to investigate the use of such systems in their own work. To this end, we have provided a list of relevant software, books and other sources of information (see Appendix (A1)). Software which is available both commercially and in the public domain can be executed on any computer from a PC to a Cray supercomputer. In addition, for those readers who wish to examine the performance of a particular network package, we have provided the data set for the "Linear" example [43] outlined in Table 2 (see Appendix (A2)). This data set can be used to investigate various network foibles such as overfitting and local minima.

References

[1] Katz, W.T., Snell, J.W. and Merickel, M.B., *Meth. Enzymol.* **210**, 610–636 (1992)
[2] Gasteiger, J. and Zupan, J., *Angew. Chem. Int. Ed. Engl.* **32**, 503–527 (1993)
[3] Eberhart, R.C. and Dobbins, R.W., *Neural Network PC Tools,* Academic Press, Cambridge, MA, 1990
[4] Salt, D.W., Yildiz, N., Livingstone, D.J. and Tinsley, C.J., *Pest. Sci.* **36**, 161–170 (1992)
[5] Qian, N. and Sejnowski, T.J., *J. Mol. Biol.* **202**, 865–884 (1988)
[6] Bohr, H., Bohr, J., Brunak, S., Cotterill, R., Lautrup, B., Norskov, L., Olsen, O. and Petersen, S., *FEBS Lett.* **241**, 223–228 (1988)
[7] Holley, L.H. and Karplus, M., *Proc. Nat. Acad. Sci. USA* **86**, 152–156 (1989)
[8] McGregor, M.J., Flores, T.P. and Sternberg, M.J.E., *Protein Eng.* **2**, 521–526 (1989)
[9] Muskal, S.M., Holbrook, S.R. and Kim, S-H., *Protein Eng.* **3**, 667–672 (1990)
[10] Wade, R.C., Bohr, H. and Wolynes, P.G., *J. Am. Chem. Soc.* **114**, 8284–8285 (1992)
[11] Broughton, H.B., Green, S.M. and Rzepa, H.S., *J. Chem. Soc. Chem. Commun.* 1178–1180 (1992)
[12] Bodor, N., Harget, A. and Huang, M-J., *J. Am. Chem. Soc.* **113**, 9480–9483 (1991)
[13] Weigel, U.-M. and Herges, R., *J. Chem. Inf. Comput. Sci.* **32**, 723–731 (1992)
[14] Kvasnicka, V., Sklenak, S. and Pospichal, J., *J. Chem. Inf. Comput. Sci.* **32**, 742–747 (1992)
[15] Kvasnicka, V., Sklenak, S. and Pospichal, J., *J. Am. Chem. Soc.* **115**, 1495–1500 (1993)
[16] Harget, A.J. and Bodor, N., *Personal Computers and Intelligent Systems* **3**, 252–258 (1992)
[17] Lacy, M.E., *Tetrahedron Comput. Methodol.* **3**, 119–128 (1990)
[18] Zupan, J. and Gasteiger, J., *Anal. Chim. Acta,* **248**, 1–30 (1991)
[19] Gasteiger, J. and Zupan, J., *Angew Chem. Int. Ed. Engl.* **32**, 503–527 (1993)
[20] Aoyama, T., Suzuki, Y. and Ichikawa, H., *J. Med. Chem.* **33**, 905–908 (1990)
[21] Topliss, J.G. and Edwards, R.P., *J. Med. Chem.* **22**, 1238–1244 (1979)
[22] Manallack, D.T. and Livingstone, D.J., *Med. Chem. Res.* **2**, 181–190 (1992)
[23] Andrea, T.A. and Kalayeh, H., *J. Med. Chem.* **34**, 2824–2836 (1991)
[24] Tetko, I.V., Luik, A.I. and Poda, G.I., *J. Med. Chem.* **36**, 811–814 (1993)
[25] Livingstone, D.J. and Rahr, E., *Quant. Struct.-Act. Relat.* **8**, 103–108 (1989)
[26] Ford, M.G. and Livingstone, D.J., *Quant. Struct.-Act. Relat.* **9**, 107–114 (1990)
[27] Wikel, J.H. and Dow, E.R., *BioMed. Chem. Lett.* **3**, 645–651 (1993)
[28] Hudson, B., Livingstone, D.J. and Rahr, E., *J. Comput.-Aided Mol. Design* **3**, 55–65 (1989)
[29] Livingstone, D.J., *Meth. Enzymol.* **203**, 613–638 (1991)
[30] Livingstone, D.J., Hesketh, G. and Clayworth, D., *J. Mol. Graph.* **9**, 115–118 (1991)
[31] McFarland, J.W. and Gans, D.J., *J. Med. Chem.* **30**, 46–49 (1987)
[32] Good, A.C., So, S.-S. and Richards, W.G., *J. Med. Chem.* **36**, 433–438 (1993)
[33] Carbo, R. and Domingo, L., *Int. J. Quantum Chem.* **32**, 517–545 (1987)
[34] Cramer, R.D., Patterson, D.E. and Bunce, J.D., *J. Am. Chem. Soc.* **110**, 5959–5967 (1988)
[35] Weinstein, J.N., Kohn, K.W., Grever, M.R., Viswanadhan, V.K., Rubinstein, L.V., Monks, A.P., Scudiero, D.A., Welch, L., Koutsokos, A.D., Chiausa, A.J. and Paull, K.D., *Science* **258**, 447–451 (1992)
[36] Livingstone, D.J. and Salt, D.W., *BioMed Chem. Lett.* **2**, 213–218 (1992)
[37] Aoyama, T., Suzuki, Y. and Ichikawa, H., *J. Med. Chem.* **33**, 2583–2590 (1990)
[38] Klassen, M., Pao, Y.H. and Chen, V., *Characteristics of the Functional Link Net: A Higher Order Delta Rule Net,* (IEEE Proceedings of 2nd Annual International Conference on Neural Networks. San Diegeo, CA., USA, 1988), p. I-507–I-513
[39] Liu, Q., Hironi, S. and Moriguchi, I., *Quant. Struct.-Act. Relat.* **11**, 135–141 (1992)
[40] Liu, Q., Hironi, S. and Moriguchi, I., *Quant. Struct.-Act. Relat.* **11**, 318–324 (1992)
[41] Liu, Q., Hironi, S. and Moriguchi, I., *Chem. Pharm. Bull.* **40**, 2962–2969 (1992)
[42] Prakash, G. and Hodnett, E.M., *J. Med. Chem.* **21**, 369–373 (1978)

[43] Lewis, D. F. V., *J. Comp. Chem.* **10**, 145–151 (1989)
[44] Coats, E. A., Genther, C. S., Dietrich, S. W., Guo, Z.-R. and Hansch, C., *J. Med. Chem.* **24**, 1422–1429 (1981)
[45] Denny, W. A., Atwell, G. J. and Cain, B. F., *J. Med. Chem.* **22**, 1453–1460 (1979)
[46] BIOPROP is available from S. Muskal, Laboratory of Biodynamics, University of California, Berkeley, CA 94709. e-mail: "smuskal@sb1.cchem.berkeley.edu"

Appendix

A1. Neural network software and other information sources

This list is by no means exhaustive, but is representative of the software that is available commercially and in the public domain for the construction of neural networks. We have also provided some electronic bulletin board addresses and hope that this information will prove useful to those who wish to experiment with networks.

1. Central Neural System BBS has an electronic bulletin board containing 26 Mbytes of files related to artificial neural networks. These include simulation packages, demos, source code, tutorials and other text. Most of these are suited to IBM PC compatible machines, but some are available for Macintosh and Unix machines. CNS BBS can be contacted via Wesley Elsberry, P.O. Box 1187, Richland, WA 99352, USA; email: elsberry@beta.tricity.wsu.edu. cost = free.
2. The *Neuron Digest* bulletin board reports (approx. monthly) on various aspects of neural network activities and can be accessed by contacting the moderator Peter Marvit on the following email address; marvit@cattell.psych.upenn.edu.
3. BIOPROP. Bioprop is a programmable neural network simulator which uses a command language. Several example scripts are provided with the manual which can be purchased from Steven Muskal, Laboratory of Biodynamics, University of California, Berkeley, Berkeley, CA 94709 (415)-486 4338; email, smuskal@sbl-4.cchem.berkeley.edu.
4. Rumelhart and McClelland published their network software on diskette in the Handbook of their series of books on "Explorations in Parallel Distributed Processing (1988, MIT Press, No. 3 in the series). Various programs and exercizes are outlined in the text, however, this software is not very user-friendly.
5. BrainMaker Pro. 3.0, California Scientific Software, 10024 Newtown Rd, Nevada City, CA 95959, USA. BrainMaker Pro 3.0 (Dos/Windows) $ 795; Brain Maker 3.0 (Dos/Windows/Mac) $ 195 (also a student version, quantity sales only, approx $ 38). Various add-ons, boards and support available.
6. The ARD Corporation have a software package called Propagator which is available for IBM PCs, Macintoshs ($ 199) and Sun workstations ($ 499). ARD Corporation, 9151 Rumsey, Rd, Columbia, MD 21045 USA. email: propagator@ard.com

7. MATLAB Neural Network Toolbox; this package contains software which allows both supervised and unsupervised learning rules to be implemented. A 350 page manual is included explaining, with examples, the methods available. Product and pricing information can be obtained from, The Math Works, Inc., 24 Prime Park Way, Natick, MA, 01760, USA.
8. The following is a list of books with code (some on diskette) and offers guidance on practical applications of various neural network models.

a. Korn, Granino A, *Neural Network Experiments on Personal Computers and Workstations,* Cambridge, MA, MIT Press, 1991
b. Caudill, M. and Butler, C., *Understanding Neural Networks,* Vol. **I** and **II**, Cambridge, MA, MIT Press, 1991
c. Eberhart, R. C. and Dobbins, R. W., *Neural Network PC Tools, Cambridge,* MA, Academic Press, 1990. Diskette for book from Software Frontiers, Gilbert, AZ
d. McCord, N. M. and Illingworth, W. T., *A Practical Guide to Neural Nets,* Reading, MA, Addison-Wesely, 1990
e. A diskette offered by AI Expert magazine (San Francisco, CA and Boulder, CO). This collection of programs supplements several articles published over a period of time
f. Staff Writers, *Neural Teacher,* Salt Lake City, UT, Softlabs Corp, 1989
g. Staff Writers, *Neural Works Explorer,* Pittsburg, PA, NeuralWare, 1990
h. Aleksander, I., *An Introduction to Neural Computing.* London, England: Chapman and Hall, 1990. Software for book from Adhoc Reading Systems, East Brunswick, NJ, USA
i. Orchard, G. A. and Phillips, W. A., *Neural Computation,* East Sussex, England, LEA Ltd., 1990
j. Blum, A., *Neural Networks Programming in C+ +,* New York, NY, John-Wiley, 1992
k. Muller, B. and Reinhardt, J., *Neural Networks: An Introduction,* New York NY, Springer-Verlag, 1990
l. Freeman, J. A. and Skapura, D. M., *Neural Networks: Algorithms, Applications, and Programming Techniques,* New York, NY, Addison-Wesley, 1991
m. Kosko, B., *Neural Networks and Fuzzy Systems,* Englewood Cliffs, NJ, Prentice-Hall, 1992
n. Staff Writers, *The Brain Simulator,* San Francisco, CA, Abbot, Foster and Hauserman, 1989
o. Staff Writers, *NetWurkz,* Palo Alto, CA, DAIR Computer Systems, 1989
p. Staff Writers, *Anwareness,* Vancouver, BC, Canada, Neural Systems, 1989
q. Masters, T., *Practical Neural Network Recipes in C+ +,* New York, NY, Academic Press, 1993

The following books are introductory in nature and combine the ideas behind expert systems and neutral networks.

r. Zahedi, F., *Intelligent Systems for Business,* Belmont, CA, Wadsworth, 1993

s. Gallant, S. I., *Neural Network Learning and Expert Systems,* Cambridge, MA, MIT, 1992
t. Blanchard, O. and Beard, P., *Intelligent Applications,* New Canaan, CT, Lionheart, 1992
u. Lawrence, J., *Introduction to Neural Networks and Expert Systems,* Nevada City, CA, California Scientific Software, 1992
v. Zupan, J. and Gasteiger, J., *Neural Networks for Chemists:* A Textbook, VCH Publishers, 1993

9. Other commercial software packages for neural network simulation. Unfortunately we have few details for this software.
a. NeuralWorks Professional 2+, NeuralWare, Inc., Pittsburgh, PA 15276 USA
b. We know of other NN packages which are available such as: NeuroCompiler (Neuro Informatik GmbH, Berlin), AIM, Brain Cel, Neural Desk, Neural Case, Neuro Windows, Explorenet 3000, Neuroshell (Systems Group) and DynaMind, DynaMind Developer (NeuroDynamX).

10. There is of course a wealth of freely available neural network software and a list of free software and frequently asked questions (FAQ) can be obtained from Lutz Prechelt, University of Karlsruhe, Germany, email; prechelt@ira.uka.de

A2. Example linear QSAR data set [43] (see Table 2 Eq. (3))

Compound	α_{calc}	μ	$E_{(HOMO)}$	$\log P_{expt}$
Tetrafluoromethane	2.5555	0.000	−20.3	1.18
Trifluoromethane	2.5400	1.652	−17.9	0.64
Fluoromethane	2.5556	1.654	−17.4	0.51
Ethanenitrile	4.6210	3.045	−15.9	−0.34
Propanenitrile	6.5043	3.061	−15.3	0.16
Methanenitrile	3.0178	3.584	−17.0	−0.25
Methanal	2.7452	1.972	−14.7	0.35
Propanone	6.5100	2.749	−13.1	−0.24
Butanone	8.4345	2.804	−12.7	0.29
Ethanamide	5.7137	3.920	−13.1	−1.26
Methanamide	3.8732	3.701	−13.6	−1.95
N,N-dimethyl methanamide	7.7540	3.453	−12.2	−1.01
Propan-2-ol	7.0014	2.049	−14.5	0.05
Ethanol	5.1232	1.945	−15.0	−0.31
Methanol	3.2533	1.979	−15.7	−0.77
Methoxymethane	5.2699	1.823	−14.8	0.10
Ethoxyethane	9.0251	1.843	−14.2	0.89
Dioxan	9.1490	0.000	−13.3	−0.27
Propoxypropane	12.8080	1.819	−13.9	2.03
Cyclohexane	11.3131	0.002	−13.5	3.44

Table **A2** continued on p. 318.

Compound	α_{calc}	μ	$E_{(HOMO)}$	$\log P_{expt}$
Cyclopentane	9.4330	0.067	−14.7	3.00
Pyridine	9.1539	2.048	−12.7	0.65
Nitrobenzene	12.2284	5.417	−12.8	1.85
Methylbenzene	11.1511	0.193	−12.9	2.73
p-Xylene	13.3834	0.004	−12.2	3.15
m-Xylene	13.3789	0.079	−12.5	3.20
o-Xylene	13.3885	0.265	−12.5	3.12
Ethene	3.9007	0.001	−15.8	1.13
Ethane	4.4304	0.000	−16.2	1.12
Propane	6.3180	0.006	−15.4	1.61
Butane	8.2057	0.002	−14.7	2.12
Pentane	10.0951	0.006	−14.1	2.67
Hexane	11.9821	0.015	−13.6	3.25
Isubutane	8.2141	0.007	−14.9	2.19
Methane	2.5519	0.000	−19.7	1.55
Benzene	9.5935	0.000	−13.9	2.13
Water	1.3474	2.100	−17.8[a]	−1.38

Data from ref [43].

[a] Corrected as per communication with David Lewis.

5.2 Rule Induction Applied to the Derivation of Quantitative Structure-Activity Relationships

Mohammed A-Razzak and Robert C. Glen

5.2.1 Introduction

The field of Artificial Intelligence (AI) has generated numerous innovative methods for the extraction of rules from data. We have applied techniques specifically tailored to the expert systems area which are useful in deriving simple rules from large datasets. These methods have been employed to analyze relatively small series of molecules with the objective of establishing rules which are predictive and reliable as well as providing an insight into the mechanisms of molecular behavior.

The derivation of useful relationships between computed and measured molecular properties of molecules and their physical and biological properties has been attempted using many different types of data generation and analysis methods. The objective is usually to increase the understanding of the processes involved and to establish predictive descriptions relating molecular properties to biological actions.

Classical QSAR analyzes may utilize physicochemical substituent constants describing properties of molecules, and trends within series may be discovered using statistical methods such as correlation analysis. More diverse types of data have been generated based on computational-chemistry methods [1]. Because of the large number of possible molecular descriptors which can be calculated, this approach may result in underdetermined problems where there may be a high ratio of molecular descriptors to cases. In addition, much of the generated data may be noise, which although accurately describes the molecules in a series, it may have not direct bearing on their biological activities.

To overcome some of these deficiencies, non-parametric statistics and pattern recognition methods have been developed [2] and applied to these problems. For example, principal components analysis (PCA), cluster analysis, partial least squares (PLS), nearest neighbour (*k*NN), etc. These are generally described as being either supervized (in which a model is fitted, e.g. linear regression) or unsupervized (in which the data is not fitted to a model, e.g. a non-linear map). Although these methods have been, and continue to be applied successfully to many problems, it is sometimes difficult to interpret the results and formulate the next step.

More recently in the field of Artificial Intelligence, methods have been specifically developed (AI) in the expert systems area designed to extract rules from data [3, 4]. The input is usually a number of test cases and the output is a tree-structured series of rules (a class probability tree). This is an example of a supervized learning method.

Applications are diverse and include, for example, electronics [5], agriculture [6] and engineering [7]. The attraction of these rule-induction methods, is that the rules generated during the induction phase can be easily interpreted and applied later to new cases. The performance of new examples can be tested and the reasons for success or failure can be easily deduced.

The software package EXTRAN [8] implements the rule induction used in the examples analyzed here and is based on the ID3 [9] algorithm. This algorithm is modified to allow pruning of "bushy" rule trees using a χ-squared criterion.

5.2.2 Rule Induction Using the ID3 Algorithm

First, we shall highlight some of the vocabulary used in rule induction.
Induction: the task of detecting rules in the example set.
Objects: each data point.
Attributes: the properties that describe the objects.
Decision tree: the set of rules (or one rule).
CX: a modification of ID3 to introduce a *chi*-squared test at each decision point.
The ID3 algorithm attempts to construct a simple decision tree from a number of objects. It may not be the best tree (the method is iterative), but it is usually compact and extracts the essence of the information contained in the examples.

If C contains p objects of class P and n objects of class N then an object will belong to class P probability $p/(p+n)$ and to class N with probability $n/(p+n)$ The information in a decision tree is, therefore:

$$I(p, n) = -p/(p+n)\log_2 p/(p+n) - n/(p+n)\log_2 n/(p+n) \tag{1}$$

If attribute A with values $(A_i, A_{i+1} \ldots)$ is used for the root of the decision tree then it will partition C into $(C_i, C_{i+1} \ldots)$ where C_i contains those objects in C that have value A_i of A. Let C_i contain p_i object of class P and n_i objects of class N. The expected information for the subtree for C_i is $I(p_i, n_i)$. The expected information required for the tree with A as a root is then obtained as the weighted average:

$$E(A) = \sum_{i=1}^{v} (p_i+n_i)/(p+n) I(p_i, n_i) \tag{2}$$

The information gain on branching A is, therefore:

$$\text{gain}(A) = I(p, n) - E(A) \tag{3}$$

The ID3 method examines all the attributes and then chooses the one that maximizes the information gain. The same process is then reiterated over the rest of the attributes, or until a suitable level of reliability is achieved (the rule tree may be "pruned" at any stage). The objects are then divided into subsets depending on their value of that attribute.

In the case of missing data, a null value is recorded and incorporated into the information entropy calculation. A *chi*-squared test for stochastic independence has been found to be useful for noisy data.

Suppose attribute A produces subsets (C_1, C_2, ... C_n) of C where C_i contains p_i and n_i of class P and N. If the value of A is irrelevant to the class of an object in C, the expected value p_i' of p_i should be:

$$p_i' = p(p_i + n_i)/(p+n) \ , \tag{4}$$

if n_i' is the corresponding expected value of n_i, the statistic

$$\sum_{i=1}^{v} (p_i - p_i')^2/p_i' + (n_i - n_i')^2/n_i' \tag{5}$$

is approximately *chi*-square with $v-1$ degrees of freedom. The tree-building algorithm can then be modified to reject attributes whose irrelevance cannot be rejected with high confidence.

5.2.2.1 Examples of Data Analysis

Two sets of data are analyzed here. The first data set consists of calculated molecular properties and the thin-layer chromatography (TLC) retention factors of a series of substituted benzoic acids. Pattern recognition and neural network methods have shown the relationships present in the TLC test set. The second dataset is a series of anticonvulsants showing anti-epileptic activity which was previously analyzed using regression methods.

5.2.2.2 Rule Induction on Thin-Layer Chromatography Data

Thin-layer chromatography retention times of 22 substituted benzoic acids were measured on 10×20 cm glass-backed plates precoated with C-18 bonded silica gel with an in-built fluorescent indicator. Thin-layer chromatography was performed in glass TLC tanks containing 5 mL of the mobile phase. The mobile phase consisted of mixtures of different solvent systems: acetonitrile-water (30:20), acetonitrile-water (40:60), acetonitrile-water (60:40), methanol-water (40:60), methanol-water (50:50) and methanol-water (60:40).

The retention factors (R_f values) were converted into R_m values by the following equation:

$$R_m = \log\left(\frac{1-R_f}{R_f}\right) \tag{6}$$

In order to use numeric data in rule-induction, the data must first be classified. In this case, the TLC plate was divided into 5 equally spaced regions of increasing R_f values, (0–0.2, 0.2–0.4, 0.4–0.6, 0.6–0.8, 0.8–1.0). The compounds and data are listed in Table 1.

The molecular structures for the twenty-two substituted benzoic acids were constructed using SYBYL [12], geometrically optimized using MOPAC [13] (AM1 PRECISE) and a range of physicochemical properties was calculated using an in-house

Table 1. R_m values of substituted benzoic acids.

Compound	Acetonitrile/ water (30:20)	Acetonitrile/ water (40:60)	Acetonitrile/ water (60:40)	MeOH/ water (40:60)	MeOH/ water (50:50)	MeOH/ water (60:40)
4-F	0.15	0.26	0.49	0.10	0.23	0.37
3-F	0.14	0.26	0.48	0.10	0.23	0.37
2-F	0.21	0.32	0.51	0.18	0.37	0.49
2-CF_3	0.11	0.23	0.45	0.10	0.26	0.41
3-CF_3	0.07	0.16	0.39	0.03	0.11	0.26
4-CF_3	0.04	0.12	0.37	0.03	0.11	0.26
2-F,4-CF_3	0.07	0.18	0.39	0.03	0.14	0.29
4-F,3-CF_3	0.06	0.14	0.35	0.03	0.11	0.20
4-F,2-CF_3	0.08	0.15	0.42	0.05	0.19	0.33
4-CH_3	0.12	0.17	0.44	0.05	0.19	0.33
2-CH_3	0.13	0.19	0.44	0.09	0.22	0.38
3-CH_3	0.12	0.18	0.44	0.09	0.20	0.38
4-NH_2	0.41	0.45	0.57	0.45	0.62	0.72
H	0.20	0.25	0.49	0.17	0.35	0.52
2-OH	0.16	0.22	0.48	0.13	0.27	0.43
3-OH,5-OH	0.62	0.55	0.64	0.62	0.67	0.78
2-OH,6-OH	0.26	0.39	0.51	0.25	0.39	0.55
2-OH,3-OH	0.29	0.44	0.59	0.26	0.40	0.59
2-OH,5-OH	0.37	0.48	0.62	0.38	0.53	0.66
2-OH,4-OH	0.31	0.44	0.61	0.30	0.47	0.61
3-OH,4-OH	0.57	0.57	0.66	0.56	0.66	0.78
2-COOH	0.37	0.50	0.63	0.35	0.57	0.67

Table 2. The calculated physico-chemical properties.

Property
Molecular area (AREA) [15] A^2
Molecular volume (VOL) [16] A^3
Ovality $(OVAL)^a$
Molecular weight (*MW*) [17] Daltons
Dipole moment $(DIP)^b$ Debye
Sum of the excess atomic charge on Oxygen and Nitrogen atoms $(QNO)^b$ electrons
Sum of the excess charge on Nitrogen atoms $(QN)^b$ electrons
Sum of the excess charge on Oxygen atoms $(QO)^b$ electrons
Number of multiple bonds (*NMULT*)
Molecular polarizability $(POLAR)^c$ 10^{-25} cm^3
Dipole moment of the solvent $(DIPSOL)^b$ Debye
Polarizability of the solvent $(POLSOL)^c$ 10^{-25} cm^3

[a] Ovality is the ratio of the surface area to the minimum surface area which would be found if the molecule were constrained to be a sphere.
[b] Atomic charges and dipoles were calculated by the Partial Equalization of Orbital Electronegativity (PEOE) [19–22].
[c] Molecular polarizabilities are calculated from Slaters rules for the calculation of atomic screening constants (to be published) [18].

package PROFILES [14]. The properties calculated are listed in Table 2. In addition, the same properties were calculated for the ionised species (e.g. COO^-, NH_3^+), for example, QNOC (parameters appended with –C to denote the charged species). The data set also included the squared terms for each of the parameters relating to the neutral species (to introduce non-linearity into the parameter set). This resulted in 36 descriptors for each molecule.

An earlier study [10] showed that pattern recognition and neural network methods utilizing some of these molecular properties were of use in predicting retention factors for this series. Rule induction offers a different perspective on the data, and perhaps some insight into the chromatographic behavior of these compounds.

The CX algorithm was used for induction, giving rise to the rules in Fig. 1.

All the molecules are classified using a subset of the original parameters which are the following: QNC^2, POLSOL, MW, POLAR, DIP^2, $AREA^2$, DIPSOL, $OVAL^2$.

The remaining parameters were unnecessary for complete classification. This is a particularly useful attribute of rule-induction, highly correlated parameters (e.g. molecular volume is highly correlated with the molecular area, r = 0.98) and those with little discriminatory value are discarded.

Classes 1 and 2 were combined due to the small sample size and a series of test runs performed on the data, which involved a "leave-one-out" strategy. All the molecules except one were used to derive rules which were employed to predict the reten-

```
                                                                Classes 1 to 5 and total.
QCN2                                                            2 27 54 36 13 132
<1.133 : POLSOL                                                 0  5 36 36 13  90
  |   <25.59 : MW                                               0  0  2 15 13  30
  |     |   <136.6 : POLAR                                      0  0  0  1 11  12
  |     |     |   <117.9  =5                                    0  0  0  1  1   2
  |     |     |   >=117.9 =5                                    0  0  0  0 10  10
  |     |   >=136.6 : DIP2                                      0  0  2 14  2  18
  |     |         <7.57 : AREA2                                 0  0  0 14  2  16
  |     |           |   <29073  =4                              0  0  0 12  0  12
  |     |           |   >=29073 =5                              0  0  0  2  2   4
  |     |         >= 7.57 =3                                    0  0  2  0  0   2
  |   >=25.59 : POLSOL                                          0  5 34 21  0  60
  |         <29.83 : MW                                         0  3 21 21  0  45
  |           |   <136.6 : DIPSOL                               0  0  5 13  0  18
  |           |     |   <1.4 =4                                 0  0  0  6  0   6
  |           |     |   >=1.4 : DIPSO                           0  0  5  7  0  12
  |           |     |         <1.57 : DIP2                      0  0  5  1  0   6
  |           |     |               <7.79 =3                    0  0  5  0  0   5
  |           |     |               >=7.79 =4                   0  0  0  1  0   1
  |           |     |               >=1.57 =4                   0  0  0  8  0  27
  |           |         <7.57 : MW                              0  1 15  8  0  24
  |           |           |   <137.6 =2                         0  1  2  0  0   3
  |           |           |   >=137.6 : AREA2                   0  0 13  8  0  21
  |           |           |         <29073 : AREA2              0  0 11  4  0  15
  |           |           |           |   <24715 : DIPSOL       0  0  8  4  0  12
  |           |           |           |     |   <1.40 =4        0  0  2  2  0   4
  |           |           |           |     |   >=1.40 : DIPSOL 0  0  6  2  0   8
  |           |           |           |     |         <1.57 =3  0  0  4  0  0   4
  |           |           |           |     |         >=1.57 =4 0  0  2  2  0   4
  |           |           |           |   >=24715 =3            0  0  3  0  0   3
  |           |           |         >=29073 : DIPSOL            0  0  2  4  0   6
  |           |           |               <1.40 =4              0  0  0  2  0   2
  |           |           |               >=1.4 =4              0  0  2  2  0   4
  |           |         >=7.57 =2                               0  2  1  0  0   3
  |         >=29.83 : MW                                        0  2 13  0  0  15
  |               <151.1 =3                                     0  0  9  0  0   9
  |               >=151.1 : DIP2                                0  2  4  0  0   6
  |                     <1.71 =2                                0  1  0  0  0   1
  |                     >=1.71 =3                               0  1  4  0  0   5
>=1.132 : OVAL2                                                 2 22 18  0  0  42
      <1.77 : POLSOL                                            0 12 18  0  0  30
        |   <29.38 : DIPSOL                                     0  7 18  0  0  25
        |     |   <1.46 : DIPSOL                                0  7  8  0  0  15
        |     |     |   <1.4 : DIPSOL                           0  2  8  0  0  10
        |     |     |     |   <1.36 =3                          0  0  5  0  0   5
        |     |     |     |   >=1.36 =2                         0  2  3  0  0   5
        |     |     |   >=1.4 =2                                0  5  0  0  0   5
        |     |   >=1.46 =3                                     0  0 10  0  0  10
        |   >=29.38 =2                                          0  5  0  0  0   5
      >=1.77 : DIPSOL                                           2 10  0  0  0  12
            <1.46 : DIPSOL                                      2  4  0  0  0   6
              |   <1.4 =2                                       0  4  0  0  0   4
              |   >=1.4 =1                                      2  0  0  0  0   2
            >=1.458 =2                                          0  6  0  0  0   6
```

Figure 1. Rule induction on the TLC data.

tion factor of the test compound. The results were excellent and correctly predicted the outcome in most cases for 90% of the samples.

Those compounds which were predicted to fall into the wrong class were invariably predicted to be in the adjacent class, which was due to compounds lying very close to the classification levels. Small changes in the induced rules (by adding or removing compounds) in some cases reclassified these into the correct classes and was due to a "splitting" in the decision tree at slightly different values. This behavior is usually seen in small data sets (such as this one), where there are relatively few examples in each class.

Table 3. Prediction of TLC retention times[a].

Class	pass	fail	% correct	% random[b]
1+2	6	4	60%	29/132 = 22%
3	9	1	90%	54/132 = 41%
4	9	1	90%	36/132 = 27%
5	9	1	90%	13/132 = 10%

[a] 10 runs were performed (50 in all) for each class. This used a "leave-one-out" strategy of using the training set to define a class probability tree which was used to predict the class membership of the missing member.
[b] This is the percentage expected for a random guess of the result.

Compounds with a low R_f value are not predicted so well. This may well be due to the difficulty in measuring retention factors near the base of the TLC plate. Small changes in the decision levels resulted in some compounds being classified into the next faster moving band of the TLC plate.

Analysis of the rules is simple and allows an interpretation of the results in a chemical sense. Branches having a high classification rate are particularly useful. For example, compounds which ascend the TLC plate rapidly have different properties to those which remain near the base. This has been explained previously in classical terms as the proton acceptor ability, proton donor ability and dipole interaction of the solute (the compound of interest) and the solvent [the stationary phase (silica in this case)] [23]. Since silica gel is amphoteric in nature, i.e. being both hydrophilic and lipophilic in character (due to the simultaneous existence of both silanol and siloxane groups), hydrogen bonding and dispersion interactions therefore are thought to be important in determining retention factors.

Examination of the induced rules shows some particularly strong branches, e.g. such as the following:

If $QNC^2 \geqslant 1.13$ and $Ovality^2 \geqslant 1.77$ then compounds are found near to the base of the plate (low R_f). On the other hand, if $QNC^2 < 1.3$ and $POLSOL < 25.5$ and $MW < 136$ then compounds are found near to the top of the plate (high R_f). The challenge then is to interpret the rules in a chemical sense, and one interpretation may be:

"compounds having less excess charge on the nitrogen and low molecular weight in a solvent of high polarizability, interact less (decreased hydrogen bonding capacity) with the stationary phase and interact more with the solvent (presumably via dispersion forces) and are moved up the plate"

while,

"compounds having good hydrogen bonding groups (more highly charged), which are more oval than spherical, are situated near to the base of the plate, presumably due to greater hydrogen bonding interactions with the stationary phase and lower mobility, due to the non-spherical shape".

These rules makes sense in the context of existing chromatographic theory and are an example of how underlying principles can be supported by the data.

5.2.2.3 Forced Induction and Exception Programing on Anticonvulsant Data

The anticonvulsant and CNS-depressant activities of sixteen commercialy available anti-epileptics were subject to regression analysis by Lien, Liao and Shinouda [24]. The maximal electro-shock data (MES) includes 16 compounds, 3 descriptors and the *MES* field. The data, compounds and a description of the attributes is contained in Table 4.

A simple correlation between anticonvulsant activities and $\log P$ was derived:

$$\log 1/C = 0.627\,(0.093) \log P + 2.58\,(0.16) \tag{7}$$

$$n = 16,\ r = 0.76,\ s = 0.342.$$

When diazepam, clonazepam and carbamazepine were omitted (on reports that they interact with different receptors) an improved fit of the data was obtained:

$$\log 1/C = 7.776\,(0.847) \log MW - 14.438\,(1.943) \tag{8}$$

$$n = 13,\ r = 0.941,\ s = 0.241$$

Table 4. CNS data of selected anti-epileptics.

Compound	*MES*	log *MW*	log *P*	Dipole Moment
phenytoin	4.42	−2.4	2.47	1.74
ethytoin	3.38	2.31	1.53	1.74
mephenytoin	3.56	2.34	2.09	1.74
phenobarbital	4.03	2.37	1.42	0.87
metharbital	3.19	2.3	1.21	1.13
mephobarbital	3.86	2.39	1.98	0.87
primidone	4.28	2.34	2.1	1.35
trimethadione	2.36	2.16	−0.37	1.74
paramethadione	2.82	2.20	0.13	1.69
ethosuximide	2.15	2.15	0.01	1.47
methsuximide	3.43	2.31	1.54	1.61
phensuximide	3.31	2.28	1.4	1.61
phenacemide	3.31	2.25	0.57	2.06
diazepam	4.17	2.45	2.82	2.65
clonazepam	3.56	2.51	2.41	2.33
carbamazepine	4.4	−2.37	2.18	2.41

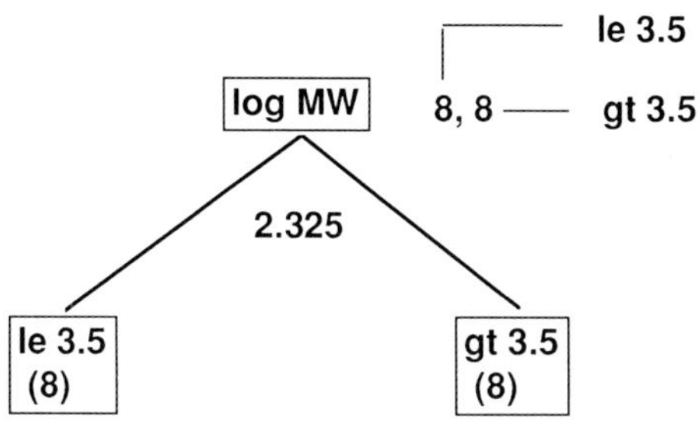

Figure 2. Class probability tree of the MES data.

The optimum lipophilicity was calculated to be 1.42 from regression (derived from the parabolic equation) with a negative dependence on μ (the dipole moment). This data set is a useful example with which to demonstrate forced induction and exception programming.

Selecting the mean of the *MES* data (3.5) and using automatic induction (CX) gives the class probability tree in Fig. 2.

Using only log *MW*, 100% classification was obtained and this was therefore, the most important descriptor. Those compounds with log *MW* value greater than 2.32 are in the most active set. This is interesting (but not surprising). The regression analysis of the full data set inferred that log *P* and dipole moment were important but that log *MW* was of lower significance. However log *P* and log *MW* were highly correlated ($r = 0.93$) for this series of compounds.

When diazepam, clonazepam and carbamazepine were omitted from the regression analysis (reports suggested that they interact with different receptors) a better fit of the data was obtained in the regression analysis resulting in the simple relationship shown in Eq. 8.

The correlation coefficient is increased and log *MW* is now the important descriptor in the regression equation. This was the result found using induction. Since induction works in a stepwise fashion, classification for the majority of compounds may be achieved early on in the analysis using only a few descriptors with outliers classified later employing additional descriptors in the set. In this way, the more important descriptors (for most of the set of compounds) are revealed.

In order to obtain more information on a diverse set of parameters, or to view the data from a different viewpoint, induction may be forced to split descriptors in a defined sequence. By forcing induction to split log *P* first, the decision tree in Fig. 3 is obtained.

This gives about 90% classification of the full data set, assuming that the tree is pruned to remove nodes 3 to 4. This can be compared with the regression equation which used log *P* as a significant contributor (where the correlation coefficient was 0.88). The induction next splits log *MW*, while the next most significant contributor to the regression equation is the dipole moment, μ.

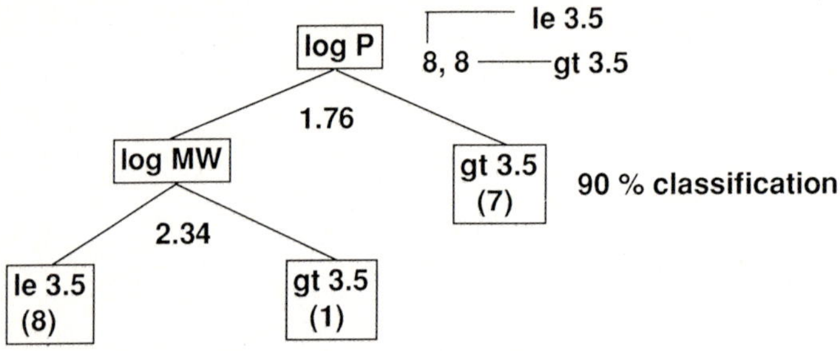

Figure 3. Class probability tree for the *MES* data (forced induction on log *P*).

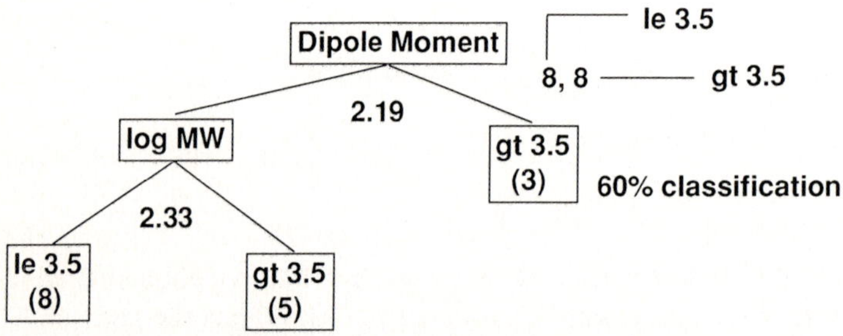

Figure 4. Class probability tree for the *MES* data (forced induction on dipole moment).

The splitting value selected for log *P* by induction was 1.76, which is similar to the optimum value found for lipophilicity by regression (1.42). So, compounds with log *P* greater than 1.42 are in the more active set. The precise value of decision values is, however, unstable in such small data sets and can change with the omission (or addition) of key compounds to the training set.

Splitting first the dipole moment gives a lower classification (only 61%, induction tree shown in Fig. 4).

This is similar to regression in which the dipole was found to be less significant. The dependence on the dipole moment was negative in the regression equation. This is not completely the case in rule induction, where a number of compounds displaying a dipole greater than 2.19 Debye are classified as more active using this descriptor. A possible explanation might be that this parameter becomes influential in cases where other descriptors have values which enable the dipole moment to become significant.

This is also an interesting example in which to attempt exception programming [25]. This method is a modification of the induction algorithm and can be used to reveal combinations of descriptors (physico-chemical properties in these applications) that are not present in the training set. It may, therefore, be useful in expanding the training set to cover more of the parameter space at minimal cost.

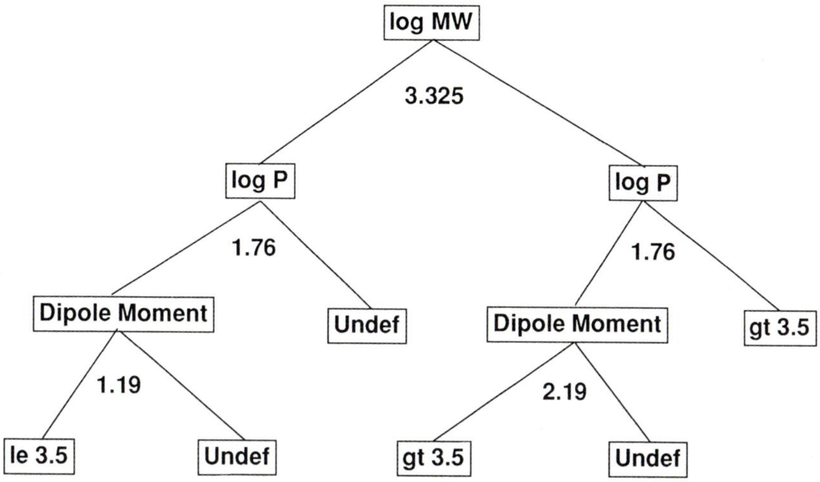

Figure 5. Class probability for the *MES* data showing exception programming.

In this method, a decision tree is generated using the automatic CX induction algorithm to determine the decision points for each of the attributes. The values of the attributes are then converted from numerical to logical values, depending on threshold values given in the trees. A new class is created (undefined) for undefined compounds or combinations. The example sets are then entered as exceptions to the general case of undefined.

A new decision tree was generated using exception programming (Fig. 5). The undefined points in the tree are cases which do not appear in the trial set (Fig. 5). For example, there is no case of a compound with a dipole $>= 2.195$ and a log $P < 1.76$ and a log $MW < 3.325$. This is a very useful feature when expanding a compound set to cover property space in the most efficient way.

5.2.3 Conclusions

Rule induction using the CX algorithm can be an alternative method for data analysis. The rules are usually compact and may offer insights into the role of molecular properties in the determining physical and biological properties. Rule induction appears to extract properties in a different way than, for example, regression in the determination of the relative contribution and importance of descriptors. The presentation of the results in the form of conditional statements is clear (from the point of view of deciding with which molecule to test next) and may be compared to, for example, regression equations or non-linear map plots which classify the objects of interest, but do not easily reveal the reasons behind the classification.

The probabilities attached to the derived rules (based on the number of correctly classified members in each branch) allows confidence to be attached to particular branches. Also, it is clear how the rules were derived and (if requested are given by the program) which examples were used to determine the rules. Also *exception programming* may be used to further explore property space economically.

Since several trees may be induced from the same data set by forcing induction on attributes of interest, we can examine the data from different angles. A number of alternative production rules may, thus, be generated from the same data set.

The results imply that in some cases it would be advantageous to use rule-induction as a complementary technique in addition to conventional statistical and pattern-recognition methods [26].

References

[1] Hyde, R.M. and Livingstone, D.J., *J. Computer Aided Molecular Design,* **2**, 145–155 (1988)

[2] Chatfield, C. and Collins, A.J., *Introduction to Multivariate Analysis,* 3rd Edn., Chapman and Hall, London, 1986

[3] *Building Expert Systems,* Hayes-Roth, F., Waterman, D.A. and Lenat, D.B., ed., Addison-Wesley, London, 1984

[4] Winston, P.H., *Artificial Intelligence,* Addison-Wesley, London, 1984

[5] Gunhold, R., Zettel, J., DIASS: *An Expert System for Diagnosis of Faults in Electronic Circuit Boards,* (Proceedings of the Artificial Intelligence and Expert Systems conference, Weisbaden 23–25 September (1986)). TCM, Expositions, Liphook, UK, (1987)

[6] Austin, D.J., Blake, P.S., Fletcher, D.A. and Garrett, C.M.E., *Chemometrics,* **5**, 53–64 (1988)

[7] Modessitt, K., *Experience with Commercial Tools Involving Induction on Large Databases for Space Shuttle Main Engine Testing,* (4th International Expert Systems Conference, London (1988), Learned Information, Oxford, UK, (1988)

[8] A-Razzak, M., Hassan, T. and Ahmad, A., *EXTRAN 7 User Manual Infolink, Decision Support Group,* 9–11 Grosvenor Gardens, London, SW1 WOBD, UK

[9] Quinlan, J.R., *Machine Learning, Vol.* **I**, Kluwer Academic Publishers, Boston, 1986, p. 81–106

[10] Glen, R.C., Rose, V.S., Lindon, J.C., Ruane, R.J., Wilson, I.D. and Nicholson, J.K., *J. Planar Chromatog.* **4**, 432–438 (1981)

[11] Liu, Q., Hirono, S. and Moriguchi, I., *Quant. Struct.-Act. Relat.* **11**, 318–324 (1992)

[12] *SYBYL* Molecular Modelling Package (1990), Tripos Associates, St. Louis, Ms. USA

[13] *MOPAC, Version 6.0,* Quantum Chemical Program Exchange, Department of Chemistry, University of Indiana, Bloomington, Indiana, USA

[14] Glen, R.C., Rose, V.S., *J. Mol. Graph.,* **5**, 79–86 (1987)

[15] Edward, J.T., *Chem. Ed.,* **47**, 261 (1970)

[16] Pearlman, R.S., *SAREA, QCPE 413*, Quantum Chemical Programme Exchange, University of Indiana, Bloomington, Indiana, USA

[17] Lide, D.R., ed., *CRC Handbook of Chemistry and Physics,* 71st edn., CRC Press, Boston, 1991

[18] Slater, J.C., *Physical Review,* **36**, 57–64 (1930)

[19] Gasteiger, J., Marsili, M., *Tetrahedron,* **36**, 3219–3288 (1980)

[20] Hinze, J., Jaffe, H.H., *J. Am. Chem. Soc.,* **84**, 540–546 (1962)

[21] Hinze, J., Whitehead, M.A., Jaffe, H.H., *J. Am. Chem. Soc.,* **85**, 148–154 (1963)

[22] Hinze, J., Jaffe, H.H., *J. Am. Chem. Soc.,* **67**, 1501–1505 (1963)

[23] Poole, C. F., Schuette, S. A., *Contemporary practice of Chromatography,* Elsevier, Amsterdam, 1984
[24] Lien, E. J., Liao, C. H., Shinouda, H. G., *J. Pharm. Sci.,* **68**, 463–465 (1979)
[25] Hassan, T., A-Razzak, M. *Exception Programming: A New Approach to Defining Specifictation Examples,* (Proceedings of the International Conference on Expert Systems, London, 181–197, 1988, Learned Information, Oxford, UK
[26] A-Razzak, A., Glen, R. C., *J. Comp. Aided Mol. Des.,* **6**, 349–383 (1992)

Index